Praise for

TIGERS AND TOMAHAWKS

"*Tigers and Tomahawks, The American Volunteer Group in China, 1941-42* by Ward Boyce belongs at the head of the long line of books about this legendary combat unit popularly known as the "Flying Tigers." The product of more than four decades of research, Boyce's book tells the A.V.G. story in great detail while adding life to the account with extensive first-person accounts by the Tigers themselves."

– Carl Molesworth, aviation historian and author of 15 World War II-related books, the latest of which is Flying Tiger Ace: The Story of Bill Reed, China's Shining Mark

"The most complete and accurate history of the Flying Tigers that I've ever read."

– Bob Andrade, A.V.G. Historian

ISBN: 978-1-943492-87-9

Cover: *Tigers Over China*. Oil on canvas painting by Roy Grinnell, used with generous permission of the artist. Aircraft depicted is the P-40B, No. 3, of the American Volunteer Group's Third Squadron, flown by Dick Rossi.

Book design by designpanacne

ELM GROVE PUBLISHING
San Antonio, Texas, USA
www.elmgrovepublishing.com

Elm Grove Publishing is a legally registered trade name of Panache Communication Arts, Inc.

TIGERS AND TOMAHAWKS

The American Volunteer Group in China 1941-1942

J. Ward Boyce

EGP

Dedicated to the 22 American Volunteer Group members who died while serving, the five who became prisoners-of-war and the additional 249 who served until the group was disbanded on 4 July 1942.

CONTENTS

PREFACE

It began with a motion picture, the 1942 Republic Pictures production of *Flying Tigers*, starring the inimitable John Wayne. Although somewhat "hokey" by today's film standards, it grossed well at the box office and was a powerful boost for American morale in the early months following the Japanese attack on Pearl Harbor. It also fired the imagination of a six-year-old lad in the small, Texas Panhandle town of Childress, and instilled a life-long interest in aviation and military history (especially that of the Flying Tigers!). Every primitive sketch of airplanes doodled on my yellow school notepads carried the leering, sharks'-mouth insignia made famous by that group.

From that time, I bought every book and magazine chronicling the American Volunteer Group that I could get my hands on. One of the first was *The Flying Tigers*, by Russell Whelan, published in 1942. *With General Chennault*, by Robert B. Hotz , followed a year later. Still an avid fan of the A.V.G. when I entered the Air Force, 17 years later, my military travels allowed me to expand my research with a visit to Stanford University's Hoover Institute, where I was introduced to the Chennault papers housed there, and later to the A.V.G. holdings at Maxwell AFB, Alabama, where I was able to obtain a copy of the Third Squadron Diary.

In 1975, while stationed at Bergstrom AFB in Austin, Texas, I learned (from a National Geographic article) that David L. "Tex" Hill lived in San Antonio. A quick call got me an invitation to come to San Antonio to meet the legendary Flying Tiger, which led to a warm friendship that lasted until Tex's death in 2007. Tex put me in touch with George "Pappy" Paxton (who just happened to be living in my hometown of Abilene, Texas!) who added greatly to my A.V.G. research material. It would be 10 years later, however, before – following a congratulatory letter I wrote to R.T. Smith for the publishing of his *Tale of a Tiger* – I was invited to attend one of their annual reunions in Ojai, California. This opened up a whole new world for me, for I took advantage of that reunion and followed with several more over the next 25 years. In addition to Tex Hill and Pappy Paxton, I was privileged to maintain friendships with Charlie Bond, Dick Rossi, Ed Rector, Buster Keeton, Chuck Baisden, J.J. Harrington, Merlin Kemph, Irving Stolet and a host of other members of the A.V.G.

In addition to the Flying Tigers I met, I was fortunate to meet two inveterate A.V.G. historians, Bob Andrade and Mark Burken. Bob worked closely with Dick Rossi for several years and Mark was the official historian of the Flying Tigers Association. Over the years both shared their resource material with me, including nearly a complete file of the A.V.G.'s individual combat reports and copies of over 27 personal diaries.

There have been literally dozens of books written about the American Volunteer Group in China, many of which have certain discrepancies, one-from-another. 35 years of researching A.V.G. material gave me an incentive to write as accurate an account of this unique combat group as possible. I have done so through access to 27 personal diaries, a vast collection of contemporary combat reports, and personal correspondence and interviews with 30 A.V.G. veterans and their Japanese counterparts.

This, then, is the story of the Flying Tigers – that group of American adventurers, idealists and yes, mercenaries, who fought the Japanese in the skies over Burma and China in the early days of World War II, and who captured the imagination of the world to such an extent that now, 78 years later, their name is still widely recognized, as is the shark's mouth insignia of their P-40s. To understand how this group came about, however, we must go back to the early 1930s in China...

J. Ward Boyce
Austin, Texas, 2021

INTRODUCTION

On February 22, 1932, Robert M. Short, a Reserve U.S. Army Air Corps lieutenant who had been hired by the L.E. Gale Company to help market American planes throughout the Orient, was in Nanking watching as the train bound for Shanghai was loading with Chinese civilians and a few Americans. Fifteen minutes later after the train pulled out of the station, three Mitsubishi B1M Type B bombers, escorted by three Nakajima A1N Type Three biplane fighters, appeared overhead. As the bombs began to fall on the train, Short raced to Shanghai's Jungoa airport, where the Chinese had a recently purchased Boeing Model 218, export version of the U.S. Army Air Corps P-12E/U.S. Navy F4B-3. Jumping into the little biplane, guns installed and ammunition belts fully loaded, he took off in pursuit of the Japanese. The Boeing was unevenly matched against the three Nakajimas, however, and, after a two-minute battle Short's aircraft fell in flames. He was the first American to fly, and the first to die, against the Japanese in support of China during the Sino-Japanese war. The Japanese aircraft returned to the aircraft carrier Kaga from whence they had come, and pilot Sasumi Kotani filed his combat report claiming victory over the Chinese aircraft before dying soon after from his own wounds from the battle.[1]

By this time, China had been at war with the Japanese for five months, since the "Manchurian Incident," when Japanese and Chinese forces clashed at the Chinese barracks in Mukden, Manchuria. On May 5, 1932, Chiang Kai-shek, head of the Kuomintang, or "Peoples' Party," and the nominal head of the Chinese government after defeating various Chinese warlords and elements of the Chinese Communist party, signed a truce with the Japanese in Shanghai.

At the time, China's Air Force consisted of some 270 miscellaneous aircraft, the remnants of those survivors who had flown from Manchuria into northern China following the attack at Mukden, and those flown from the eastern provinces of China to the far west to escape destruction by the Japanese. Still, barely a third could be considered combat-ready, and most of these were two-seat, general purpose biplanes.

Following the loss of Manchuria and defeat at Shanghai, Chiang Kai-shek approached the British government for help in rebuilding the decimated Chinese Air Force. He wanted not only aircraft, but also a first rate, professionally-run training program. The British, however, had no desire to disrupt their relations with Japan and refused Chiang's request.

At that point, another door opened for China in the guise of Capt. George C. Westervelt, U.S.N. (Ret), the leading representative for Curtiss-Wright Aircraft Corporation in China. In a letter to Soong Tse-ven, chairman of China's Commission of Aeronautical Affairs (C.O.A.A.), he wrote, "I take it for granted that the events of the last few weeks have sufficiently demonstrated the significance of air power." He went on to suggest (with an eye toward Curtiss-Wright military sales) that: (1) China should hire a high-ranking U.S. Army officer to serve as an advisor on military aviation. (2) China should ask the U.S. government to arrange for training pilots and technicians in the United States. (3) A small aircraft factory be built in Shanghai. (4) Commercial air service be expanded throughout China; and (5) A modern aerodrome be built in Shanghai.[2]

Soong accepted the proposals enthusiastically. Educated at Harvard and Columbia Universities, he spoke English fluently and was one of the most forward-thinking members of China's wealthiest and most powerful families. Referred to by Americans as T.V. Soong, he was Chiang Kai-shek's brother-in-law. Indeed, each of his three sisters married into wealth and/or power – the youngest, Mei-ling, was wed to Chiang Kai-shek; the eldest, Ai-ling, was the wife of China's Minister of Finance, Dr. H.H. Kung; and the third sister, Ching-ling, married Dr. Sun Yat-sen, first president of the Republic of

China. He would have a see-saw career in and out of Chiang's service over the next 17 years, going from prominence to political exile and back, as his relationship with his volatile brother-in-law waxed and waned, and would later figure prominently in the establishment of the American Volunteer Group.[3]

Soong had reservations on just one of Westervelt's proposals: Instead of sending Chinese students to the United States for flight training, China would seek to bring American instructors to China. He turned first to Dr. Arthur N. Young, an American economist who was serving as a financial adviser to the Chinese government, placing him in charge of the project. Young in turn went to Mr. Edward P. Howard, U.S. Trade Commissioner in Shanghai for the Department of Commerce, and Julian Arnold, commercial attaché, who approached American Ambassador Nelson T. Johnson with the proposal. Would the United States government favor a petition for an American mission to establish and operate a school for military aviation in China? Johnson and his military attaché supported the idea. It would certainly help sell American airplanes to the Chinese, and would boost the American aircraft industry, hard hit by the Depression now in its third year. T.V. Soong emphasized that the mission should be headed by a high-ranking military officer with a "prestigious background in military aviation." [4]

Ambassador Johnson cabled the request to Washington, along with his own endorsement for support. If the request was refused, he argued, other nations would certainly move in – probably France and Italy – and an American trade opportunity would be lost.[5]

The State Department and War Department wanted no part of it. Given the state of hostilities that existed between Japan and China, in the view of the State Department, Japan would surely declare war on the United States if such a mission were sent to China. The War Department went a step further. Anticipating a Chinese attempt to hire U.S. Army aviators, they specifically prohibited active-duty Army Air Corps officers from accepting employment, instructing, or even giving advice to the Chinese while on leave in China. Washing their hands of the problem, the State Department passed the buck to the Commerce Department, who looked much more favorably on the proposal. The establishment of an American training program would open up an aviation market worth millions.[6]

Receiving the go-ahead from his Commerce Department superiors in Washington, Howard went to work. He drafted a training program in early April calling for twelve aircraft and eight instructors to train fifty pilots a year. After gaining approval of the basic program, he then turned to the task of finding a "high-ranking officer with a prestigious background in military aviation." As a long-time aviation enthusiast – he had conducted an extensive survey of the "Northeast Air Force" in Manchuria in July of the previous year – Howard had many acquaintances in aviation circles. He promptly contacted "just the man for the job."[7]

CHAPTER 1

From Jouett to Chennault

Lieutenant Colonel John Hamilton "Jack" Jouett had recently been retired from the U.S. Army Air Corps, a victim of field-grade officer cutbacks dictated by the Depression economy.[1] A West Pointer, he was an energetic and capable officer – and a fine aviator. He had been a squadron commander in World War I and later served as U.S. Army Air Corps Chief of Personnel during the 1920s.[2] Although he was the director of the aviation division of Standard Oil of New York at the time, he wasted no time in accepting Howard's offer.

With his contacts in the Air Corps personnel office, he obtained access to 200 records of recent Air Corps flight school graduates who had gone into the Reserves. He selected nine of the best – Gerrardus B. "Jerry" Clark, Christopher Mathewson, Jr. (son of baseball great, "Christy" Mathewson), Morris B. "Reed" Knight, Lucius Roy Holbrook, Ellis D. Shannon, W.C. "Foxy" Kent, Harry T. Rowland and Roy L. Sansbury. He hired four mechanics – James A. Coughlin, William A. "Al" Spencer, W.H. Pennington and H. Ponceti. A flight surgeon, Doctor Howard Cooper, and a secretary, Alma Wade, rounded out the party. Each had received a letter of instructions informing them they should meet him in San Francisco at the Sutter Hotel before noon on June 16, 1932, for a departure the following day. He further advised them that, "...As to clothing, it is advised that you take over such civilian clothes as you now possess, purchasing the necessary linen suits and other clothing which is peculiar to that country and Shanghai. A civilian overcoat is recommended, as it gets cold in San Francisco and out on the water...." He and his group boarded the *President Hoover* in San Francisco on June 17, 1932, and three weeks later they stepped off the boat in Shanghai.[3]

The following year two more instructors joined the group – John Schweitzer and Harvey Greenlaw. Greenlaw, West Point Class of 1920, would become Claire Chennault's executive officer in the American Volunteer Group eight years later.[4]

Shien Chiao airfield, east of Hangchow, had been scheduled as the site for the "Central Aviation School" as it was to become known. After securing quarters in Hangchow (there was no housing available on base), Jouett and his staff set up shop. Looking around the field, Jouett found a dismal sight – deteriorating hangars and a handful of obsolete British and Russian aircraft in varying states of disrepair. There was no drinking water, electricity, machinery, or repair shops. Jouett's first task was to dig a water well, get started on a new hangar and administration offices, and erect a radio station. Later would come an operations office, woodworking shop, aircraft maintenance shops, and an auditorium for lectures.[5]

Jouett's plan was to pattern the school after U.S. Army Air Corps training methods – a Chinese Randolph Field. He had brought with him a complete set of U.S. Army flight manuals and training aids. The course would last twelve months and would consist of four, three-month stages – primary, basic, and advanced. Aircraft for the school initially consisted of fifteen Consolidated Fleet trainers bought by the Chinese and delivered on the same ship that brought Jouett and his people. One-hundred men would follow over the next two years.[6]

Within sixty days, Colonel Jouett had established the Central Chinese Aviation Academy and begun the task of rebuilding the Chinese Air Force with the some 200 pilots on the rolls. With the full support of Chiang Kai-shek, Jouett began refresher courses for the Chinese pilots, both to determine their real qualifications and to develop a cadre of Chinese instructors.

Starting with a group of fifty on September 17, he soon found that individual flying ability of these "qualified" pilots ranged from dismal to fair. The officers, most of whom were field grade, had come from wealthy families, had received minimal training, and looked upon the Air Force as one large social club. Of the initial fifty, six were unable to pass Dr.

Cooper's physical exam, and only twenty-seven of the remaining forty-four passed the course, each requiring an average of four hours and twenty minutes to solo.[7] Jouett subsequently dismissed over half of the previously "trained" pilots for incompetence and reduced the remainder two grades in rank – but kept their pay at the same level, lest they "lose face." To boost morale after these blows to Chinese egos, however, he raised their pay from $30 to $100 (Chinese) a month, and added a $50 bonus for flight pay.[8]

Aviation cadet training began on October 7, 1932, with a class of 88 highly-motivated and intelligent young men. Less than half completed the initial course, but those that did could measure up to their counterparts at Randolph Field! About one-hundred cadets reported for each class, with a new class reporting every six months.[9]

One aspect of the American-style instruction irritated Chiang. Chinese aviation cadets had come from families of the socially elite. The stringent standards demanded by Colonel Jouett resulted in a fifty percent wash-out rate. Chiang was being constantly reminded of this "loss of face" by irate parents of the washed-out cadets. He was pleased with the school otherwise and managed to attend every class graduation.

The school soon expanded to include bombardment training at Nanking, equipped with the Italian Fiat BR-3s, and a course for mechanics and engineers. Also, by September, 1933, China had received fifty Curtiss Hawk II biplane fighters, all of which were brought to Shien Chiao.[10] The Hawk II, Curtiss's export model of the U.S. Navy's XF11C-2 "Goshawk," was China's first modern fighter and would be the mainstay of its fighter force for the next four years.

Chiang Kai-shek was pleased with Jouett's progress and proud of the Central Aviation School. However, three factors converged during 1933 that eventually brought an end to the Jouett mission. The first was Japanese diplomatic pressure. Japanese reconnaissance planes had been invading Chinese air space at will, and could frequently be seen in the skies above Shien Chiao.[11] It wasn't long after that Japan lodged a formal protest through diplomatic channels claiming illegal American intervention. The Jouett mission, the Japanese claimed, along with Robert Short's attack on the Japanese bomber formation one year earlier, was a clear indication of continued and illegal American involvement in Sino-Japanese affairs.[12]

Meanwhile, the Japanese were continuing their campaign to nibble away at China. Early in 1933 Japanese forces based in Manchuria crossed the Great Wall and invaded the northern Chinese province of Jehol. Facing little resistance, they soon took the entire province. The campaign ended with the Tangken Truce, signed in May 31, 1933.

There had been little Chinese resistance in Jehol because it was Chiang Kai-shek's avowed policy to bring the warlords and Communists under control first and then fight the Japanese. At the time of the Tangken Truce, rumors began to reach Chiang of a rebellion building in Fukien province – largely in opposition to his reluctance to fight the Japanese. Rumor became fact, and on November 20, 1933, the rebels proclaimed a "Peoples' Revolutionary Government" at Foochow.

Chiang immediately ordered a general campaign and flew to western Fukien to take command personally. He ordered Colonel Jouett to support the campaign with his instructors and aircraft from the Central Aviation School. Jouett refused, emphasizing that he and his mission had been hired to train pilots, not to engage in combat. Infuriated, Chiang ordered other elements in from Chinese Air Force units. Operating from airfields at Wenchow and Chuchow, they provided some air support against the Fukien rebels. Within two weeks, Foochow was occupied by Nationalist Chinese Marines and the rebellion was put down. Chiang was victorious – but with no help from the Americans.[13]

Earlier in the year, Italy had been making overtures to the Chinese. Dr. H.H. Kung, T.V. Soong's brother-in-law (who had replaced Soong as head of the Commission on Aeronautical Affairs in late 1932), went to Washington in early 1933 to try to obtain credit for surplus American military equipment and aircraft parts. The State Department refused on the grounds that the executive branch had no authority to extend credit. Rebuffed, Kung returned home – by way of Italy. Mussolini welcomed him with open arms, offered to build an aircraft factory in China, set up a much cheaper school than the Americans, and offered the long-term credit that the American government had refused. As a result, the Chinese opened the door to the Italians, purchasing twenty Fiat BR-3 bombers to be assembled at Hangchow.[14]

In early 1935, the Italians moved into China full of promise. General Silvio Scaroni (World War I ace with 26 aerial victories) and Colonel Roberto Lordi, Regia Aeronautica, brought the Italian air mission, 40 pilots and 100 engineers and mechanics, opening a flying school at Loyang with Italian aircraft – graduating all cadets who survived the course, in contrast to Jouett's school at Shien Chiao. Scaroni became Chiang Kai-shek's air advisor, setting up operations from their headquarters at Nanchang in Kiangsi Province. They promised to build not only an air force but an aviation industry as well, and they would train engineers along with pilots.

Mussolini's government took full advantage of their position and over the next two years supplied hundreds of Italian aircraft – Fiat CR.30s and 32s, Breda Ba.27s, Caproni Ca.101s, and Savoia-Marchetti S.M.72s. By 1937, Chinese air strength had swelled to over 600 combat planes of multi-national origin, but mostly Italian.

American presence at Shien Chiao continued in aircraft maintenance as well. On April 14, 1934, the Central Aircraft Manufacturing Company (CAMCO) was incorporated to "assemble and repair aircraft and to manufacture parts" as a subsidiary of the New York-based Intercontinent Corporation. CAMCO was the brainchild of 38-year-old William D. "Bill" Pawley, sales representative for Curtiss-Wright, Douglas and Vought aircraft companies in China. A very successful entrepreneur, Pawley had been made president of the Douglas-backed China National Aviation Corporation (C.N.A.C.) and was also president of the Intercontinent Corporation, an exporting company headquartered in New York. CAMCO was subsequently formed as a subsidiary of Intercontinent. Pawley installed his brother, Edwin, as CAMCO president and named Paul A. Zimmerman as manager. Although it was staffed with Curtiss-Wright engineers, CAMCO was jointly operated by the Nationalist government and under contracts jointly signed by Curtiss-Wright, Vought, and Douglas aircraft companies. Bill Pawley and CAMCO would figure prominently in the establishment of the American Volunteer Group seven years later.[15]

It was no surprise that Jouett's' contract wasn't renewed when it expired in June 1935. Chinese aviation cadets had come from the families of the socially elite. The stringent standards demanded by Colonel Jouett resulted in a fifty-percent washout rate. Chiang Kai-shek was being constantly reminded of the "loss of face" by irate parents of washed-out cadets. Jouett had been successful in what he was hired to do, however. He left the Chinese Air Force in much better shape than he found it – a nucleus of 355 well-trained pilots, with 258 more in training greatly improved facilities, and a fine training system on which to build. He had developed a competent cadre of Chinese instructors, and he had established a first-rate flying school modeled on the U.S. Army Air Corps standards.[16] And the Commerce Department was happy: By 1934 the Chinese had bought $9,000,000 worth of American aircraft, parts, and supplies, including Fleet Model 5s and 10 trainers, Curtiss Hawk IIs and IIIs, Northrop Gamma light bombers, Vought V.65 Corsair reconnaissance bombers, and Douglas O-38 observation planes. Nevertheless, Japanese pressure on the U.S. State Department, Chiang's displeasure with Jouett over the Fukien Rebellion incident, and the emergence of the Italians as an attractive alternative, brought about Jouett's departure.[17]

Although the Jouett mission had come to an end, the Chinese government still valued American expertise and requested Washington's approval to hire another military officer to run the school at Shien Chiao and to act as senior air advisor to the Chinese Commission of Aeronautical Affairs (C.O.A.A.). The State Department was cool to the idea (as ever). Secretary of State Cordell Hull was specific in his position that the American government desired "no special bond" between the United States and China.[18]

Lacking support by the U.S. government, American influence in China waned. The Central Flying School stayed in operation, however, with the support of Chinese instructors and three American holdovers – Ray Holbrook, Foxy Kent, and Al Spencer. Harvey Greenlaw and his wife Olga left the school but remained in China. Eventually, in the summer of 1936, China was able to hire another American superintendent, George E.A. Reinberg, West Point Class of 1930. Reinberg served for six months until he died of a heart attack in December, 1936.[19]

Impressed with the success of Jouett's mission, Chiang's rival in Canton, Marshall Chan Chi-Tong, had negotiated with the U.S. Commerce department for his own mission in 1933. With little fanfare, Mr. Edward Deeds, of United Aircraft, arrived in Canton with six instructors. This was followed by the purchase of several Boeing aircraft and the eventual establishment of an aircraft factory in 1934 at Shinochow, 125 miles north of Canton.

The Nationalist Air Force received an unexpected prize on July 13, 1936, when the Cantonese Air Force defected en masse to Chiang. Chiang had moved against the Cantonese warlord, Chan Chi-tong, two weeks earlier. Opting for the winning team, Cantonese Air Force commander, General Wong Kwang-yui, took his 90-plane force – by some sources the best-trained and -equipped in China – and joined the Nationalists. Chiang reciprocated by making him commander of the Nationalists' Air Force. Chiang also inherited Marshall Chan's Shinochow Aircraft Works, recently contracted out to Bill Pawley. As the assembly plant for Chiang's Curtiss Hawk IIIs, by October the factory had become the Nationalists' First Aircraft Manufacturing Factory, under the Commission on Aeronautical Affairs.[20]

Although the American-run school at Shien Chiao had been eclipsed by the Italians at Nanchang, the Chinese continued to purchase American aircraft in large quantities – helped, no doubt, by the aggressive Bill Pawley. To fly and maintain these planes properly, China needed American-trained pilots and mechanics; in January 1935, Chiang sent a delegation of the Commission on Aeronautical Affairs, headed by Col. Mao Pang-tsu, to the United States to explore means of getting the additional aviation equipment and to recruit more American instructors.[21] As the guests of the ever-enterprising Bill Pawley, the delegation attended the All-American Air Races in Miami, Florida, in mid-January. A feature of the show was the U.S. Army's aerobatic team, a flight of three Boeing P-12Es from the 12th Pursuit Squadron at Maxwell Field, Alabama. Dubbed "Three Men on a Flying Trapeze," the team put on a spectacular demonstration of precision flying.

When the team landed, Pawley and Col. Mao were introduced to the team captain and his two wingmen. The leader was a slender, black-haired Army Air Corps captain named Claire Lee Chennault. His tanned, heavily-lined face reflected thousands of hours flying in open cockpits, as well as his advancing age – at 41, he was a very senior captain! His two wingmen, although sergeants in the (Regular) Air Force, wore the gold bars of (Reserve) second lieutenants and introduced themselves as William C. "Billy" McDonald and John H. "Luke" Williamson. Pawley told the trio how impressed they had been with their flying exhibition and invited all three to join them on his yacht in the Miami harbor for some relaxation.

Aboard Pawley's yacht that evening, Pawley relayed to Chennault the delegation's invitation for Chennault and his wingmen to come to China as flying instructors at the Hangchow flight school. All three declined at the time, but the seed was sown.

Roy Holbrook, still at Shien Chiao, became Chiang's technical advisor to CAMCO and Chiang's chief recruiter, hiring twenty-two Americans during 1936, many of whom held reserve commissions. Holbrook used his former Army Air Corps contacts for potential candidates, among whom was Capt. Chennault, at that time still assigned to Maxwell Field, Alabama. Holbrook had served with Chennault at Brooks Field, Texas, during the late 1920s when they were both flying instructors. Holbrook needed, he wrote Chennault, about a dozen pilots and mechanics. Could Chennault recommend some good men who might like to come to China for high wages and adventure? You bet he could – Luke Williamson and Billy McDonald![22]

Both had flown with Chennault for three years. With McDonald and Williamson (who had earlier replaced Lt. Haywood Hansell), Chennault had organized the Army's first aerobatic team at Maxwell Field – the Army's answer to the Navy's "Helldivers." Following the Miami airshow, and nearing the end of their enlistments, they both had applied for commissions in the Regular Army. As two among four-hundred applicants vying for 52 commissions, however, their lack of formal education kept them from being accepted, despite the fact that they were among the top three in total flying time and were clearly the best in flying ability. Bitterly disappointed, the two sergeants decided to leave the Air Corps for more profitable flying jobs, as both had airline transport ratings.[23] Their farewell performance was flown in December 1935 at the Eighth Annual All-American Air maneuvers in Miami.

In addition to Williamson and McDonald, Chennault recommended a third pilot, Sterling Tatum; an armament specialist, Rolf Watson; and two mechanics, John Holland and C. Biggs "Seebie" Smith, whom he had known since 1931 when they met at Maxwell Field. All were on active duty with the Army Air Corps; however, obtaining honorable discharges presented no problem. They departed out of Victoria, British Columbia, on July 11, 1936.[24]

Chennault had been corresponding with Roy Holbrook ever since he had recommended McDonald and Williamson for instructor jobs. He had relayed his own conditions for employment with the Chinese, when and if he should retire. Just a week after the departure of the five, Chennault received a cablegram on July 20 from the Chinese Commission on Aeronautical Affairs offering him a two-year, renewable contract for $12,000 a year to be in complete charge of China's advanced pursuit training, developing training manuals, textbooks and tactical directives. Additionally, he would be a major participant in selecting future purchases of pursuit aircraft and would help in organizing an early-warning system. China would pay his transportation to China and his return to the United States at the end of the contract. Finally, in the event of his death, his family would receive an additional $12,000. As a further incentive, they included the option of a three-month trial, should he accept. Re-evaluating his 19-year career in the Army, and considering his future prospects, he quietly cabled the Commission on Aeronautical Affairs of his acceptance of their offer, conditional on his retirement from the Army.[25]

Chennault's Army Air Corps career had peaked. He had joined the Army in 1917 after the United States declared war on the Central Powers and applied for flight training. Various circumstances kept him on the ground until the end of World War I, but he was initially accepted as an aviation cadet in November 1918 and reported to Kelly Field, just outside of San Antonio, Texas. Despite a personality clash with his first instructor (he could be just as ornery and stubborn as the most hard-nosed flying instructor), Chennault was awarded his wings in the spring of 1919. Mustered out as a Reserve officer in the spring of l920, he applied for a Regular commission and was back in the Air Corps six months later.[26]

Chennault had flown fighters (known as pursuit ships at that time) most of his 20-year career – first in SPADs and SE-5s with the First Pursuit Group at Ellington Field, then to Hawaii in the Thomas-Morse MB-3, where he took command of the 19th Pursuit Squadron at Luke Field. Following Hawaii was a stint as director of primary flight training at Brooks Field, Texas (where he met Roy Holbrook), then on to Maxwell Field, Alabama, where he was senior instructor in pursuit tactics at the Air Corps Tactical School.[27]

It was during this time that Chennault's early career began to fade. He and four other officers went to Washington, D.C., to testify before the Howell Committee on air power. Chennault's blunt testimony on how he felt air power had been misused during the Army maneuvers of 1934 was ill-taken by one General C.E. Kilbourne of the War Department General Staff. Kilbourne had been one of the senior planners for the exercise and resented being openly criticized by an upstart captain. As a result, Chennault's name was removed from the list of nominees to go to the Army's Command and General Staff School at Fort Leavenworth, Kansas. Without the school, Chennault knew he had little hope of further promotion.[28]

Chennault was to pursuit aviation what Brig. Gen. Billy Mitchell had been for bomber forces. The only problem was that pursuit aviation was totally eclipsed by the bomber crowd in the U.S. Army Air Corps of the 1930s. The general consensus could be summed up by a comment made by Air Corps General Walter (Tony) Frank: "...due to increased speeds and limitless space it is impossible for fighters to intercept bombers and therefore it is inconsistent with the employment of air force to develop fighters."[29]

During his five years at Maxwell, Chennault championed the cause of the pursuit, writing scathing comments regarding the misuse of air power during Army maneuvers. One such blast at one Lt. Col. Henry H. Arnold, then commander of March Field, California (and later commander of the U.S.A.A.F. in World War II), brought the response from Arnold, "Who is this damned fellow Chennault?" He also wrote a text on pursuit tactics, "The Role of the Defensive Pursuit." Although never officially accepted as a classroom textbook, it outlined the concept of an early-warning net that Chennault would put into practice with great success in China.[30]

Blunt and outspoken, Chennault made few friends at Maxwell, nor did he succeed in converting the bomber crowd. In early 1936, the Tactical School discontinued the pursuit tactics course and Capt. Chennault was made commander of a services squadron on the base. Two months later Chennault got back into the pursuit business briefly when he was assigned to Barksdale Field, Louisiana, as executive officer for the 20th Pursuit Group. In the fall of 1936, however, the bronchitis that had been bothering him for months gradually got worse, and he spent the winter in the hospital at Hot Springs, Georgia, recuperating.[31]

An Army Air Corps medical board met on February 25, 1937, to review Chennault's case. Chronic bronchitis (Chennault was an inveterate, three-packs-a-day chain smoker), low blood pressure and impaired hearing all counted against him. He was 43 years old, still a permanent captain (although he had been promoted to temporary major), and he was grounded – never again to fly the pursuits he loved so much. His was a "voice crying in the wilderness" so far as the official Air Corps was concerned. When the board recommended retirement, Chennault, tired of fighting the system and perhaps looking forward to better things in China, accepted. His retirement orders were issued in late March for an effective date of April 30, 1937.[32]

Chennault spent April preparing for his departure to China. He moved his family into a house on Lake St. John, near Waterproof, Louisiana, and arranged to have his retirement checks, along with an additional sum from his salary in China, sent to his wife, Nell, through the Chase Manhattan Bank in New York City. On May 1, the day after his official retirement, he kissed his wife and seven children goodbye and left for San Francisco, bound for China.

After a short layover in Hawaii, Chennault re-boarded the Dollar Lines' SS President Garfield and headed for Kobe, Japan. Here he was met by his old flying buddy, Billy McDonald, and took a several-day layover to survey the cities of Osaka, Kyoto, and Kobe, filling notebooks of potential military intelligence.[33]

One month to the day after leaving his home in Waterproof, Louisiana, Chennault stepped off the gangplank onto the docks at Shanghai. Chennault and McDonald were met by W.H. Donald, a ruddy-faced Australian and ex-journalist who served as confident and advisor to Chiang Kai-shek. Donald was to prove one of Chennault's most valuable friends in China.[34]

Chennault didn't meet the Generalissimo and Madame Chiang right away. First came a trip to Shien Chiao, a round of introductions, and a few welcoming parties. Finally, one sultry afternoon, Roy Holbrook picked Chennault up at his hotel and drove him to meet his new boss, temporarily quartered in the high-walled French Concession of Shanghai. As the slim and attractive Madame walked into the room where Chennault and Holbrook were waiting, Chennault was visibly surprised, as he had expected someone at least ten years older.

Soong Mei-ling (or westernized, Mei-ling Soong), one of three daughters of Charles Jones Soong, a wealthy and influential Chinese merchant who insisted on his daughters obtaining western educations, had married Chiang Kai-shek in 1923. Pretty and feminine, she could nevertheless be hard as nails and provided strong support for Chiang. Since Chiang knew but few words of English, she frequently acted as an interpreter for him. Having received her education at Wellesley, she spoke English fluently and with a soft, southern accent. She completely captivated Chennault, who recorded in his diary that night, that she "will hereafter be 'The Princess' to me." Madame Chiang quickly filled him in on what was required. She needed facts on the Chinese Air Force: number and condition of the aircraft, capability of the pilots, improvements needed – and she needed them quickly.[35]

Chennault was given two aircraft, Douglas O2MC-4s, for his survey. Starting at the Hangchow Flying School in late June, he selected Billy McDonald to fly the second Douglas. His mechanic from Maxwell Field, "Seebie" Smith, would fly with Chennault, and the assigned interpreter, Colonel P.Y. Hsu, flew in McDonald's back seat.[36]

Over the next several days, the four covered the airfields at Nanking, Canton, and Nanchang in the South, up through Hangkow in Central China, and into the north at Sian. Chennault was at first shocked, and then disgusted, at what he found. The Italian-run flying schools at Nanking and Loyang pinned pilot wings on every cadet who survived the course.

Chiang Kai-shek was happy; he was gaining hundreds of pilots (never mind how well-qualified) and the cadets and their wealthy families were happy; no one had to endure the "loss of face" that came with washing out.

The Italian assembly plant at Nanchang was turning out large quantities of the aging Fiat CR.30, which Chennault deemed "a fighter that proved to be a firetrap in combat..." The Savoia-Marchetti bombers that were produced there were so obsolete that they were good only as transports. Finally, as Chennault took inventory of Chiang's aircraft, he counted only 91 aircraft as being fit for combat – not quite 20 percent of the claimed 500 aircraft on the rolls. The Chinese Aero Commission, with the encouragement of their Italian advisors, never removed an aircraft from the official inventory, whether wrecked, cannibalized for parts, or so obsolete that it couldn't be used for combat.[37]

On paper the Chinese Air Force had an impressive number of aircraft; in actuality, they would be hard-pressed to defend themselves against a determined war lord, much less the progressing military might of the Empire of Japan. It was a dismal picture – an air force made up largely of poorly-trained pilots and obsolete, inferior aircraft. The only bright spot was the pool of American-trained pilots graduated by Jouett. Later, those who went through the school at Hangchow operated under the American system of flight training.

Chennault was inspecting the Italian school at Loyang when he learned of the Japanese attack at the Marco Polo Bridge. Japanese military units had been garrisoned in the Peking (Peiping) area ever since the Boxer Rebellion was put down in 1900. The Boxer Protocol of 1901 provided for foreign signatories to station troops between Peking (Peiping) and the sea. On the night of July 7, 1937, a Japanese company was conducting maneuvers at Lukouchiao, near the historic Marco Polo Bridge, fifteen miles southwest of Peiping, when the incident occurred. Chinese and Japanese versions of just what actually happened differ. Chinese accounts insist that the Japanese demanded to enter the nearly walled village of Wanping in search of a missing soldier. When refused by the Chinese garrison there, the Japanese opened fire on the city, occupying it at 4:30 a.m. on the 8th. However, historian John Toland, in his scholarly history of modern Japan, *The Rising Sun*, maintains that the Japanese troops under Gen. Gun Hashimoto were fired upon by the Chinese garrison as they were ending their field exercise.[38] Whichever account is correct, the incident soon became the spark that ignited the Sino-Japanese War.

Despite attempts by the local commanders to settle the incident, within six days Tokyo dispatched 20,000 troops to Peiping, reinforcements from both Manchuria and Japan. Chiang Kai-shek, at a conference at Kuling, proclaimed, "If we allow one more inch of our territory to be lost or sovereign rights to be encroached upon, then we shall be guilty of committing an unpardonable crime against the Chinese race... China's sovereign rights cannot be sacrificed, even at the expense of war, and once war has begun there is no looking back."[39]

Eight days later, on July 25, Chinese and Japanese troops clashed again, this time at the railroad station at Langfang, fifty miles south of Peiping. The skirmish soon turned into a major conflict. Japanese reinforcements poured into the area, the following morning Japanese planes bombed the Chinese barracks, and on the 26th Japanese troops occupied the town. The war that Chiang Kai-shek had been able to delay since 1931 was on.

Upon learning of the fighting that had broken out between the Chinese and Japanese troops, Chennault sensed that war was imminent and immediately wired Chiang Kai-shek, volunteering his services in any capacity. Two days later he received the Generalissimo's reply, "Your voluntary offer of services gratefully accepted. Proceed to Nanchang. Direct final combat training fighter groups there." [40]

William D. Pawley, head of the Central Aircraft Manufacturing Company (CAMCO), later played a central part in the hiring of Chennault by the Chinese.

The "Flying Trapeze" team. Left to right: Sgt. W.C. "Billy" McDonald, Capt. Claire Lee Chennault and Sgt. J.H. "Luke" Williamson. (via W.C. McDonald)

Above: Chiag Kai-Shek and Lt. Col. John Jouett on reviewing stand. Front row, left to right: Jouett, Chang Hsueh-liang and Chiang Kai-shek. (via Malcolm Rosholt)

Left: George E.A. Rhineberg replaced Col. Jouett in 1935. (via Malcolm Rosholt)

CHAPTER 2

Italians, Russians and the 14th International Squadron

When Chennault arrived at Nanchang, commanded by recently-promoted Brigadier General Mao Pang-tsu (anglicized to P.T. Mow, or Peter Mow), he found three fighter groups – one equipped with the Curtiss Hawk III biplane fighter, principally used as dive bombers, another group of Hawk II fighters and the third with two squadrons of Fiats and one squadron of Boeing P-26s.[1] Of the lot, only the Boeings showed any promise of success against the Japanese.

Chennault brought Billy McDonald with him to Nanchang. The two soon found they had their work cut out for them. Although some of the pilots were graduates of the Jouett school, and therefore very competent, the majority had passed through the farce of Italian training and were, in Chennault's words, "a menace to navigation." [2]

A few days after arriving at Nanchang, Chennault and Mao were summoned to Chiang Kai-shek's summer palace in Kuling for a conference with the Generalissimo. As the Chinese leader and his wife stepped out onto the sun porch of their one-story bungalow, Chennault had his first glimpse of the Generalissimo – the first opportunity since his survey of the Chinese Air Force to tell Chiang of his findings. The Generalissimo was a bit shorter than Chennault's height of five-feet, ten inches, with a shaved head and clipped mustache. A slight smile played on his lips. His pressed, khaki uniform bore no insignia of rank. Chennault was anxious to report the findings of his survey, but Chiang posed the first question to General Mao: "How many first-line fighters are ready to fight?" Mao shifted uneasily and replied: "Ninety-one, your Excellency." Immediately enraged upon hearing this surprising report, the Generalissimo stared hard at a trembling Mao and then turned to Chennault. As the Madame translated, Chennault confirmed the number, explaining how the Aero Commission's figure of 500 had been inflated with aircraft that had crashed, were totally obsolete, or otherwise unsuitable for combat. He went on to outline the details of the dismal proficiency of his pilots and the poor conditions of the air bases. In short, the Chinese Air Force wasn't ready for war.[3]

The Japanese General Staff anticipated that the Chinese campaign would take no longer than three months.[4] They would seize North China and force Chiang and the Nationalists into economic cooperation. After all, their real threat came from Russia, not China. With Manchuria already under Japanese control, the provinces of North China would provide an increased buffer to the colossus to the north.

City after city fell, as the Japanese forces pushed back the Nationalists' 29th Army. By late July, they were at the gates of Peiping. Chiang ordered the city's evacuation on July 28th, moving his capital to Nanking as the Japanese approached. On the 30th they took Tientsin and on August 8 they moved into Peiping as fighting spread to Shanghai.

Chennault was at the Nanking Military Academy for a war council meeting on Friday, August 13, when Madame Chiang asked his recommendations for defending Shanghai. He replied that dive and high-level bombing attacks on the Japanese warships at anchor off the harbor at Shanghai would certainly help.[5] The next thing he knew he was being asked to plan the attack. He and Billy McDonald worked far into the night, mission planning for an attack the following day on the Japanese heavy cruiser *Idzumo*, flagship of the Japanese fleet.

The *Idzumo* lay at anchor in the Whangpo River just opposite the Japanese consulate at the edge of the International Settlement. August 14 was a grey day with a low overcast. As the two Chinese Northrop "Gamma" light bombers approached their target area, they dove to get under the overcast. Trained to drop from an altitude of 7,500 feet, they were forced to descend to 1,500 feet. to make their bomb run. Picking up speed in their dive, they were also faster than their accustomed bomb release airspeed. In the excitement of the moment, the inexperienced crews forgot to adjust for the lower altitude and higher airspeed and all dropped their bombs far short of the *Idzumo*, hitting instead the British Sector of the International Settlement. Two 1,100-pound bombs hit on Nanking Road, the main thoroughfare, resulting in 950 people killed and 1,150 wounded.[6]

The first aerial attack on the Japanese launched by the Chinese Air Force had ended in a tragic fiasco. Later in the same day, another Chinese bomber, obviously aiming for the Japanese Sector, again dropped two bombs on the British Sector, this time killing or wounding over 3,000. Clearly, the inept Chinese Air Force bomber crews were doing more damage to Shanghai than the Japanese.

Throughout the latter part of August the Chinese held their own in the air. On August 14 (the same day that the Chinese Air Force made their disastrous attack on the *Idzumo*), three Chinese Hawk III squadrons of the 4th Fighter Wing attacked a force of Mitsubishi G3M2 "Nell" bombers en route to bomb Hangchow. Six Nells were shot down without a single Chinese loss. Two days later eighteen Japanese bombers returning from a bombing raid on Nanking were intercepted by Chinese Boeing Model 281s (P-26 exports) of the 17th Squadron. This time eight bombers fell, one each credited to Sun-Shui Wong and Pan-Yang Wong, both Chinese-Americans (the former born in Los Angeles and the latter in Seattle – both would achieve "ace" status before the A.V.G. was organized).[7] On August 20, six Nells attacked Nanking's Chuying airfield. Eight Boeing fighters rose to meet them and in less than five minutes shot down all six. During the five-day period, August 16th through the 20th, the Japanese made four raids with unescorted bombers – at the cost of 48 bombers.[8]

The Japanese fighters encountered by the Chinese at first were mostly the obsolete carrier-based Nakajima A2N1 and A4N1 biplanes, which the Curtiss Hawk IIIs and Boeing P-26s were able to match handily. It was quite another story when the Japanese 12th and 13th *Sentais* arrived, equipped with the Mitsubishi A5M2 Claude. Despite its fixed-gear and open-cockpit configuration, the little fighter was equipped with a 610-hp Kotoki radial engine, giving the light-weight fighter a maximum speed of 265 mph. With its elliptically-shaped wing (similar to the British Spitfire), it was highly maneuverable. On one day alone, the new Claudes shot down eleven of sixteen defending Chinese fighters.[9] Meanwhile, the war continued to expand.

As Chiang's German-trained 87th and 88th Divisions battled the Japanese to a standstill in and around Shanghai, the Japanese Air Force attacked sixty-two cities and towns in eight provinces throughout China.[10]

Following the attack on Shanghai, Chiang Kai-shek called upon the western democracies for aid. Unfortunately, the Western world was in a period of isolation and appeasement. The United States made a lame protest that it "considered Japan's behavior in China to be contrary to the Kellogg-Briand Pact of 1928 and the Nine-Power Treaty of 1922" – but there was no material aid offered to China.[11] To the contrary, there was constant pressure on those Americans who had gone to China as instructors and advisors – and even airline pilots – to leave the country. The League of Nations also condemned Japan for their aggression in China, but did nothing.[12] Britain did sell China a few aircraft – 38 Gloster Gladiator single-seat fighters – but withheld armament for them. Only one country agreed to come to the aid of China – Soviet Russia. On August 21, 1937, they signed a non-aggression pact with China which was followed by an economic agreement, whereby the Soviets granted China credits for $100,000,000, in return for Chinese tungsten, wool and tea.[13] A substantial military mission came with the deal – 500 Soviet Army "advisors" and five wings of Soviet aircraft, complete with aircrews, to bolster China's flagging air defense.

Certainly there was no love lost between Stalin and Chiang Kai-shek – Chiang had been fighting the Chinese Communists for ten years – but Russia clearly saw the advantage to helping China. So long as Japan was tied up in China, they wouldn't be likely to take an interest in extending their conquests north of Manchuria.[14]

Back at Nanking, Chennault continued to plan offensive combat missions, defensive maneuvers, and set up air raid warning net covering the Shanghai-Hangchow-Nanking triangle. He had first devised the system years earlier while he was at the Air Corps Tactical School.[15] Four years later the warning system would serve the American Volunteer Group to great advantage.

In addition to Chennault, Madame Chiang relied heavily on the services of four other American airmen in the early months of the war. Royal Leonard and Julius Barr, both of whom had flown commercially in China for several years and who had been personal pilots for the Generalissimo, were placed in charge of all Chinese Air Force bombardment squadrons, which consisted of approximately 40 light bombers. Billy McDonald, Chennault's long-time friend, took charge of observation aircraft and Sterling Tatum, another American pilot who had instructed at Shien Chiao, supervised fighter operations.[16]

In August, Madame Chiang purchased a Curtiss Hawk 75, the fixed-gear predecessor of the later P-36 of the U.S. Army Air Corps, for $55,000. The first bought by China, she presented it to Chennault for his personal use. A clean, all-metal monoplane with an enclosed cockpit, the Hawk 75 was powered by a 1,100-hp Wright R-1820 Cyclone radial engine and carried one .30-caliber machine gun in each wing. Chennault and Billy McDonald flew it for ten months, using it for numerous reconnaissance flights and to observe air battles with the Japanese. The aircraft was finally damaged beyond repair when a ham-handed American test pilot ground-looped it while he was taxiing out for a test hop.[17]

Chennault's use of the Hawk 75 has led to much speculation regarding Chennault's actual participation in air combat during his pre-A.V.G. days in China. Some sources claim that he shot down between 30 and 40 Japanese aircraft. No official Chinese records support such a claim, however, and, as late as 1987, retired Nationalist Chinese Air Force Gen. Lo Yuing-teh maintained that Chennault did not engage in combat. Even Chennault himself, in his autobiography *Way of a Fighter*, says nothing of even firing at a Japanese aircraft, much less shooting one down.

There are some clues, however, that might indicate otherwise. He wrote:

> *"...It was in the Hawk Special [Hawk 75] that I got my first taste of Jap flak and fighter tactics, and I learned some of the lessons that saved many an American pilot's life over China." He also mentions that the Hawk Special survived "ten months of combat and operational flying,"* and *"My years of air combat in China before the United States entered the war gave me an excellent opportunity to observe the combat pilots of the major Pacific powers."* And finally, *"...Many times I went aloft in the Hawk to umpire air battles over Nanking, Nanchang and Hankow and found that umpire baiting was as popular in China as it is in Brooklyn. Again, the baiters paid a price."*[18]

Certainly, an argument can be made for either position. On the one hand, one wonders how Chennault could have gained the knowledge of Japanese tactics that he displayed later in his "chalk talks" in front of the A.V.G. ground school classes at Toungoo, unless he had personally engaged in combat with the Japanese. On the other hand, any early victories by an American in the Sino-Japanese War (much less the 30-to-40 claimed) would not have long gone unnoticed by inquisitive journalists. The fact remains that the Nationalist Chinese to this day deny that Chennault engaged in combat and certainly there has been no documented evidence brought forward to the contrary.

By mid-October China had only a dozen combat capable fighters left, and many of the experienced pilots graduated by the Jouett school were dead. At this time, the Italians pulled out of China, due in part to Chiang Kai-shek's displeasure over Mussolini's urging that Chiang accept Japanese surrender terms, as well as the fact that Italy had their hands full with the invasion of Abbysinia.[19] By October 19, 1937, Japan enjoyed complete air supremacy over Shanghai, Nanking, and Hangchow.

The battle for Shanghai lasted three months to the day – from August 8 to November 8. The Japanese had not anticipated such fierce opposition; indeed, they had anticipated the entire war would be over within ninety days. The end for Shanghai came when the Japanese, landing at Hangchow Bay, out-flanked the Chinese troops south of Shanghai and advanced swiftly toward Nanking. On December 7, two days before the Japanese arrived, the Generalissimo and Madame Chiang left Nanking and set up a temporary capital at Hankow.[20]

What followed became known throughout the world as the "Rape of Nanking." As General Matsui and his army entered Nanking they embarked on an absolute orgy of slaughter, rape and pillage. Over a two-month period, more than 200,000 residents of Nanking – men, women, and children – were indiscriminately massacred. The wholesale atrocities committed by Matsui's troops brought worldwide condemnation on the Japanese and resulted in the General Matsui's recall to Japan.[21]

The Russians began arriving in China in late September 1937, just as the Chinese Air Force had virtually ceased to exist. The "advisory force" brought with them four fighter squadrons equipped with the Polikarpov I-152 biplane (two squadrons would be replaced with the I-16 monoplane fighter in early January) and two bomber squadrons flying the Tupolev SB-2 light bomber. All aircraft carried the twelve-pointed star insignia of the Nationalist Chinese Air Force.[22]

Commanded by a General Asanov, the Russians made their base of operations at Lanchow in northern Kansu Province. From Lanchow, they dispersed to air fields at Nanchang, Sian, and Hankow.[23] They also set up training schools at Lanchow, Nanchang, and Hami.[24] By the end of the year they had established an assembly plant at Lanchow, which was eventually capable of turning out ten I-16s or three Russian bombers per day, opened a supply route from the Soviet Union into China, and re-equipped the Chinese Air Force with some 220 aircraft, the first deliveries of a 400-plane contingent.[25]

Concurrently, Madame Chiang resurrected the idea of an international volunteer squadron to augment her rapidly-diminishing pool of experienced pilots. Dr. H.H. Kung, China's Minister of Finance, had suggested it as far back as 1931 following the Japanese attack on Shanghai, but Jouett had nixed the idea as being detrimental to the expansion of the Chinese Air Force. When Madame Chiang brought it up to Chennault, he was no more enthusiastic over the proposal than Jouett had been.[26]

At that point, however, CAMCO head Bill Pawley came forward with an offer of a group of Vultee V-11 bombers – a slow, old bird, but one that boasted a 2,000-mile range. Chennault, anxious for a long-range bomber force, finally bought the idea. Julius Barr and Royal Leonard, who had been in charge of bomber operations for Chiang, were given the task of selecting the best available from the hodge-podge of volunteers who were cropping up in response to rumors that an international squadron was being formed. Although a few had been trained by the military, most had only civilian pilot rating, and some not even that.[27]

In his autobiography *Way of a Fighter*, Chennault stated:

> *"We hired a weird collection of four Frenchmen, a Dutchmen, three Americans, and a German and filled out the squadron with six of the best Chinese bomber pilots."*[28]

There were considerably more than the nine foreigners indicated by Chennault, however. In his well-documented, *Flight in the China Air Space, 1910-1950,* Malcom Roshalt lists five Europeans – Frenchmen A. Boulingre, William La Bussiere, Marcel Floueir, W. La Roce, and a Dutch, Jan Rouffaert. The several Americans included Vincent Schmidt, a World War I and Spanish civil war veteran from Mineola, New York; Elwyn H. Gibbon, an ex-Army Air Corps pilot; George H. Weigle; and Lyman Woelpel and James W.M. Allison, two commercial pilots from Seattle. Allison, a professional soldier-of-fortune, had flown against the Germans and Italians in Spain. There was no mention of a German, although Chennault may have considered Schmidt, whom he designated as the squadron commander, to be German.[29] Other names listed as passing the flight check by Julius Barr in the Douglas DC-2 included R.G. Whitehead and Harold Welch, although no further mention is made of them. Sixteen Chinese air gunners and mechanics rounded out the squadron.[30]

Officially formed in October 1937, the group was given the designation "14th Volunteer Bombardment Squadron" and equipped with a combination of the Vultee V-11s provided by Pawley, Martin 139s (export version of the U.S.A.A.C. B-10), and Northrup Gammas. Operating out of Hankow, they flew mainly harassment missions against the Japanese in northern China, hitting rail centers and knocking out a few bridges. The entire squadron was often grounded for weather, however, and was severely hampered by lack of spare parts.

Sometime during the winter of 1937, the 14th Squadron pilots were issued the Chinese prototype of what became known as the "blood chit" by successive American pilots flying for China in the American Volunteer Group and later in the Fourteenth Air Force.

Initially called a "rescue patch," the approximately 8"x11" piece of cotton-backed, white silk cloth bore the Nationalist Chinese flag at the top, with five vertical rows of Chinese characters accompanied by an official government "chop," roughly translated as, "This foreign person has come to China to help in the war effort. Soldiers and civilians, one and all, should rescue, protect, and provide him with medical care." Worn on the front of flight suits by the 14th Squadron pilots, it was usually sewn on the back or inside of the leather flight jackets of the A.V.G.[31]

As more and more Soviet equipment poured into China, the Chinese Air Force began to take shape once again. By January 1938, Chinese pilots had completed intensive training on the Soviet aircraft and had formed new squadrons – seven fighter squadrons of eight I-15s or I-16s each, and five bomber squadrons, each equipped with six SB-2s. Although the fighter squadrons were manned primarily with Chinese pilots, Soviet "volunteer" pilots and bombardiers were sent to man the bombers. By mid-February, their training requirements completed, the Russian and Chinese squadrons moved forward to Lanchow and Nanchang, ready to take on the Japanese.[32]

Unlike the ever-cautious American government – who discouraged American involvement in even the instruction of combat flyers in China (let alone flying combat missions!) – the Russian government sent their pilots and aircrews to China with the expectation that they would be flying combat. They were rotated every four months, which, over the three-year period that the Soviets were in China, provided combat experience to a large percentage of the Soviet Air Force.[33]

The Russians, led by Feodor Polynin, launched their first raid on the Japanese on February 23, 1938. Flying 28 SB-2s, they attacked the Japanese-held airfield at Taipei, Taiwan, with complete surprise, destroying forty parked aircraft.[34]

Meanwhile, despite his misgivings, Chennault was still trying to make some use of the 14th Volunteer Bombardment Squadron, sending them in their Vultee V-11s to Pengpu on their first bombing raid on February 7, 1938. Never more than lukewarm about the idea of an international squadron, Chennault didn't find much to change his opinion, as he later wrote:

> *"The big troubles of the International Squadron were not in the air but on the ground - especially in the unsavory thoroughfare of Hankow called Dump Street and inhabited chiefly by barkeepers, opium peddlers, and prostitutes. The International aces established their field headquarters on Dump Street and subsisted almost entirely on high-octane beverages. Well lubricated, they would tell any and all who would listen all they knew about Chinese Air Force operations and a great many things they didn't know."*[35]

Chennault's fears proved correct when, in mid-March 1938, he planned a mission against a large troop-movement center at Tsinan in the Shantung Province. The Volunteer Squadron had their aircraft loaded with bombs and gassed-up the afternoon before the planned early morning mission on March 15 and then departed for their favorite haunts on Dump Street. The Japanese agents in the bars and brothels duly passed the word, and just at dusk Japanese bombers appeared over the field and laid their bombs across the flight line where were parked all of the mission aircraft – wingtip to wingtip. The bombers went up in flames, one exploding bomber igniting another.[36]

Six days, later, Vincent Schmidt and six of his pilots arranged a meeting with Madame Chiang to demand a pay raise. Chennault was standing beside the Madame's desk as the group filed in. Schmidt was rude and demanding, charging that, "You want us to go on missions your Chinese pilots cannot accomplish because of their incompetence and their lack of courage." That was the proverbial straw that broke the camel's back. The next day, March 22, 1938, the 14th Volunteer Bombardment Squadron, Chinese Air Force, was disbanded.[37] Chennault ordered the remaining Vultees and the two Martins to Chengtu to be used as trainers. Gibbon and Allison were retained as instructors at Chengtu; the remainder were paid off through April and released.

As the Russians installed themselves more firmly in China, Chennault's influence began to wane, and he was relegated to primarily a training role. He did, however, participate in some joint activities with the Russians, one of the most notable being the "Emperor's Birthday Celebration." Following the fall of Nanking, Hankow became the main target for the Japanese air attacks. Anticipating that the Japanese would welcome an opportunity to deliver a decisive blow to that city on the birthday of Emperor Hirohito, Chennault met with the Russian and Chinese commanders to plan a clever ruse. On April 28, the day before the Emperor's birthday, Chennault directed that all the Chinese and Russian fighters take off from the Hankow airfield, circle the city, and set a course for Nanchang. As soon as they saw the fighters depart, Japanese intelligence agents in Hankow reported that the city was defenseless. What they didn't see was that two hours later, just at sundown, the fighters returned at treetop level to the Hankow airfield.[38]

Chennault had thirty-five I-16s of the Russian squadrons airborne the next day, patrolling thirty miles east of

Hankow on the expected route of the Japanese coming out of Wuhu. Nine I-15s of the Chinese Air Force 4th Group orbited over Tsingshan. The twenty-one remaining I-15s and I-16s of the 3rd and 14th Groups remained on the ground, standing by on alert.[39]

At 1310 the first call from the warning net came in – 15 bombers flying at 12,000 feet, escorted by 24 pursuit planes at 16,000 feet, had just passed over Lukiang on a heading toward Hankow.[40] Twenty-one Chinese pilots rushed for their aircraft and were soon airborne, miraculously avoiding collisions during the scramble. The bombers managed to drop their loads on Hengyang, one of the three cities in the Wuhan area, and on the airfield south of Wuchang before the Chinese fighters attacked. Although two bombers fell, they turned for home with the formation relatively intact.[41]

Diving out of the sun, the Russian fighters caught the retreating Japanese completely by surprise. The first attack separated the fighters from the bombers. While one group worked over the bombers, the other chased down the fighters, already low on gas and unable to turn and fight. Elements of the Chinese 4th Group also managed to join the Russians as they decimated the Japanese formation. The official report at the end of the day claimed eleven bombers and ten fighters downed by the combined Chinese/Russian force.[42] American intelligence sources later estimated only three of the original thirty-nine Japanese aircraft returned to Wuhu.[43]

The Chinese lost four pilots and nine planes, the Russians two planes and no pilots. It was the greatest single air victory recorded by the Chinese. Undaunted, the Japanese radio announced later that fifty-two Chinese aircraft had been shot down over Hankow in celebration of the Emperor's birthday! [44]

The largely Russian victory brought an almost immediate protest from Mamoru Shigemitsu, Japanese ambassador to Moscow. He charged that the Soviets had supplied the Chinese with 500 planes and 200 "advisory" personnel since August, 1937, and cited the Russian pilots found in the aircraft downed by the Japanese. Shigemitsu finished by warning Soviet Foreign Minister Maxim Litvinov that Tokyo would hold the Soviet Union responsible for any confrontation which might arise between the two countries as a result of Soviet interference in the war in China.[45]

As could be anticipated, the warning fell on deaf ears. Aid continued to flow and Russian airmen continued to mount impressive air victories over the Japanese. Meanwhile the Japanese juggernaut continued to roll. Following the fall of Nanking, the main Japanese forces had moved northward along the Tientsin-Pukow Railway while other elements from North China moved southward to take the vital communication center at Hsuchow in late March. A stubborn defense mounted by the Chinese thwarted their advance at Taierhchuang, near Hsuchow in southern Shantung, and the Japanese sustained 30,000 casualties. Superior numbers of Japanese troops finally prevailed, however, and Hsuchow fell on May 19. The Japanese also took Amoy and Soochow in May and Kaifeng and Anking in June.[46]

While Soviet pilots were being rotated every few months, constantly bringing in new blood and replacements for those lost in battle, the Chinese were not so fortunate. By mid-year, 1938, most of their experienced pilots were either dead or in the hospital. However, the Chinese did manage to launch an ambitious propaganda mission to Japan on May 19. Flying two Martin 139s loaded with leaflets to "call up the self-consciousness of the Japanese people," the Hankow-based bombers staged out of Lo-She Airfield at Ningpo, and departed for Nagasaki just before midnight. Three hours later they were 10,000 feet over Nagasaki. After they dropped leaflets on that city, the two bombers separated, each proceeding to different areas and dropping leaflets all across the island of Kyushu before they turned for home. After nearly twelve hours in the air, they landed back at Hankow.[47] It had been a clear demonstration of China's ability to penetrate Japanese air space – although the Martin 139s would never have been able to fly that far with a full bomb load.

The last major air battle that the Chinese Air Force participated in was on May 31, when eighteen I-15s and I-16s of the 3rd and 4th Groups joined twenty-one Russian fighters as they rose to attack thirty-nine Japanese bombers over Tsinshan. The defenders successfully turned back the bombers, knocking down nine for a loss of two of their own.[48]

During the summer of 1938, the remnants of the Chinese Air Force pulled back into the interior of China, leaving the conduct of the air war to the Soviets, who would remain in China until April 1941, when Russia signed a non-aggression pact with Japan. Chennault was ordered to Kunming, in the province of Yunnan, to set up another training program in the hope of rebuilding Chiang's air force.[49]

Going from chief air advisor to "chief instructor" didn't set well with Chennault, and his initial reaction was to write to Washington requesting return to active duty in the U.S. Army Air Corps. His application was promptly rejected, however, with the comment that there were "no funds available for the return of retired officers to active duty"[50] – at least no funds to recall retired Captain Chennault who had been such a burr under the saddle of Air Corps officialdom. So Chennault settled down to the task, which he frankly viewed with a great deal of pessimism.

Chennault brought with him to Kunming a group of Air Corps reserve officers who would serve as his instructor staff. There was C. Briant "Skip" Adair, who would later help Chennault with recruiting for the American Volunteer Group; Boatner Carney, who remained with Chennault until after the A.V.G. was formed; Billy McDonald, who had been with Chennault for most of the past eight years; Frank Higgs, who was the prototype for artist Milton Caniff's "Dude Hennick" in the early *Terry and the Pirates* comic strip; Jim Bledsoe, who would later command a P-38 squadron under Chennault in the Fourteenth Air Force; Willie Heston, Jr., who had been a star football player at the University of Michigan as well as Johnny Preston, Emil Scott, Harold Johnson, Bill Cherymisin, and Harold Mull.[51]

Using the same techniques that Jouett had used seven years earlier, Chennault began giving re-qualifying check flights to the Chinese pilots. Recently equipped with Curtiss Hawk 75s, the Chinese had wrecked two-thirds in accidents in the first month. Chiang Kai-shek ordered the Hawk pilots to fly the remaining aircraft to Kunming for refresher training. As the thirteen aircraft came in to land, seven cracked up while Chennault watched in amazement. He immediately ordered each Hawk pilot to undergo a check flight in a basic trainer. General Chow, commander of the Kunming training command, protested to Chiang, only to be told that Chennault was acting under the authority of the Generalissimo. Chennault washed out half of them and sent the remainder to re-qualify in basic trainers.[52]

Chennault, Madame Chiang Kai-shek and the Generalissimo in pre-A.V.G. days. (via Chuck Baisden)

The Curtiss Hawk 75 purchased by Madame Chiang Kai-shek for Chennault's use in reconnaissance flights and observing Japanese air battles. (via Wikipedia)

Coming to the aid of China in 1937, the Russian contingent brought four squadrons of Polikarpov I-15 biplane fighters.

Over six weeks, starting on 13th December 1937, in what became known as "The Rape of Nanking," soldiers of the Imperial japanese Army slaughtered over 200,000 Chinese citizens of Nanking, not even sparing small children.

Above: George H. Weigle and Elwyn H. Gibbon in hankow, 1938. Weigle is wearing the early "blood chit" issued to the 14th Bombardment Squadron during the winter of 1937-38. (via San Diego Aerospace Museum)

Right: French volunteer Andre Boulingre, probably taken at Hankow, China, in 1938. (via Malcolm Rosholt)

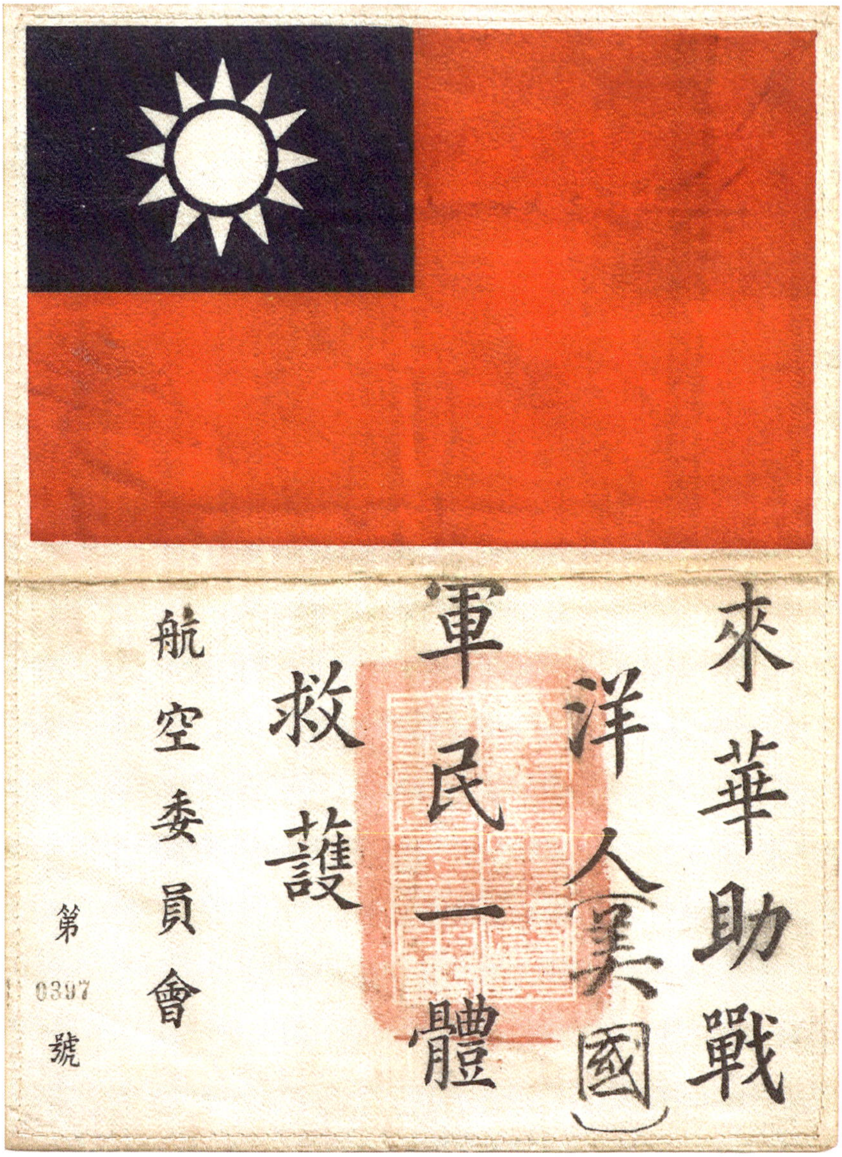

The "rescue flag," later to become known as the "blood chit," issued to members of American Volunteer Group – Flying Tigers. (Courtesy R.E. Baldwin)

Vultee V-II, as provided to Nationalist China's air force through the Central Aircraft Manufacturing Company (CAMCO), and subsequently used by the 14th International Bombardment Squadron.

CHAPTER 3

Chiang Gains American Support

As the southwestward advance of the Japanese continued, millions of Chinese refugees clogged the roads in one of the largest mass migrations in history. On October 21, 1938, after successive air raids on the coastal seaport, Canton fell to the Japanese. Four days later they moved into the provincial capital of Hankow, forcing Chiang to move his seat of government once more, this time far into the interior of China to Chungking, in Szechwan Province.[1]

The loss of Canton underscored the importance of China developing an alternate supply route. Discussions regarding a "back door" to China had been going on for months, but now, as Japan was threatening to cut off China's seacoast, it became imperative. Indeed, in November 1937 Chiang had ordered General Lung-yun, Governor of Yunnan Province, to begin construction of what would soon become known as the "Burma Road." It would stretch from Kunming, China, to Lashio in Burma, where it would connect with the Rangoon-Lashio rail line. Although a straight-line distance of only 320 miles, a continuous series of hairpin curves winding through mountains, around precious rice fields and sacred burial grounds, increased the road distance over two-fold, to 717 miles. The Chinese would build all but 117 miles of the route, to Wanting on the China-Burma border, a feat completed in one year by thousands of Chinese peasants using primitive tools and their bare hands, and which had not been seen since construction of China's Great Wall. The British agreed to construct the remainder from Wanting into Lashio.[2]

The capture of Hankow marked the end of what may be considered as the first phase of the undeclared Sino-Japanese War. Chiang had chosen the only strategy that seemed feasible at the time; as Tsar Alexander I had done when faced with Napoleon's invasion of Russia, he emphasized the strategy of "strategic retreat," drawing the enemy further and further into China's vast interior. Avoiding direct confrontation, he was able to preserve his forces intact while exhausting the Japanese and stretching their supply lines to the breaking point.[3]

By the end of 1938, the Japanese could claim control of 1,500,000 square kilometers of territory with a population of 170,000,000.[4] At this point, however, the Japanese forces stalled out and a stalemate ensued. While the Japanese occupied most of the cities and lines of communication throughout the eastern half of China, the Chinese pursued a hit-and-run strategy of guerilla warfare. The Japanese held the cities, but the Chinese held the rural areas, a pattern that would be repeated in Vietnam and Afghanistan years later. The second phase of the war, from December 1938 to December 1941, would be largely one of attrition.[5]

As the Japanese ground campaign ground to a halt, their air activity picked up. Just as Hankow had been a major target before, the new capital of Chungking now commanded Japanese attention. With the Japanese using Hankow as a base of operations, Chungking was within easy reach of their bombers.

The first raid of the New Year came on January 15, when twenty-seven Japanese bombers flew over the city. Chungking in the winter is often shrouded in dense fog, as it was on this particular day. Unable to identify a clear aiming point, the bombers dropped their bombs harmlessly twenty miles away.[6]

As spring cleared away the winter mists, the Japanese intensified their bombing campaign. The coastal city of Foochow suffered at east fifty raids by Formosa-based bombers, while the Japanese training school at Canton used the cities of Liuchow, Nunning, Kweilin, Kienow and Kanchow as practice targets for their student aircrews. Kunming also proved to be a popular target as Chennault related:

"The Japanese had too keen an appreciation of air power to allow the Chinese to hatch a new air force unmolested. During good weather the Japanese bombers were over Kunming almost every day. We had only training planes on the Yunnan fields and a few short-range anti-aircraft guns at Kunming. Most of our flight training had to be done in the early morning or late afternoon when the probability of air raids was smallest. Enemy bombs killed Chinese cadets in their barracks, demolished my house in Kunming, and slaughtered thousands of civilians in the city. A near miss splattered my office with shrapnel and tile fragments." [7]

Japanese bombers returned to Kunming with a vengeance on May 3. Twenty-seven bombers flying in "V" formation dropped their loads. This time the village was clearly visible and the bombs fell with devastating effect. An estimated 10,000 Chinese died in the raid. The next day the Japanese returned to repeat their performance.[8]

By mid-1939 the Russians moved many of their units further into the interior to airfields around Chengtu. There they joined Chinese squadrons equipped with I-15s and I-16s, bringing with them two Soviet fighter squadrons and one bomber squadron of twelve SB-2s.

Then, on August 23, 1939, in a move that shook the governments of both China and Japan, Stalin signed a non-aggression pact with Adolf Hitler. Nine days later, German troops invaded Poland, signaling the start of World War II in Europe. Surprisingly, this event didn't affect the level of support to China. To the contrary, the Soviet government urged Chiang to accept an increased number of Soviet advisers. This was viewed by alarm to Chennault, who claimed the Russians were trying to assume control over all the advisory and training functions and whose aim was to "get the American out." American diplomats were similarly concerned and so advised the State Department.[10]

Russian aid to China continued at a steady level into the early months of 1940. The flow of materiel slowed somewhat during the Russo-Finnish War in 1940, but Russian flyers continued to provide spirited defense for Chungking, Chengtu and Kweilin. According to Japanese reports, forty Soviet fighters were encountered over Kweilin. The Japanese claimed to have shot down fourteen and destroyed another nine on the ground.[11] As newer Japanese equipment came on the scene, the Russians were finding that their older I-15s and I-16s were rapidly being outclassed.

Discouraged by the lack of advances on the ground, the Japanese General Staff, during secret meetings in early April 1940, declared that, unless they could achieve total victory within a year, they would gradually withdraw Japanese forces from China, leaving only a contingent of troops in northern China as a buffer against the Soviets.[12]

Their pessimism was whisked away six weeks later, when, on May 10, Hitler launched his blitzkrieg against France, Belgium and Holland. Holland surrendered on the 14th and France fell on June 21. Encouraged by Germany's lightning victories, the Japanese military leaders began setting their sights for the oil and other natural resources of Southeast Asia. Support for the long-discussed Tripartite Pact with Germany and Italy gained momentum and, on September 27, the agreement was signed.[13]

Reaction by the United States was swift. Branding the three as "gangster nations," the U.S. added scrap metal of every kind to the list of embargo materials announced in July. Russia denounced the treaty as "further aggravation of the war and an expansion of its realm." [14]

Chungking continued to bear the full brunt of Japanese air attacks. Indeed, by the end of 1940, the city shared with London the questionable distinction of being the most- bombed cities in the world. The bombing of Chungking was clearly a reprisal by Japan for China's intransigence. It was by no means a strategic target. There was no railway station, no airfield, and no heavy industry – merely cottage industries and small factories.[15] But it was Chiang Kai-shek's capital.

During 1940, the Japanese stepped up their attacks on Chungking eight-fold – 168 daylight and 14 night bombing raids. During the summer, the raids intensified. Between ninety and one-hundred bombers flew over virtually unopposed each day, pounding the city unmercifully. The frequency of the raids increased until they were coming an average of once every five hours.

As the Japanese Navy's new Mitsubishi A6M2 (the vaunted Zero) began to appear in the early Fall, they swept away what few Russian I-15s, I-16s and Hawk IIIs that were left to defend the capital. The new Zero claimed 101 Chinese aircraft destroyed from September 12 to December 10, 1940.[16] To demonstrate their complete mastery of the air, the

Japanese fighter pilots actually landed their planes on the Chengtu aerodrome, jumped out of their idling machines and set fire to the remaining Russian fighters and training planes camouflaged under rice straw. While a covering flight circled above, the marauding pilots climbed back into their cockpits and took off, completely unscathed. Bomber pilots would make repeated low passes over target cities in eastern China, their bomb bay doors open, before actually dropping their bombs – to impress upon the Chinese their utter helplessness.

The raids on Chungking led to the development of a highly efficient air raid warning system. Using Chennault's system of a net of radio and telephones between the Japanese bases and Chungking, the Chinese would receive word as soon as the Japanese aircraft took off. A tall pole and hollow paper balls were used as an air raid warning. As word was received that the enemy bombers were airborne, one ball was hoisted to the top of the pole, signaling the coming air raid, or *jing bao*. When the bombers were half way to the city, a second ball was raised, which meant that the people must hurry to the shelters throughout the city. A few minutes before the engine noise was heard, the third ball went up. By that time, everyone was to be in a shelter. In tribute to the construction of the air raid shelters and the efficiency of the air raid warning system, only five people died during the raids on Chungking in 1941, in contrast to the 10,000 people killed in the raids on May 3-4, 1939.[17]

Japan moved quickly to take advantage of the German invasion into Western Europe. Not long after France capitulated on June 21, the Japanese demanded that France close the railway from Hanoi to China. As Britain, beaten at Dunkirk, retreated to the home islands, Japan threatened war with Britain unless they immediately closed the Hong Kong border, as well as the Burma Road. Backed against the wall – for Churchill could get no assurance from Roosevelt that the United States would involve themselves in war at that time for Britain – on July 12 Britain agreed to close the Burma Road for three months to give China the opportunity to come to terms with Japan. China was now completely isolated, save for the Silk Route supply line coming down from Russia.[18]

Although the Russians continued to supply China with war materiel, trucking it over 3,000 miles from Alma Ata to destinations in the Szechwon Province, aircraft deliveries had slowed markedly. By October 1940, only eighty Soviet aircraft had been brought in for the entire year – not even enough to replace those lost in combat. By that time, also, the Soviet "volunteers" had been recalled to Russia, influenced in no small measure by the expanding war in Europe.[19] The British had just staved off a German invasion in the Battle of Britain in September, and Rumania had joined the Axis.

As Soviet aid began to diminish, Chiang Kai-shek increased his efforts to obtain American assistance. Chungking had submitted a plan to Washington earlier in the year, seeking to increase the size of the Chinese Air Force to 500 planes. The plan was flatly rejected by the War Department. But with Soviet support fading, and with British aid completely out of the question, America stood, literally, as China's last hope. Chiang turned once again to his brother-in-law, T.V. Soong. Soong, at this particular time the Chairman of the Board of Directors of the Bank of China, was dispatched to Washington in late June 1940 to serve as China's Foreign Minister to the United States and to act as a lobbyist on China's behalf.[20]

Chiang had made an excellent choice. Harvard educated, Soong moved easily in Washington circles and proved to be a polished diplomat. Within two weeks of this arrival in Washington on June 26, he had arranged a luncheon with Treasury Secretary Henry Morgenthau. On July 12, he sent Morgenthau a note suggesting that the United States loan the Nationalists $370 million – $50 million to support the grossly inflated Chinese yuan, $250 million to improve the Burma Road, China's sole remaining supply link to the sea, and $70 million for military aid. As part of the military aid package, Soong asked for 300 pursuit planes and 100 light bombers with spare parts – a compromise reduction of their earlier request for a 500-plane force.[21]

Throughout the summer, Soong cultivated the influential of Washington political society – presidential aide Lauchlin Currie; Thomas "Tommy the Cork" Corcoran, close friend and confidant to President Roosevelt; and newspaper columnist Joseph Alsop, who happened to be a distant cousin to both the president and Mrs. Roosevelt. Soong worked with Arthur Young, an American on Chiang's payroll as his economic adviser, to refine China's shopping list, preparing tables of equipment for the air force they hoped the United States would supply. Negotiations went slowly, despite the obvious sympathetic support that Soong had been able to elicit for China's desperate situation. Feeling that he had reached an impasse, in late September Soong cabled the Generalissimo that, "It would assist in convincing authorities here if programs transmitted were supported by Colonel Chennault." [22]

Acting on T.V. Soong's suggestion, on October 12 Chiang Kai-shek summoned Chennault to the provisional capital of Chungking. Although he had no idea of what the Generalissimo wanted to talk with him about, the summons had said "urgent," so Chennault wasted no time in boarding the Beechcraft D-18 that Chiang had placed at his disposal and headed for Chungking. Chennault had just landed at the airstrip outside Chungking when, as if to underscore the desperate situation, the air raid sounded. Chennault was forced to take off to avoid the falling bombs and landed at a field 150 miles away, where he remained until the raid was over.[23]

Hours later he was able to return to Chungking. Smoke hung low over the city as he made his way through streets piled high with the debris of Japanese bombing attacks. When he finally arrived at the Generalissimo's quarters, Chiang and the Madame were waiting. It was obvious that the Chinese leader was agitated and desperately worried. Speaking through his wife, Chiang emphasized China's deteriorating hope and told of how the constant, large unopposed Japanese bombings were wearing down the will of the Chinese people. He feared that surrender was not far away. He then asked Chennault what he thought of the idea of buying modern American aircraft and hiring American pilots to fly for the Chinese. Chennault didn't think much of it, obviously recalling his experience with the International Squadron. He pointed out that American aircraft factories were even then straining to supply Great Britain in their battle with Germany and to meet the expanding needs of the American armed forces. Then there was the problem of getting State Department approval for those who would volunteer for such a scheme. Chiang listened impassively to Chennault's arguments and then concluded their meeting.[24]

The concept of American volunteers flying for China wasn't new. Dr. H.H. Kung, Chinese Minister of Finance at the time, was probably the first to suggest it, when he brought the subject up to Col. Jack Jouett in 1931. Later came Madame Chiang's insistence on recruiting an international squadron that led to the abortive experiment of the 14th Volunteer Bombardment Squadron in 1938. Bill Pawley also got into the act, maintaining years later that he originated the idea as early as 1937 after observing Japanese bombing attacks on Chinese cities, and broaching the subject to Chinese officials at the time.[25] A letter Chennault wrote in late 1938 to Elwyn Gibbon lends some credence to his claim. Gibbon, a former member of the 14th Bombardment Squadron, had written Chennault for another job. Chennault responded:

> *"Dear Gibbon:*
>
> *I don't think there's a chance... for reasons with which you are acquainted. If I took you on for the school even there would be a storm of opposition from Lauchow to Kunming – via Chengtu. However there is a project under discussion which might afford you a chance later. I can't divulge any of the details by letter and am not sure that the thing will materialize. If it does I assure you that you will like it and that I'll count you in the outfit. Don't hold too much hope of it, however, we ought to know definitely within two months."* [26]

A few weeks later Gibbon received a Christmas card from Chennault with a note that read:

> *"Write W.D. Pawley, Intercontinent Corporation, Hong Kong, for job as a pilot in a special squadron. Also, write Intercontinent Corporation, New York City, the same proposition. C.L.C."*

Although no further information has surfaced, it certainly points to tentative plans for another volunteer squadron and connects William D. Pawley to these early plans.

Still later, U.S. Navy reserve commander Bruce S. Leighton, vice-president of CAMCO's parent Intercontinent Corporation, during a meeting on January 17, 1940, with Chief of Naval Operations Admiral Harold Stark (probably speaking for Bill Pawley), suggested that a force of 100 pursuits, 100 bombers, and 10 transports be provided to China to be flown by 50 American pilots operating with a Chinese staff. The pilots could be augmented by U.S. Army and Navy fliers if government approval could be secured. Looking at the venture as a business proposition, Leighton went on to say that the Intercontinent Corporation should handle the whole job under commercial contract with China and "without any direct participation by the United States government."[27]

On October 21 Chiang Kai-shek called Chennault back to Chungking. There was little discussion this time. Chiang merely handed him a set of orders directing him to report to T.V. Soong in Washington. His brief instructions to Chennault: "Do what you can to get American planes and pilots." [28] Unbeknownst to Chennault, Chiang Kai-shek had met with the U.S. Ambassador to China, Nelson T. Johnson, just three days earlier to enlist his support for American aid. China's situation was clearly desperate – without American aid China might soon collapse. The war had devastated the Chinese economy; it was imperative that China be given a stabilization loan. Chiang added that 500 to 1000 aircraft were also needed, with American pilots, to defend the Burma Road and key Chinese cities. Johnson forwarded Chiang's message to Roosevelt, along with his own strong endorsement. Support for China was growing.

Accompanied by Major General Mao Pang-tsu, Director of Operations of the Chinese Air Force, Chennault left on October 22 for Hong Kong. After a two-day layover there, they boarded a Pan American Clipper bound for the United States, the two arrived in Washington on November 1 and reported to Dr. Soong. Soong wasted no time in introducing Chennault to the influential around Washington. On the evening of his arrival, Chennault met two journalists who had been following the Sino-Japanese War with great interest – Edgar A. Mower, of the *Chicago Daily News*, and Joseph W. Alsop, Jr., of the *New York Herald Tribune*, who would later join Chennault's staff in China. Both agreed with Chennault's original estimate what first-rate fighters would be hard to come by.[29]

Chennault set up shop in a red brick building at 1601 "V" Street, Northwest, later to become the headquarters of China Defense Supplies, Inc. There he began formulating his plans for an American Volunteer group. Working with Soong, Mao and Young, often until late at night in Young's apartment, Chennault hammered out an initial proposal which Soong submitted to the White House on November 25. More grandiose even than Soong's own proposal of July 12, it called for 100 of the new Curtiss P-40s and 250 radial-engine Navy fighters, 150 North American basic trainers and 10 Douglas DC-3 transports – along with 350 American pilots and crews to fly them. Also included was a twenty-percent spares stockpile, material for 14 major airfields and 122 additional landing strips, and ammunition and bombs for a year's combat operation.[30]

President Roosevelt received the "aviation program," as Arthur Young described it, as the 1940 presidential election was drawing to a close. FDR had been elected to an unprecedented third term, and could be assured of the popular support of the people and of Congress. He had had a life-long interest in China through his family – he often told of how the Roosevelt fortune was built on the Shanghai opium trade. He had known both T.V. Soong and his brother-in-law, H. H. Kung, for years, and proudly displayed a small wooden model of a Chinese junk on his desk, a personal memento of their friendship.[31]

But Roosevelt had other, more practical reasons for leaning toward China. At the present time, 750,000 Japanese troops were tied down in the Sino-Japanese war. If China were forced to capitulate, as Chiang Kai-shek was broadly hinting through Ambassador Johnson and in personal correspondence with the President, Japan would be free to move south into Indochina and the Netherlands East Indies to satisfy her need for oil and rubber – perhaps even to link up with her Axis partners across India. Chiang's messages to FDR were speaking more and more of Japanese peace overtures and of German offers to guarantee Japanese peace terms.

President Roosevelt was convinced of the need – the problem revolved around how to support the need. The war in Europe demanded U.S. priorities; the Axis held most of Western Europe and England was making a valiant lone stand against Germany. Britain's survival was vital to U.S. security – and Japan was clearly threatening British possessions in the Far East.

Following the November election, in one of his radio "fireside chats," Roosevelt announced his lend-lease plan – a plan to provide war materiel to any nation whose defense was deemed vital to American security.[32] Passage of the bill through Congress was still weeks away, however, and aid to China couldn't wait that long. At the suggestion of presidential advisor Lauchlin Currie, a front organization was set up to coordinate the military aid for China the framework for aid in place, Roosevelt directed Henry Morgenthau to provide a $50 million "stabilization loan" to the Chinese – a sum that was subsequently raised to $100 million after negotiations between Morgenthau and Soong.[33]

With the money available, and a conduit through which to spend it, the next step was to find the airplanes to purchase. Earlier in November, Chennault had visited the Curtiss-Wright plant in Buffalo, New York. Burdette Wright, vice-president and general manager of the company's airplane division, had known Chennault from years before when they were based together in the Army Air Corps. Wright pointed out that his company was producing a 630-plane order for the H81-A2 – the export model of the P-40B "Tomahawk" – for Britain's Royal Air Force, which had taken over a Swedish order following the fall of France.[34] The British had just received the first few aircraft of their order, and deliveries would continue until March 1942. Wright suggested that, if an order could be placed immediately, his company could set up an extra assembly line and produce 300 additional aircraft by late spring.[35]

The order was placed, and Curtiss-Wright set up a seventh assembly line. But China still wasn't guaranteed the aircraft. Other countries were also clamoring for planes and promises of support had been made. On December 23, Secretary of State Cordell Hull convened a meeting in his office to allocate the 300 extra aircraft. Present were three other cabinet members – Morgenthau, Navy Secretary Frank Knox, and Secretary of War Henry L. Stimson – plus assorted other military and civilian advisers, including Army Chief of Staff George C. Marshall and Chief of Naval Operations Admiral Harold Stark.[36]

Getting British compliance proved to be more difficult than expected. They weren't anxious to give up a hundred "birds in the hand" for the promise of more down the line. However, General Marshall intervened on the side of the Chinese, suggesting that if the British didn't transfer the planes now, they might lose the chance to get the improved aircraft.

The conferees carried on a lively discussion. Hull, mindful of a promise that Roosevelt had made earlier to the Greeks, suggested 150 for China, 120 for South America and 30 for Greece. Secretary Knox countered by saying that at least 150 should go to the British. In the end, South American and Greek interests were set aside and the group agreed to a fifty-fifty split between England and China. Then someone suggested, in order for China to get some aircraft immediately, that Britain be asked to relinquish 100 planes coming off the assembly line – 50 in January, 25 more in February, and 25 in March. In return, England would get the entire run of 300 later in the spring. As an added incentive, the added run of aircraft would be of an improved model, the H81-A3, or the export version of the P-40E. Morgenthau said that he would get in touch with the British to gain their concurrence.[37]

Later on, Secretary Morgenthau added his weight also, and on January 10 the British agreed to the proposition. China Defense Supplies made the purchase on behalf of the Chinese, and for $4.5 million China bought a new air force.

The first thirty-six of the H81-A2s – the export model of the Curtiss P-40 – were completed in January and each one fitted into two crates – one for the fuselage and engine and one for the wing assembly. From Buffalo they went by rail to New York Harbor, where they were loaded aboard a Norwegian freighter headed for Burma by way of South Africa and the Indian Ocean. The first batch left New York on February 19.[38]

Meanwhile, Chennault immersed himself in the details of preparing for the volunteer group. He later described the depth of planning required:

> *"Many long and dreary days were spent at my desk on "V" Street, ploughing through the myriad of details for these plans. There was no precedent for this kind of an air force. Every policy and detail had to be thoroughly planned in advance. I made a deal with Jerry Huang to have his War Area Service Corps house and feed all American airmen in China for one dollar a day. The WASC had no experience with the personal habits of Occidentals, so I had to plan the hostels to be built for them in great detail, down to the size of the rooms required for the pilots, the amount of hot water required for showers, how many latrines necessary for each hostel, and quantities of cattle, pigs, eggs, and chickens the meat-eating Americans would consume per month... I had to prepare requisitions for every piece of equipment we needed – how many rounds of ammunition, how many pounds of oxygen, flying suits, paper clips, six-ton trucks, machine-gun barrels, ink, candy, sun goggles, shaving cream, razor blades, medical supplies, carbon paper, typewriters, and all the other strange gear with which an air war was fought..."* [39]

Working closely with Dr. Soong, Chennault soon met the members of the small but powerful "Friends of China" group – Knox, Morgenthau, Currie and Corcoran. The first two, in particular, had come on line earlier in support of China. Secretary Knox had previously provided the Chinese with information concerning aircraft maintenance that the War Department refused to give them.[9] Secretary Morgenthau had tried earlier (in vain) to work a three-way deal with the Soviet Union, whereby the -strategic goods to Russia who would then provide war materials in turn to China.[40]

One of the first that Chennault met was Frank Knox. Knox invited Chennault to a meeting with the Navy's top admirals, who listened attentively to Chennault's briefing on the Chinese situation. Chennault pressed the need not only for first-rate aircraft, but for first-rate pilots, as well – a point that was not lost on Knox. He explained that there was not time for an extensive training program. Therefore, he could accept only trained, combat-ready pilots and ground crew-men. To get this ready-made air force, Chennault argued, he would have to go directly to the only source possible – the existing Army, Navy and Marine squadrons.[41]

At this point, Bill Pawley threw a large monkey wrench into the works. As sales representative for Curtiss-Wright in China, Pawley claimed that by virtue of his contract he was entitled to a 10% commission on the sale of the 100 P-40s, or $450,000. Curtiss-Wright refused his demands, maintaining he had nothing to do with the sale. Weeks passed as valuable time was lost with negotiating. Finally, on April 1, just as Curtiss-Wright was about to cancel the Chinese purchase and sell the aircraft back to the British, Secretary of the Treasury Morgenthau called a meeting to resolve the issue. Pawley finally agreed to a cash settlement of $250,000 and a contract for CAMCO to assemble, test-fly and service the P-40s in Burma and China.[42]

Securing the aircraft was just the first hurdle – spare parts, guns and ammunition were the next major problems. To expedite the shipment, the P-40s had been purchased without guns, ammunition, radios or spare parts. Most did come equipped with British .303-caliber wing guns, but this merely increased the problem of getting ammunition because the American standard was .30- and .50-caliber. While the P-40s were en route to Rangoon, Mr. Archie Lockhead of the Universal Trading Corporation began to tackle the problem. He initially bought .30- and .50-caliber machine guns from the British and a few 7.92-caliber guns from Colt. He negotiated radios from RCA (commercial models designed for light, civilian aircraft), and he was working with Bill Pawley to obtain spare parts from Curtiss.[43] Spare parts proved to be a problem, partly because Curtiss had ceased production of the P-40B. By August 25, Curtiss announced delivery schedules were still 30-to-90 days away. Bowing to Army pressure, however, the aircraft company accelerated their production of spare parts and shipment was ready for delivery 30 days ahead of schedule.[44]

Buying ammunition was even more difficult. The Chinese had requested 2,500,000 rounds of .50-caliber ammunition and 8,000,000 rounds of .30-caliber. General Marshall stated that the Army couldn't supply the ammunition, but pointed out that "the British agreed to supply the guns and ammunition for the 100 airplanes allotted to China."[42] Although the notion was held by many, including. T.V. Soong and members of the Universal trading Corporation, this later proved to be a mistake, since no one had bothered to inform the British of the agreement. The deadlock was broken through the efforts of Mr. Lauchlin Currie.

Working through General Gerow of the War Plans Division, one lot of 900,000 rounds was purchased from the Raritan Arsenal in New Jersey, and sent from New York on the S.S. *Warrior* on August 14. A second lot of 600,000 rounds was sent from Manila on September 16 on the steamer Iran. Although far short of China's stated needs, ammunition was on the way.[45]

As difficult as procurement of aircraft, spare parts, and armament proved to be, the most difficult task by far was obtaining American military aviators. The first hurdle was the legality of the operation. In July 1941, the War Department had sent warnings to all retired officers that the Constitution specifically forbade their being hired by foreign governments to instruct their military personnel.[46] Inquiring into his own status, Chennault was told that those so employed must have the consent of Congress; otherwise, they would be liable for court-martial. Chennault solved his problem by becoming an employee of the Bank of China, rather than the Chinese government.[45] The same scheme was adopted for the volunteers – the men would be hired by a civilian agency, not by the Chinese government. Once again Bill Pawley came forward and volunteered the services of CAMCO They were already in China in the aviation business and could provide a believable front.[47] Although his legal counselors advised him that recruiting the men would be in violation of the neutrality laws, Pawley agreed to take the chance.

With the mechanism for hiring the personnel in place, Chennault then went head-on into the halls of the military. Accompanied by presidential adviser Lauchlin Currie, who had just returned from a six-week fact-finding tour of China, Chennault went to see Major General Henry H. "Hap" Arnold, chief of the Army Air Corps, in early April. Arnold was adamantly opposed to giving up any of his pilots. He told Chennault and Currie that he "couldn't spare a single staff officer without endangering the Air Corps expansion program," and that he would "oppose vigorously any diversion of Air Corps strength to any other country." [48] It was "that damned fellow Chennault" attitude all over again.

The U.S. Navy, in the voice of Rear Admiral Jack Towers, Chief of the Bureau of Aeronautics, raised similar objections.[49] He was overruled, however, by Secretary Knox, who wrote Secretary of State Cordell Hull, arguing that "the government should support a program to allow a considerable number of American aviators to volunteer their services by combat flying in China." [50] As a precedent, he cited the American volunteers in the Eagle Squadrons of the British R.A.F., who were even then flying combat against the Germans, and suggested that the State Department could handle volunteers for China in the same manner.[51]

War Department officials continued to protest. A mission such as this, they warned, might provoke an attack against American possessions in the Pacific, and those Americans who were hired by China could lose their American citizenship.

The tide of support building for China swept away these objections, however, and Tommy Corcoran, speaking for the pro-China group, took their arguments to President Roosevelt. The President didn't need much further convincing. He had long held the belief that Japan had long-range plans for the conquest of Asia and the Pacific,[52] and Lauchlin Currie's report following his trip to China vividly described Chiang's desperate plight. Over the protests of his military chiefs, Roosevelt verbally authorized reserve personnel of the U.S. armed forces to be recruited for duty in China.[53] That finally opened the door for Chennault to get his people.

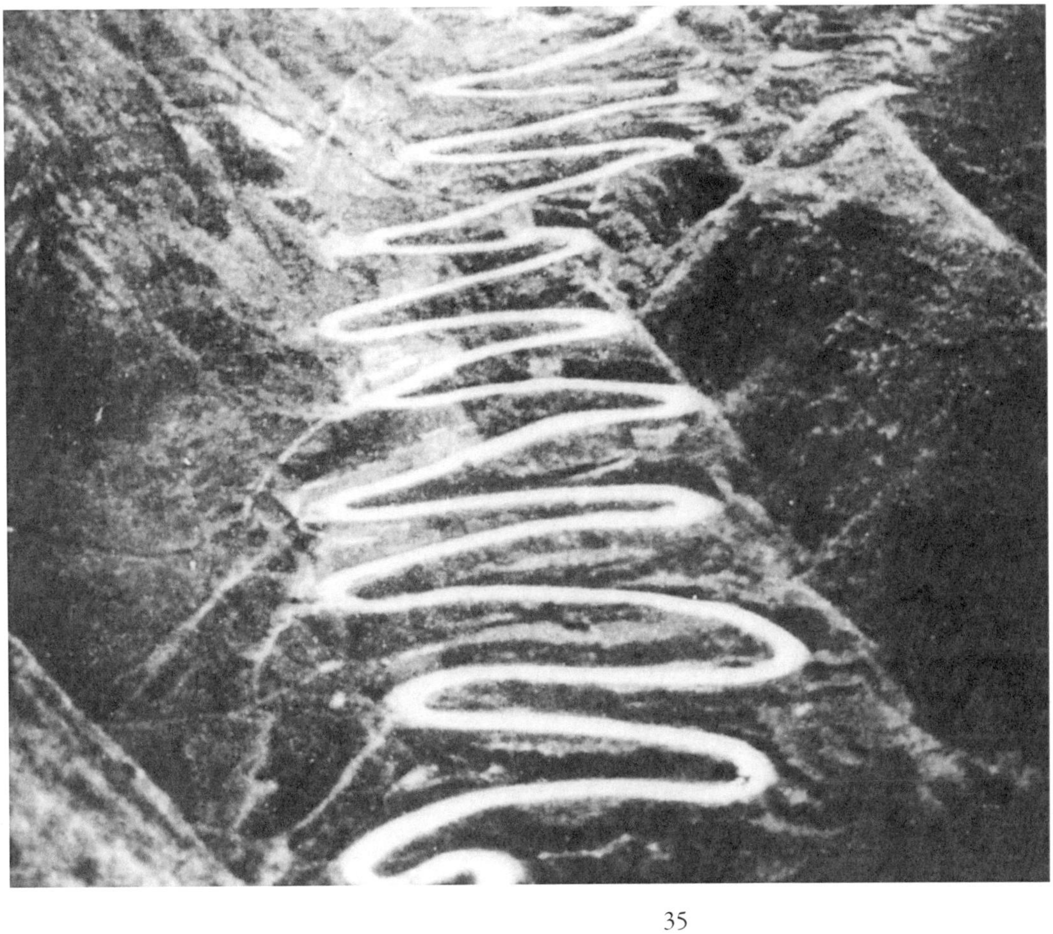

Above: Burmese and Chinese laborers, using simple hand tools, carved out 600 miles of the Burma Road.

Left: Aerial photograph of the Burma Road. (National Archives)

Beginning in 1938, combined units of the Imperial japanese Army Air Service and Imperial Japanese Navy Air Service laid waste to Chiang Kai-shek's provisional capitol of Chunking.

T.V. Soong, shown here with several Chinese generals, was one of China's most effective diplomats in lobbying for American support.

CHAPTER 4

Recruiting

Although Bill Pawley had agreed in February for CAMCO to be the recruiting agent for the American volunteers, it wasn't until April 15, 1941, that a formal contract was signed, by T.V. Soong on behalf of the Chinese government and by Bruce G. Leighton for CAMCO The document was a model of ambiguity and double meaning, calling for CAMCO to hire personnel for "three advanced training and instruction units" which would operate "active trainers" under the immediate direction of an "American supervisor." Further, they would "furnish technical assistance for assembling, maintaining and repairing equipment of the units, receive material shipped to Rangoon, and send technical personnel to assist in operations of the training and instruction units." CAMCO would pay the volunteers from a revolving fund of $700,000 set up with the Bank of China, which would be replenished when it fell below $250,000.[1]

Pawley hired a retired Army Air Corps officer, Colonel Richard Aldworth, to head CAMCO's recruiting team. Aldworth, however, did little of the actual recruiting since he spent much of his time in Walter Reed Army Hospital, suffering from a kidney ailment brought on by an airplane accident earlier in his career. Most of his duties were therefore passed to C. Briant "Skip" Adair, another ex-AAC pilot who had most recently served as an instructor in the Kunming flying school run by Chennault. Two other former Army pilots, Harry Claiborne and Senton L. Brown, and a retired Navy commander, Rutledge Irvine, rounded out the team.[2]

Beginning in late March, the recruiters fanned out to military units from coast to coast. Army bases visited included Bolling Field, Washington D.C.; Selfridge Field, Michigan; McDill Field and Eglin Field in Florida; Mitchell Field, New York; Langley Field, Virginia; Hamilton Field and March Field in California; Maxwell Field and Craig Field in Alabama; Barksdale Field, Louisiana; and Randolph Field, Texas. Naval air stations visited were Norfolk, Virginia; Pensacola and Jacksonville in Florida; and San Diego, California. Quantico Marine Air Station, Virginia, was the only Marine installation visited.[3]

The Department of the Navy supported the recruiting team by providing letters of introduction to naval air station commanders authorizing the CAMCO representatives to recruit fliers on naval air stations. They further provided for volunteers to tender their resignations without prejudice, allowing them to return to active duty with the U.S. Navy when their contract expired. General Arnold, despite his misgivings, followed suit and sent directives to all Army air fields to be visited, instructing commanders to let the CAMCO recruiters talk to their people.[4]

Rutledge Irvine walked into the pilots' ready room at San Diego's North Island Naval Air Station in early April. The room was full of young naval aviators curious to hear what Irvine had to say. With little formality he launched into his pitch. An American volunteer group was being organized – pilots and ground crews – to fly P-40s for the Chinese. Pilots were to be offered a minimum of $600 per month. Those experienced enough to qualify as flight leaders were to receive $675, while squadron leaders were promised $750. Volunteers would be provided room and board, travel expenses, and a 30-day leave at the end of the one-year contract. Ground personnel were to receive from $150 to $500 per month, depending upon their individual skills, while pilots would receive from $600 to $675 per month – a doubling of the military salaries of $300 for U.S.A.A.C. lieutenants or U.S. Navy lieutenants, junior grade. Irvine assured them that they would be free to rejoin their respective services, if they wished, at the end of their contract without loss of either rank or service time. Irvine also mentioned a $500 bonus for every Japanese plane shot down, "although," he added, "you'll probably never see one."[5]

Lt(jg) George Paxton, a PBY pilot from Patrol Squadron 12, raised his hand and asked, "How are we going to get out of the Navy, under the present circumstances?" Irvine replied, "That will be taken care of in due time. First your application to the Central Aircraft Manufacturing Company must be approved. Take no further steps until that is formalized. Only then should your resignation from the Navy be submitted. We will handle any difficulties that may arise." [6]

Ensign Joe Rosbert later wrote:

> *"That sent a thrill up and down my spine. I had always wanted to be a fighter pilot but had been stuck in a patrol bomber squadron for the last year and a half. Listening to Irvine, I thought, 'This is right up my alley!'"* [7]

Although Commander Irvine cautioned that they only needed fighter pilots at this time - a bomber group would be organized later – he did allow all to submit applications. From this group came C. Joseph Rosbert, George L. "Pappy" Paxton, Thomas J. Cole, Jr., Robert J. "Catfish" Raines, Henry G. Gilbert, Carl K. Brown, Edwin S. Conant, Leo J. Houle, Morris P. Bohman, and Richard J. White – all patrol bomber pilots.[8]

Charlie Bond learned about the American Volunteer Group from an Air Corps buddy who was an old friend of Skip Adair. Bond was a "frustrated fighter pilot" who had been assigned right out of flight school to the 2nd Bomb Group, the first Air Corps unit to get the new B-17. As the European war got hotter, he was assigned to the new U. S. Army Air Corps Ferrying Command in Long Beach, California, ferrying Lockheed Hudsons to Montreal for the British Royal Air Force. When his friend told him that Adair was hiring pilots to fly P-40s, he jumped at the chance. He called Adair the next day and was given the number of a "Col Green" in Washington. Adair told him, "If you want to join the A.V.G., just call that number and leave your name and duty station." Bond later recalled:

> *"It didn't take long for me to decide. The lure of adventure in a foreign country on the other side of the world was exciting… the opportunity served to satisfy my dreams – a chance to get back into fighters, a chance for combat experience which might help me secure a regular commission, and a chance to earn fast money which would put me in a position to buy my parents' home."* [9]

Bond talked to two more of his Ferrying Command buddies, pilots Jim Cross and George Burgard, and within twenty-four hours Ferrying Command headquarters received a wire that three of their pilots were resigning. [10]

Eriksen E. Shilling was another bomber pilot who welcomed the opportunity to get into fighters. Stationed at Langley Field, Virginia, flying B-18s with the 41st Long Range Reconnaissance Squadron, he heard rumors that China was looking for fighter pilots to protect the Burma Road. A pilot friend passing through Langley confirmed the rumor and gave him the address of CAMCO's recruiting office in New York City. Wangling a B-18 for a weekend cross-country, he flew into Mitchell Field, took the Long Island train into New York and from there caught a ride to the Chrysler Building, where he met Skip Adair. Although Adair wasn't very impressed with Shilling's previous time in P-36s, he was subsequently hired.

R.T. Smith found out about "tall, bronzed American airmen... quietly slipping away from east- and west-coast ports making their way to Asia..." to fly P-40s for the Chinese Air Force in the June 23, 1941, issue of *Time* magazine. The article went on to say that Army Air Corps pilots had been allowed to resign their commissions for duty in China.[11] Another would-be fighter pilot who was stuck with instructing at Randolph Field, near San Antonio, Smith jumped at the chance to get into fighters. He and his instructor roommate, Paul J. Greene, contacted a Kelly Field lieutenant who knew something about the volunteer program and shortly after dispatched a telegram to C.B. Adair in Washington, D.C. Adair wired back that he'd be at the Gunter Hotel in San Antonio on June 30.[12]

The two eager young flyers met Adair almost as soon as he checked into his hotel. After a brief interview, however, it was obvious that they didn't meet the qualifications that Chennault had set down. Although they each had a thousand hours flying time, it was all trainer time – no fighter time, no aerial gunnery and no training in fighter tactics. Both had been given assignments as flying instructors right out of pilot school. Hell, they hadn't even seen a P-40. "Damn," Adair said, "that's too bad – I mean about your experience. I'm sorry gents, but I'm afraid we won't be able to use you."

Smith and Greene were dismayed at this turn of events, but, in a "no hard feelings" gesture, Adair offered to take them out to dinner. Over dinner Adair tried to explain that "the old man"– meaning Chennault – had given him and the other recruiters specific instructions to hire only pilots with fighter experience. Only in rare instances would he allow an exception to be made. [13]

The next day, Smith and Greene worked out their strategy to qualify as "rare exceptions." They showed up at Adair's room with a new quart bottle of "I.W. Harper" and a new supply of resolve. Two hours of friendly conversation, punctuated by frequent trips to the room's little desk bar, resulted in Adair saying, "Okay, gents, you've convinced me. I'll probably catch hell, but it won't be the first time." He wouldn't regret it, and neither would Chennault.[14]

Second Lieutenant Albert E. "Red" Probst, a flying instructor at Craig Field, Alabama, met Skip Adair by accident. Adair walked into his office looking for Lieutenant Albert J. "Ajax" Baumler, who had flown for the Loyalists in Spain during the Spanish Civil War two years earlier. Ajax was interested in signing up for another foreign adventure. Since Baumler was out on a gunnery training mission, Adair made his pitch to Lt. Probst. As Probst recalled:

> *"I didn't know anything about Chennault or anything about the A.V.G. until Skip Adair walked into my office. He talked to me a few hours. I was head over heels in debt and he was going to pay me six hundred dollars a month... Kuykendall, Sandell, and I decided then and there that we would all go. We were buddies."* [15]

The three filled out their application papers and took the forms to the base commander for his required signature of approval. The colonel took one look at the papers and exploded. "I don't know who you think you are," he said angrily. "War is imminent and it is with Germany! We're training fighter pilots and you're a fighter pilot and if you think you're going to resign, you're out of your mind!" Two days later he called Lt. Probst back in. Shaking his head, he said, "I don't know who you are, but I have a telegram that you are resigned. You are to be processed off this base and separated from the Air Corps within twenty-four hours." This scenario was to be repeated several times at other bases across the country as incredulous commanders were made painfully aware of the President's order. [16]

Navy carriers in port gave up a fair share of recruits. From the *Saratoga* came fighter pilot Frank Lawlor and dive bomber pilots Robert H. Neale, Henry M. Geselbracht, Jr. and Fritz E. Wolf; William E. Bartling left the *Wasp*, David L. "Tex" Hill came off the *Ranger*, while the *Yorktown* gave up John V. "Jack" Newkirk. Marine Lieutenants Thomas C. Haywood, Jr., Kenneth O. Jernstedt, and Charles H. "Chuck" Older signed up at Quantico, while Lieutenants C. H. "Link" Laughlin and Gregory "Pappy" Boyington were picked up at Pensacola. [17]

As the recruiters went through the Army bases, they signed up Robert P. "Duke" Hedman, William McGarry, Fred S. Hodges, Robert A. Smith and Robert Moss from the 1st Pursuit Group at Selfridge Field, Michigan. The 8th Pursuit Group at Mitchell Field, New York, lost even more: pilots Arvid E. Olson, Jr., Parker S. Dupouy, Robert L. Little, Charles W. Sawyer, Neil G. Martin, Peter W. Atkinson, Elmer J. Cook, Robert H. Walroth, John P. Kelleher, and Robert R. Brouk, as well as armorers Charles N. Baisden, Joseph A. Poshefko, Paul J. Perry, Clarence W. Riffer, and Roy G. Hoffman, and crew chiefs Irving J. Stolet and Joseph N. Peeden. Down at Eglin Field, Florida, the recruiters picked up two experienced test pilots from the 23rd Proving Ground Group – George B. McMillan and Frank Schiel, Jr. [18]

While the pilots and ground personnel came from Army, Navy and Marine flying fields, the recruiters had to look elsewhere for their medical staff. They hired all their physicians and the one dentist from the infantry post at Fort Bragg, North Carolina – Doctors Thomas C. Gentry, Lewis J. Richards, Samuel B. Prevo, and Everett W. Bruce. Nurse Jo Buckner Stewart came from Dallas, Texas and Nurse Emma J. "Red" Foster was hired at State College, Pennsylvania.[19]

Initial recruiting went slowly. Chennault had emphasized that he wanted experienced P-40 pilots, or at least pilots with a minimum of 200 hours on pursuits. With such criteria, the recruiting team had only accepted eighteen of seventy-three applicants by May 3.[20]

It became obvious that the standards would have to be relaxed somewhat if the recruiting goals were to be met. Working on that premise, they were able to hire fifty-five pilots by the end of May, but then there was a lull, and only five more were recruited by June 15.[21]

Lauchlin Currie asked the War Department's permission to expand the recruiting effort to P-40 bases in Puerto

Rico, the Panama Canal Zone, the Philippines, and Hawaii, as well as to release the twenty American instructors at Kunming. The War Department refused on all counts – no instructors would be released and no recruiting would be allowed in Hawaii or the Philippines.[22]

To complicate the situation further, the Navy began to hold up resignations of their volunteers, on the basis that more Navy pilots were being recruited than those from the Army. The complaint wasn't valid, since no agreement of balanced recruiting had ever been made. The fact was that more Navy pilots wanted to sign up than did their Army counterparts. Lauchlin Currie and Admiral Nimitz finally reached an agreement to limit Navy recruiting to sixty-three pilots. A total of fifty-nine Navy pilots were finally recruited. [23]

Meanwhile, the first P-40Bs had reached Rangoon in mid-May. Several had already been delivered to Mingaladon Aerodrome, where they were still to be uncrated, and nine more were on a cargo ship in the harbor waiting to be unloaded when Walter Pentecost, Allison engine specialist, stepped off the gangplank of the S.S. *Karda* on May 23. He had answered an ad in the *Los Angeles Times* looking for "Aviation people, for work overseas..." Leaving his job with North American Aviation in Los Angeles, he signed up with the Intercontinent Corporation and left San Francisco on April 21. His job when he reached Rangoon, he was told, would be to take charge of assembling 100 P-40s for the Chinese at as they arrived from the States. He would join two other Americans at Rangoon: Andy Sargeant, field manager for CAMCO and Byron A. Glover, a Curtiss-Wright test pilot.[24]

Through an agreement with the British, CAMCO was being permitted to assemble the P-40s at Mingaladon Airfield. The Royal Air Force had a group of hangars and administrative buildings on the field, but CAMCO had nothing but an "A" frame out in the open, plus dozens of Burmese laborers and 165 Chinese mechanics brought down from CAMCO's factory in Loiwing, China. Each crate was removed from the trucks by sheer manpower. After the fuselage was uncrated, it was pushed (again by manpower) under the "A" frame where it was hoisted to receive the wings. Another group of laborers then walked each wing beneath the fuselage and bolted it in place. The landing gear was lowered, and the aircraft could then rest on its wheels.[25]

Next the aircraft was pushed into the assembly hangar for additional rigging, sub-assemblies, and hydraulic checks. More Burmese and Chinese hands pushed it out into the open once again to a two-bay shed next to the taxiway where the propeller was installed. On the first few P-40s to be assembled, the workers painted the twelve-pointed Chinese star above and beneath the wings of the aircraft. Later, as work became more rushed, only the necessary assembly work was done at Mingaladon and aircraft markings were added at Kyedaw.[26]

Thirty-two more P-40s arrived on July 12, along with radio equipment and about half of the required machine guns, bringing the total to sixty-eight aircraft. Assembly went slowly at first, as the workers struggled with a new aircraft. After the first ten had been completed, however, their pace picked up and they were even finally able to turn out one aircraft a day. [27]

While Skip Adair and his recruiting team were signing up pilots and ground crew, Chennault was beating the bushes for an administrative staff. He recognized his own weakness in that area and knew he'd need some help. He had tried in vain to recruit some qualified staff personnel through Air Corps chief, General "Hap" Arnold. He initially asked for six trained staff officers to set up his headquarters. Arnold rejected that out of hand. Chennault cut his request in half. "No," replied Arnold. Finally he argued for the release of just one staff officer. Arnold adamantly refused, telling Chennault that "the loss of even one staff officer would cripple the Air Corps expansion program." [28] He had been forced to let his pilots and ground crew go, and now, by damn, he was going to stand his ground!

Forced to look elsewhere for his staff, Chennault decided to use Skip Adair as his supply officer, as soon as his recruiting duties were finished; and he had a few other "old China hands" in mind that he would contact when he got back to China. American vice-consul to China John Davies recommended a candidate for a chaplain to Chennault: Paul W. Frillman, a Lutheran minister who had been a missionary in northern China for five years, had just returned from China. Davies had introduced Frillman to Chennault at a 4th of July party in Hankow in 1938, but Chennault didn't remember him. Nevertheless, Frillman was at loose ends and was willing to sign on for another tour in China.[29]

Chennault had originally intended for the group to train at Kunming. It offered excellent facilities and was also strategically located to serve as a base of operations for defending the Burma Road. However, delays brought about by the conflict with Pawley, slow recruiting, and difficulty in getting spare parts for the P-40s pushed Chennault's training schedule into the summer – and the summer, as Chennault well knew, brought monsoon rains that would soon turn Kunming's grass runways into a sea of mud. Preparations were already being made for hard-surface runways.

The British in Burma had been quietly working with the Chinese government for a joint defense effort, and in April, 1941, had agreed to send material to Kunming to build all-weather airfields around the city, in return for Chinese support for Burma, should the need arise.[30] At the same time, they agreed for the P-40s of the volunteer force to be assembled and test flown in Burma, but "under no circumstances could actual combat training be permitted under the British flag."[31] Time had run out on preparing the Kunming base for summer training, however, and Chennault had to look elsewhere for a training site that could be used immediately.

Chennault had arrived back in Rangoon on July 23, less than a week before the first contingent of volunteers was to arrive. Gathering General P.T. Mao, Bill Pawley and his brother Ed, Chennault arranged a conference with Sir Reginald Hugh Dorman-Smith, governor of Burma; his military commander, Lieutenant General D.K. McLeod; and the senior R.A.F. officer in Burma, Group Captain E.R. Manning. A cable to the British War Office brought permission for the volunteers to use Kyedaw Aerodrome, the small Royal Air Force auxiliary field slashed out of the Burmese jungle on the Sittang River. Bill and Ed Pawley went to work immediately contracting Burmese workers to cook and clean for the men, and Chennault brought down Boatner Carney, one of his American instructors from the Chinese flying school at Kunming to oversee the preparations at Kyedaw.[32]

In early June, the first group of volunteers gathered at the Jonathan Club in Los Angeles, where they'd been told to report to assemble for their departure to the Orient. There were the four armorers from Mitchell Field – Charles N. "Chuck" Baisden, from the 57th Pursuit Group, and Paul J. Perry, Clarence W. Riffer, Joseph Poshefko (who came down with appendicitis and didn't depart with the first group) – and Roy G. Hoffman from the 8th Pursuit Group. Ex-Army crew chiefs made up nearly half of the Group – Leo J. "Red" Schramm, Daniel W. Keeler, Henry L. Olson, Robert A. Smith, Carl E. Schur, Charles V. Misenheimer, John E. Fauth, Edward L. Stiles, Glen O. Blaylock, Charles R. Engle, Robert W. Rogers, Walter J. Dolan, Charles D. Kenner, John J. Uebele and sheet metal mechanic Joseph Gasdick. Most of these men had one thing in common – experience on the P-40. This would be the advance group to get the planes ready by the time the pilots arrived. Rounding out the group were two communications specialists, Joseph H. Sweeney and Harvey G. Cross; finance clerks Harold G. Wylie and Joe T. Jordan; engineering clerk Sharon L. McHenry, and photographer James E. Regis.

Into this group walked the new chaplain, Paul Frillman, feeling very self-conscious. He was the only full-fledged civilian in the group. He tried to keep a low profile, but Bill Pawley's brother, Gene, who was there to see the men off, let the cat out of the bag. "Holy smoke!" he said, in a loud voice when he was introduced to the chaplain. "You can't be the Frillman we're expecting!" For the next several days the men would refer to the hapless chaplain as "Sky Pilot" and "Holy Joe."[33]

As the lobby of the Jonathan Club began to fill, the men began to compare notes on where they'd come from, what they did, and to speculate on where they were going. They found that each of their passports carried a bogus occupation. For instance, Chuck Baisden was listed as a "metal worker," while P.J. Perry was an "installation mechanic." The State Department was still very cautious about Americans becoming involved with the war in Asia – hence the attempted subterfuge. Within a day or so of his awkward introduction to the group, Frillman managed to get to know the men a little better and observed that:

> *"Most men were escaping from frustrations or looking for adventure, as perhaps I was. They hoped an unknown future in unknown places would somehow give them a second chance. One of the oldest was a tough former sergeant-major about forty-three, irreconcilably divorced. One of the youngest was a boy of nineteen who had enlisted in the Army, then got right out again for this junket; he was longing for adventures with lots of shooting, perhaps because he was small for his age."*[34]

After a few days in Los Angeles, the group boarded two buses for the ten-hour ride to San Francisco – not exactly the first-class transportation they had been told to expect. However, a couple of laundry tubs full of beer and ice, plus a few jugs of liquor helped to ease the discomfort of the hot and crowded bus.[35]

Things weren't much better when they arrived in San Francisco Their "luxury liner" may have been one once, but had since been converted to an Army troop transport, its former name, the S.S. *President Pierce*, painted over. The volunteers would share space with a load of soldiers on their way to the Philippines. There would be no private staterooms, not even a private locker or toilet. Their quarters would be rows of cots set up in the former ballroom on the top deck. [36]

The men boarded the ship on June 6, and it got underway about 3:00 PM. Shortly afterward, a soldier brought Frillman a message from the ship's radio room – Chennault had assigned him as leader of the group during the voyage. "Some job," he thought, "having to maintain order and discipline over twenty-nine young Air Corps troops for several weeks packed into a crowded Army troopship. But to give the job to a chaplain?!" He was already having misgivings about signing on with Chennault; after all, his family had told him he was "a fool to go off to die with a bunch of harem-scarum adventurers." [37]

After eighteen days of having to endure the drills and inspections imposed by the Army troop commander, they finally reached Manila, where the Army disembarked. By this time the volunteers had stopped calling Frillman "Holy Joe" and "Sky Pilot" and had developed a grudging respect for their new chaplain. The now nearly-empty troop ship departed Manila for Hong Kong and the group was told that their final destination was not China, but Rangoon, Burma.

After an overnight stop in Hong Kong, where they stayed at the Peninsula Hotel, at 10:00 AM on July 3 they transferred to a small Dutch freighter called the H.M.S. *Van Houetz* that made the six-day trip to Singapore.[38] There they had a two-week layover at the Raffles Hotel awaiting a shipment of heavy machinery for Rangoon, much to the dismay of the British authorities. Frillman was constantly having to break up fights, and to take care of paying for damages for bars wrecked by his exuberant flock. In one instance, two young mechanics practiced their golf game from their room in the Raffles Hotel. The managers' temper was shattered, as were lamps, vases, and windows.[39]

Much to the relief of their British hosts, the group departed Singapore at 5:30 P.M. on July 22, this time on a third Dutch freighter, the *Hong King*.[40] Leaving Singapore, the freighter moved slowly up the western side of the Malay peninsula, made its way up the Rangoon River, and finally dropped anchor at the Rangoon docks on July 28. Chennault, Bill Pawley and his brother Ed were there to meet the first contingent. Paul Frillman later recalled his impression of Chennault on that hot and steamy day in Rangoon:

> *"That day on the docks he was wearing some slapdash, adventurous costume as usual – mosquito boots, officer shirt with Chinese insignia, beat-up Air Force cap – which emphasized his gamecock look. Watching him for only a few minutes, anyone would get the impression of informality and lack of military pomp, plus a quick, sure air of decisive authority. I don't think any of the men on our ship had seen him before, and I looked down the rail where they were lined up, staring silently at him. I could see that, for the time being, anyway, Chennault had them all in his pocket."* [41]

John E. Fauth had the distinction of being the first volunteer off the boat, followed by P.J. Perry. After they all disembarked, Chennault had them gather around for a short briefing. They weren't being sent to China right away, they learned. They would spend some time at a Royal Air Force auxiliary field called Kyedaw Aerodrome, about 170 miles to the north, near the village of Toungoo. There the group would train, preparing for operations in China later on. Chennault continued to say that sixty-eight P-40s had already arrived and were in the process of being assembled at Mingaladon Airfield, just outside Rangoon. The mechanics in the group would go immediately to the airfield to help with the assembly. In the meantime, the remainder would get a night's rest in Rangoon and then take the train up to Toungoo.

As the group broke up, Chennault called Paul Frillman over and handed him a scribbled paper – a long list of "shopping items," including such things as auto parts, tires, electrical and telephone supplies, screen doors, typewriter, hunting rifles, and musical instruments. "A few other things I thought we might need," he explained to the puzzled chaplain. "Get the money from the Pawleys and bring the stuff up to Toungoo when you come."[42]

After they left the docks, Chuck Baisden recalled:

> *"We were put up in the Minto Mansions. Not the best in town, but okay. The Chinese Embassy had a big party for us where we met Chennault for the first time. The next day Paul Frillman informed me that Chennault wanted me to take the hold baggage to Toungoo via car (why me I never knew). Anyway, I took off with a Burmese driver and Studebaker station wagon with the thing crammed full of trunks. Arrived at Toungoo late that same afternoon. I was the first person in the A.V.G. to arrive and was not very much impressed."*[43]

The first contingent – less Chuck Baisden (who had left for Toungoo by car), Paul Frillman (who would spend days in Rangoon trying to fill Chennault's shopping list) and those mechanics directed to stay at Mingaladon – gathered at the Rangoon train station on July 29 to catch the 4:00 P.M. Mandalay Express to Toungoo.

The little narrow-gauge train looked like a holdover from the 19th century – a small, coal-burning engine and six European-style passenger cars strung out behind. The men's "first class" accommodations allowed them to travel in the last two cars on the train, away from most of the soot and sounds from the engine and with the added privilege of having no pigs, ducks, or chickens traveling in their compartments, as the third-class passengers up front were subjected to.[44]

The 170-mile trip to Toungoo took five-and-one-half hours, through the Irawaddy Delta and along the Sittang River. It was a quick introduction to an Orient that most had seen only in books and National Geographic magazines – endless rice paddies occasionally broken by villages of thatch-roofed houses surrounded by palm and banana trees.

The trip was broken up by a seemingly endless succession of stops at each village along the way. The heat was stifling. July wasn't the best time of the year to visit exotic Burma. P.J. Perry found a way to beat the heat, however – he simply stripped off down to his drawers. He was in that comfortable mode when the train pulled in to yet another whistle stop. He opened the outside door and stepped out into the nearly isolated jungle for a breath of fresh air. Just as he did, one of his traveling companions in the six-man compartment shut the door and locked it. Someone tossed his sports jacket outside and yelled, "Put this on!" As the train began to move again, Perry had visions of having to wait at the train station in his skivvies and sports jacket. One of his "buddies" unlocked the door at the last minute, though, and he jumped back in, muttering, "You can't trust anybody."[45]

At 1000 that evening the train pulled into the little station at Toungoo and the men climbed out of their six-passenger compartments onto the dimly-lit station platform, glad for the opportunity to stretch their legs. Despite the time of day, the temperature still hovered around ninety degrees and the humidity made them feel like they'd stepped into a Turkish bath. Looking down the one main street of the village, they could see rows of shops on one side while open market stalls lined the other. Even at that time of night, traffic along the street was heavy with trucks bound for Lashio and the Burma Road.[46]

Studebaker station wagons were waiting to take them out to Kyedaw, and they made the last eight miles of their journey in less than fifteen minutes. As they tumbled out of the station wagons, Chuck Baisden showed the group to their barracks, a long, thatch-roofed teak building with woven bamboo walls and a veranda that stretched the length of the building on the west side. Two dim bulbs, about which swarmed a cloud of insects, were suspended from the ceiling and provided just enough light to see the interior – two long rows of teakwood beds with a straw mattress laying on a lattice-work of rope. Hanging from the rafters above each bed was a wooden frame for the mosquito netting that had to be dropped over the bed and the edges tucked under the mat tress upon retiring. This proved to be an absolute necessity, since mosquitoes "the size of wasps" were permanent residents at Kyedaw. The barracks had large open windows for ventilation, but no screening of any kind. It wasn't the Raffles Hotel, nor even the Minto Mansions, but the first contingent of the A.V.G. had arrived at Kyedaw, and they would call it home for the next four months.[47]

Bus taking the first 30 volunteers from Los Angeles to San Francisco, June 1941. (via Baisden)

The first contingent of A.V.G. personnel as they prepare to depart on the SS President Pierce. *Front row, left to right: Habler, McHenry, Metasavage, Jordan and McKinney. Second row: Pawley, Schurr, Fauth, Baisden, Cross, Wylie, Blaylock, Wyke, Perry, Stiles and Rogers. Third row: Engler, Kenner, Gasdick, Misenheimer and Riffer. Fourth row: Hoffman, Dolan and Kennedy. Fifth row: Bugler, Keller and Frillman. Sixth row: R.A. Smith and Sweeney. (via Baisden)*

Waving goodbute on the SS President Pierce. *Left to right: Frillman, Keller, Sweeney, Kenner, Schurr, Stiles, Bugler, Baisden, Perry, Gasdick, McHenry, Wyke, Engle, Hubler and Riffer. (via Baisden)*

En route to China, outside of Diamond Head, Hawaii, June 1941. Standing, left to right: Misenheimer, Keller, Sweeney, Schramm, Hubler, Olson, Stiles, Dolan and Baisden. Kneeling: Metasavage, Ubele, Blaylock, Kenner and Wylie. (via Baisden)

Party at the Raffles Hotel in Singapore on the way to Rangoon. Left to right: Joseph H. Sweeney, Danny Keller, unknown, Robert A. Smith, missionary woman, Paul Frillman, guest, and Harold G. Wylie. (via Baisden)

Walter Pentecost with unidentified CAMCO employee at Mingaladon. (via John Williams)

Assembling one of the first H81-A2 Tomahawks arriving at Mingaladon in July 1941. (via Hotz, with General Chennault)

First off the assembly line, A.V.G. fuselage code "1", fin number P-8126, was test-flown on July 25, but wasn't delivered to Toungoo until August 28, 1941. (via Lee Burgard)

The last of the 99 Tomahawks, fin number P-8101, being fueled at Mingaladon prior to delivery to Toungoo on November 28, 1941. (National Air & Space Museum)

Chennault at the bat! (via John Williams)

CHAPTER 5

Training

The next morning most of the men slept in. Those that went over to the mess hall, however, were treated to a breakfast of warm fish and rice, hardly what they'd expected. Then, stepping outside into the blazing sun and dripping humidity of the Burmese summer, the more curious began to explore their new surroundings. As they slogged through the mud left by the most recent downpour, they could see the decrepit R.A.F. barracks and operations shack, next to which was a spindly, open-air control tower. There were two runways, a 4,000-foot asphalt strip running north/south, and a second unpaved strip running east/west. In addition to their newly-constructed barracks, mess hall and cookhouse, there was an armament shop, hospital building and dental hut, and a large hangar, with enough space to hold at least six P-40s.[1]

As the temperature climbed to near the 100-degree mark, most drifted back to the barracks to try to sleep a bit or over to the mess hall for a few beers – anything to escape the oppressive, sticky heat. Before the sun set on their first full day at Kyedaw, many were doubting their good sense in signing on for their "great adventure."

The morning calm on August 3 was shattered by the throaty roar of an Allison engine. Those outside were treated to the sight of a brown and green P-40 flashing low across the field. The aircraft arced up in a climbing turn that brought it on a downwind leg to the macadam runway. The men watched as the gear and flaps came down and the pilot started a smooth turn toward final approach. Moments later the big fighter was on the ground, taxiing in toward the hangar.

By the time the big three-bladed prop wind-milled to a stop, a small crowd had formed on the parking ramp. Byron Glover climbed down from the cockpit and shook hands with the nearest in the group that was now crowding around the aircraft. The first of the P-40s had arrived! Even those who had worked on the bird back in the states felt excited about the event.

The P-40 was painted in the R.A.F. camouflage that had been applied at the factory – dark green and tan on the upper surfaces, with light gray below. The twelve-pointed star of the Chinese Air Force had been freshly painted on the top and bottom of each wing. The tailfin bore the stenciled serial number P-8124 in white. On closer inspection of the aircraft, the men found that there were no guns installed, nor was there any radio or oxygen equipment. These had been left to be added by the ground crews at Kyedaw.

Glover didn't stay long at the field. After filling out the necessary paperwork, he got a ride into Toungoo and was on the next "down-mail run" on the Mandalay Express back to Rangoon. He was back again on August 7th, however, with P-40 number P-8108, and two days later delivered the third aircraft, P-8126. [2]

The aircraft machine guns – two-hundred .50-caliber guns for the nose and four-hundred .30-caliber wing guns – arrived at the railway station in Toungoo on August 6. Chief armorer Roy Hoffman and a group of hired Burmese laborers loaded the guns on trucks and brought them out to the field. The armorers spent the next several days cleaning the cosmoline off the packed guns. It wasn't until ten days later that P.J. Perry and Joe Poshefko installed the first guns on aircraft number P-8108.[3]

Meanwhile, the second contingent of volunteers was two days away from Manila, heading west. They had departed San Francisco on July 10 on the Dutch Java Pacific Line's *Jaegersfontein*. This was more in keeping with the sales pitch given the men. After the hurriedly assembled first contingent departed, CAMCO contracted with the Dutch V.N.S. (Java Pacific) Line to carry the remainder of the volunteers to Rangoon.[4]

The second group was to be the largest sent – 121 men (and two female nurses) - and included 37 pilots, most of whom were from the Air Corps' 1st, 8th, and 20th Pursuit Groups. With that in mind, during the latter part of June, Lauchlin Currie suggested to President Roosevelt that some protection be given to the *Jaegersfontein* when it sailed in July. Roosevelt agreed and passed the word to Secretary Knox and Chief of Naval Operations Admiral Stark, which resulted in two U.S. Navy cruisers acting as escorts for the Dutch liner.[5] Radioman Robert M. Smith recorded the incident in his diary:

> *"...yesterday afternoon (July 16) at one-o'clock we left Honolulu and headed for the Phillipines. We didn't know it, but we were following a United States cruiser. This morning we saw two cruisers on the horizon. They signaled and the M.S.* Jaegersfontein, *out of Batavia, Dutch East Indies, stood by. The warships pulled alongside. They are the U.S.S.* Salt Lake City *and the U.S.S.* Northhampton. *The latter dropped a boat and sent an officer, a lieutenant (junior grade), and two seamen aboard. The lieutenant had a star on his shoulder indicating a staff officer. The seamen are signalmen and stayed on the* Jaegersfontein *after the officer left. They brought their sea bags and apparently will stay here for some days... I guess the cargo on this ship must be valuable!"* [6]

The two American cruisers stayed with the *Jaegersfontein* for over two weeks, from Honolulu to the Torres Straits, where they passed the escort duties to a Dutch gunboat. A week later, on August 11th, they pulled in to Singapore, the scheduled end of the line for the *Jaegersfontein*. It had been sixteen days since they had left Honolulu, and everyone was anxious to get off the boat to see exotic Singapore.[7]

Shortly after they arrived, Charles D. Mott, who was in charge of the group, was told that there was no transportation available to Rangoon in the foreseeable future – and to make matters worse, the men were being told that there were no hotel accommodations. Mott grabbed the ship's captain and the two went to see the American Consul. A brief discussion with that gentleman worked wonders. The experience with the first thirty volunteers to pass through Singapore three weeks earlier was still fresh in his memory, and under no circumstances did he want 123 more wild Americans running around Singapore waiting indefinitely for transportation. Arrangements were hastily made with Java Pacific Lines to continue on. The boys were rounded up, put back on board, and the *Jaegersfontein* sailed for Rangoon.[8]

The ship arrived before dawn on August 16, and the men were up at 0530 for coffee. At 0900 they boarded a lighter that took them to the Rangoon docks. After a breakfast at the "Silver Grill," a restaurant that was to become a favorite haunt of the Americans, they boarded the Mandalay Express for the last leg of their journey to Toungoo.[9]

The pilots were dropped off at a large tin-roofed building of teak with woven bamboo walls that turned out to be the pilot's mess. There was a bar in the corner, well-supplied with scotch and gin, and the men took full advantage of a little relaxation before being shown their sleeping quarters.

The mechanics, armorers, and radiomen started to work early the next morning, Monday, the 18th. Robert M. Smith (one of four Robert Smiths in the A.V.G. – the other three were pilots Robert T. and Robert H., and crew chief Robert A.) was assigned the chore of ripping out all the British radio wiring from the P-40s. Being built for the British, the aircraft were wired for British radios, which were not supplied in the deal. Instead, China Defense Supplies, unable to obtain military equipment, purchased the only alternative available, the RCA-7-H, a light-weight commercial transceiver designed to be used in light sport planes. 10 From Smith's diary, we get an idea of his task:

> *"Knocking out radio wiring is hot and difficult work. All I have is a very large screwdriver and a ball peen hammer to work with. I have to reach into the plane through a small door in the tail to get at the joints where the wiring is bolted to the airframe."* [11]

With a pool of pilots now available, Byron Glover was able to get some assistance in ferrying assembled aircraft from Mingaladon to Kyedaw. Nearly half of the pilots who had come over on the *Jaegersfontein* had come from P-40 squadrons in the states. The 8th Pursuit Group at Mitchell Field alone lost nine pilots to the A.V.G. who were fully qualified in the aircraft. Six former 8th Pursuit Group pilots: Arvid E. Olson, E.J. Cook, Peter W. Atkinson, R.H. Walroth,

Neil G. Martin and Robert L. Little, made the long train trip back down to Rangoon and returned, three at a time, on August 18th and 20th with six new Tomahawks, as the British had dubbed the early model P-40s. [12]

It had taken Walter Pentecost and his crew three weeks to assemble the first P-40. As the Chinese mechanics' experience grew, however, and with the added help of the newly-arrived former Air Corps mechanics, they were turning out one P-40 a day. By the end of August, twenty-two aircraft were on the ramp at Kyedaw. [13]

Chennault arrived at Kyedaw on August 21, after spending most of the month in conferences with the Chinese and British. One of his more important discussions took place in Singapore with Air Chief Marshall Sir Robert Brooke-Popham and his aide, Air Vice-Marshall Pulford, where he obtained permission to fire the P-40s guns at ground targets near Toungoo for strafing practice. Lower British officials in Burma had objected to such training on the grounds that "firing would stir native unrest" – and in any case they couldn't allow it without approval from Singapore. It turned out that Brooke-Popham was just as anxious to have a combat-ready American force as was Chennault. He proved to be one of Chennault's strongest British supporters.[14]

When Chennault arrived at the training field, he was dismayed at what he found. Only thirteen P-40s had been delivered, although the first thirty-six had arrived in Rangoon almost three months earlier – and only a third of his pilots had arrived. He had initially hoped to have been well into the training program by this time. Although most of the personnel had been at Kyedaw for less than a week when Chennault arrived, they were already grumbling about the food, the primitive living conditions, the insects, and the climate. Chennault was forced to agree with them in that regard. Having seen Kyedaw, he knew why the British agreed so readily to turn the field over to the A.V.G. The R.A.F. found it totally unusable during the three-to-four-month monsoon season. European sensitivities couldn't accept such abominable conditions. Chennault later described the field as:

> *"... a shocking contrast...to a peace time Army or Navy post in the United States. The runway was surrounded by quagmire and pestilential jungle. Matted masses of rotting vegetation carpeted the jungle and filled the air with a sickening smell. Torrential monsoon rains and thunderstorms alternated with torrid heat to give the atmosphere the texture of a Turkish bath, Dampness and green mold penetrated everywhere. The food provided by a Burmese mess contractor was terrible, and one of the principal causes of griping."* [15]

Clearly it was time for Chennault to have a talk with the men. Gathering them together, he reiterated what they were there for, reminding them that the defense of the Burma Road was vital to China's survival – it was the last remaining supply link to the outside world. He went on to announce that, as of August 1st, the First American Volunteer Group had been officially constituted for service with the Chinese Air Force, and, because of their peculiar status as foreign volunteers fighting for China, it would be necessary for each to sign up as a volunteer for the Chinese Air Force to give them some protection in the event they were captured by the Japanese.[16]

He minced no words about what they could expect, as was noted by crew chief William E. Schaper:

> *"He told us exactly what we would run into in a couple of months from now, and I might say it looks pretty exciting, with the Japs probably giving us a warm reception."* [17]

Later, Chennault had individual meetings with those who had personal complaints or who were known trouble makers. One of the most welcome things he did, however, was to change the mess hall menu to items more recognizable to Americans. Secondly, he established a more realistic work schedule. Work would be done early and late, with the men knocking off during the middle part of the day, when the heat was the most oppressive. Then, training began in earnest. [18]

At 0600 on the morning of the 22nd, Chennault gathered the newly-arrived pilots into the makeshift classrooms in the operations building and began what he termed his "kindergarten." First came an overview of the war in China, a brief lesson in the geography of the country and then a briefing on the air raid warning net. Basic ground school in flying the P-40 – aimed at the majority of the pilots who had flown only multi-engine bombers or single-engine dive bombers –

evolved into detailed lectures on how to fight the Japanese.

Also on that first day, Chennault issued General Order Number 1, organizing the American Volunteer Group into three squadrons and a headquarters element. Twelve pilots formed the nucleus of each squadron. Fourteen or fifteen crew chiefs, four or five armorers, four-to-six radiomen, and five-to-seven clerks made up the remainder of the squadron. The chaplain, medical personnel, photographers, parachute riggers and assorted specialists fell under the headquarters section. Boatner Carney, Chennault's only in-place staff member, was listed as "Chief of Staff and Executive Officer."

The First Squadron was made up mostly of ex-Army pilots and would be commanded by Robert J. Sandell, who had instructed in P-40s at Craig Field, Alabama. Slightly built, dark and intense, "Sandy" Sandell hailed from St. Charles, Missouri. He would turn out to be a martinet with his men and was the least-liked of the three squadron commanders. Chennault put most of the ex-Navy pilots in the Second Squadron and selected John V. Newkirk, who had flown dive-bombers off the *Yorktown*, as commander. A native of Scarsdale, New York, "Scarsdale Jack" Newkirk was tall, blonde, and wiry, and was described by one as "perhaps the most outstanding personality among the A.V.G. pilots... he never missed a chance for action and always pressed his men for as much as they could give or a little more." The Third Squadron would turn out to be the composite squadron, with pilots from the Army, Navy and the Marines. Arvid E. Olson, another experienced P-40 pilot from the 8th Pursuit Group at Mitchell Field, New York, (with 450 hours in the airplane), was chosen to lead this group. "Oley" Olsen, tall and slim, had just turned twenty-five. Chennault would later call him "one of the best tacticians in the Group."

In an attempt to put together a staff, Chennault had to rely on whomever was available in the Orient. Of necessity he called upon those with whom he had worked before. Boatner Carney, one of his instructors from Kunming who had been in China since 1938, was the first to arrive. He helped prepare Kyedaw for the arrival of the volunteers and later served as Group Transportation Officer until April, 1942, when he was returned to the United States for killing an Army man during a drunken brawl. Although Carney was listed as "Group Executive Officer" on General Order Number 1, he was replaced by Harvey Greenlaw when Greenlaw arrived with his wife, Olga, at the end of August.

Harvey Greenlaw, West Point Class of 1920, had come over as an instructor with the Jouett mission in 1933. In 1936, he and his wife went back to the States for a little over a year, then returned to China when Harvey found a job as an aircraft salesman for North American Aviation. Chennault, who had met the Greenlaws in China earlier, located him in Hong Kong during the summer of 1941 and offered him the job of operations officer with the A.V.G.[19]

Harvey, large, heavy-set, with a somewhat round face, wasn't at all popular with the men. He tended to be a bit on the officious side, a characteristic that many of the volunteers had sought to escape. Olga, with long, dark hair and several years Harvey's junior, was a very pretty woman with the figure of a movie star – indeed, she was often described as looking like Barbara Stanwyk. She would prove to be as interested in the boys as they were in her and Chennault would have occasion to regret ever inviting Harvey into the outfit. They didn't live on the field, preferring to rent a two-story bungalow in the village of Toungoo – a house that Olga described later as the "Greenlaw Hotel." [20]

As Chennault surveyed the new arrivals, he noted some young bodies that showed the results of the good food and easy living of their six-week cruise. He immediately put the jeeps, sedans and trucks off limits except for strictly official business. The men would at least get some exercise walking around the base, or even into town. Daily calisthenics were started, led by Chaplain Frillman, and nearly every afternoon, when it didn't rain, the group gathered for a baseball game. Chennault joined in with the men most of the time, always playing pitcher.[21]

The men didn't mind the calisthenics, and enjoyed the ball games, but hated having to walk everywhere they went. Then someone discovered a supply of cheap Japanese bicycles in a shop in Toungoo. Within hours, the shopkeeper had sold all his stock. For weeks afterward, nearly every "up-mail" train from Rangoon brought more bicycles.[22]

Many afternoons were spent going into Toungoo to shop the bazaars, often starting with a somewhat slow-motion bicycle race into the village. Everyone had to have tailor-made khakis and custom-made knee-length "mosquito boots." The unofficial uniform of the A.V.G. then became a khaki bush jacket, khaki shorts and mosquito boots, topped by a cork pith helmet. This is not to say that the men were "uniform" in any sense of the word – there were absolutely no dress standards in the organization. Although Chennault and most of his staff wore their U.S. Army uniforms (with Chinese insignia), the pilots and ground crew could be found in any combination of dress – blue jeans, khakis, tee-shirts, cowboy boots, sandals, or whatever. Later on, the soft-billed caps worn by the Chinese army became popular with some. Only on special occasions, such as funerals, banquets and award presentations, would the men bother to put on a service uniform. Toward the end of their tour in China, some had an "A.V.G. uniform" made by an Indian or Chinese tailor – a light grey or tan garment based on the U.S. Army uniform, but with A.V.G. buttons and insignia. Some had such a uniform made on their way back to the States. [23]

Day after day, Chennault drilled the pilots with tactics from his several notebooks, filled over the last four years through personal experience. For textbooks he used captured Japanese flying and staff manuals, translated into English by the Chinese. Chennault had observed the Japanese in the air and had read transcripts from many hours of interviews with captured Japanese airmen. He told his pilots:

> *"You will face Japanese pilots superbly trained in mechanical flying. They have been drilled for hundreds of hours in flying precise formations and rehearsing set tactics for each situation they may encounter. Japanese pilots fly by the book, and these are the books they use. Study them, and you will always be one jump ahead of the enemy." "They have plenty of guts but lack initiative and judgment. They go into battle with a set tactical plan and follow it no matter what happens. Bombers will hold their formations until they are all shot down. Fighters always try the same tricks over and over again. God help the American pilot who tries to fight them according to their plans. The object of our tactics is to break up their formations and make them fight according to our style. Once the Japanese are forced to deviate from their plan, they're in trouble. Their rigid air disciplines can be used as a powerful weapon against them."* [24]

He briefed them in detail on the flying characteristics and capabilities of every Japanese aircraft he expected them to encounter. He passed out mimeographed sheets with specifications and drawings of the new Japanese Navy Zero, as well as other fighters and bombers. Going to the blackboard at the front of the room, Chennault drew diagrams of the enemy aircraft, then circled each vulnerable spot – gas tanks, oxygen storage, oil coolers, and bomb bays – with colored chalk. Then he'd erase the circles and have one of his pilot students step up and redraw them from memory.[25]

Regarding tactics, Chennault emphasized:

> *"You must use the strong points of your equipment against the weak points of the enemy. Each type of plane has its own strength and weakness. The pilot who can turn his advantages against the enemy's weakness will win every time. You can count on a higher top speed, faster dive, and superior firepower. The Jap fighters have a faster rate of climb, higher ceiling, and better maneuverability. They can turn on a dime and climb almost straight up. If they can get you into a turning combat, they are deadly.*
>
> *Use your speed and diving power to make a pass, shoot and break away. You have the edge in that kind of combat. All your advantages are brought to bear on the Japanese deficiencies. Close your range, fire and dive away. Never stay within range of the Japs' defense firepower any longer than you need to deliver an accurate burst.*
>
> *You need to sharpen your shooting eye. Nobody ever gets too good at gunnery. The more Japs you get with your first burst, the fewer there are to jump you later. Accurate fire saves ammunition. Your plane carries a limited number of bullets. There is nothing worse than finding yourself in a fight with empty guns."* [26]

Time and again he repeated the tactics that he had devised for the heavier, slower P-40 to beat the faster, more maneuverable Japanese:

> *"Fight in pairs. Make every bullet count. Never try to get all the Japanese in one pass. Hit hard, break clean and get position for another pass. Never worry what's going to happen next, or it will happen to you. Keep*

looking around. You can lick the Japanese without getting hurt if you use your heads and are careful. Follow them home. They are usually low on gas and ammunition when they break off and head for home. If they maneuver or open full throttle, they will not get back." [27]

The "fire and dive away" tactics were new to the men. Those who had trained previously in fighters were more familiar with the tail-chasing dogfight tactics of World War I, and were more than a little doubtful of this new approach the "beat-up, old captain" was putting out. The R.A.F., upon hearing of Chennault's tactics, was horrified at such hit- and-run antics and even went so far as to post a notice on the bulletin boards of their Brewster Buffalo squadrons in Rangoon that "any R.A.F. pilot seen diving away from a fight will be subject to court martial." [28] The question of tactics was permanently decided four months later when the two groups fought side-by-side over the skies of Rangoon.

Every pilot who arrived before September 15 was put through seventy-two hours of ground school and sixty hours of flight training in the P-40, with the ex-Navy types getting priority on the schedule. Each morning following ground school, while it was still relatively cool and before the daily rain began to fall, the pilots filed out for specialized flight training. Colonel Chennault would climb the rickety, open-air tower beside the operations building, settle into a bamboo chair and spend the next couple of hours observing the pilots sent aloft in the new P-40s through a pair of binoculars. He maintained radio contact through a field radio and gave each pilot a running commentary on his flying, as he dictated additional notes to his secretary. When the pilots landed, they reported to Chennault for an additional critique from his notes.[29]

Single-ship flying progressed to pairs, and then to squadron formation flying. While one would take the role of the bomber, the others would make successive passes in pairs, the wingman always protecting his leader's tail. In the same fashion, one would be the "enemy' fighter while the other attacked.

It was in just this "one-on-one" combat practice that the A.V.G. lost their first pilot. Two 2nd Squadron pilots, John D. Armstrong, of Hutchinson, Kansas, and J. Gilpin Bright from Reading, Pennsylvania, had taken off on the morning of September 8 for individual dog-fighting. The sky was clear and blue that Monday morning, as the two climbed out together, Armstrong in P-40 number P-8112, with Bright in P-8117. Both were former Navy pilots but had transitioned quickly into the P-40. Armstrong had even ferried a P-40 back from Rangoon on August 30th, just two weeks after his arrival at Kyedaw. Chennault was away in Chungking that morning and no one was in the tower to observe the dogfights. By eleven-thirty, all aircraft up for the morning flying should have landed. After the planes had landed and taxied in to their parking spots, there were still two missing – Bright and Armstrong. While the pilots were in the mess hall having lunch, a call came in from a village north of the field with the report that two aircraft had collided and had gone down in the jungle near the Mandalay road. Harvey Greenlaw, in charge during Chennault's absence, immediately directed three pilots to take off to locate the wreckage and then organized a crash convoy. A short time later, one of the pilots radioed back the location of the crash and the convoy sped off, led by chief flight surgeon, Doctor Tom Gentry.[30]

As they drove down the dusty dirt road toward the crash location, they came upon Gil Bright, riding back on a bicycle that he'd borrowed from a native. He was bruised and shaken up, but otherwise unhurt. As the group gathered around Bright, he recounted the accident. After climbing out together, the two of them separated, turning back in to one another when they were about two miles apart. The normal procedure was to make a head-on pass, followed by each pilot trying to get on the other's tail. The two P-40s, their closing speed better than 600 miles per hour, rushed at each other. Neither pilot gave way until the very last moment. Bright said he rolled right and down, but Armstrong's plane sheared his right wing off just at the fuselage. He had barely managed to get out of the wildly spinning aircraft. He looked for another chute, but didn't see one, and thought that Armstrong had gone in with the airplane.[31]

The crash party finally found the wreckage of the planes, their engines buried ten feet into the soft jungle ground. Jack Armstrong's body was still in his cockpit – the first fatality for the A.V.G.[32]

There were no caskets in Toungoo, so Harvey Greenlaw called Ed Pawley at the CAMCO offices in Rangoon to have one sent up. The casket arrived by special truck the next morning, accompanied by mounds of flowers. The wives of all the British officers in Toungoo also sent flowers, and two of the wives came to the Greenlaw house to help Olga make wreaths. The service was held that afternoon at four o'clock in the pilots' mess hall at the field. Paul Frillman read the service, after which Jack Newkirk led the six pallbearers who accompanied the flag-draped casket to the cemetery. Armstrong was the first American to be buried in the little St. Luke's Church of England cemetery in Toungoo. [33]

A week after the midair collision, the third contingent of volunteers arrived. Fifteen pilots and ten ground personnel had left San Francisco aboard the Java-Pacific Line's Motor Ship *Bloemfontein* on July 24. They made up nearly a third of the passenger list, many of whom were missionaries traveling to the Far East. It made an incongruous mixture, and even became a little embarrassing, since a few of the volunteers had "missionary" listed as their occupation on their passport.

Edgar T. Goyette, a former Navy lieutenant and fighter pilot, was placed in charge of the group. Although eleven of his charges were former Army pilots, he was selected because he was four-to-five years senior to most of the others. His Navy compatriots on the voyage were David L. "Tex" Hill, Allan B. Christman and Ben C. Foshee. Tex Hill, a slender, six-foot-two dive bomber pilot from the Lone Star State, had been born in China where his father was a missionary. He would become the second highest-scoring ace in the A.V.G. and one of the best-known of the Flying Tigers. Bert Christman had been a comic strip artist before he joined the Navy, working initially as an assistant to Milton Caniff, originator of the adventure strip, *Terry and the Pirates*. Later he had his own adventure strip, *Scorchy Smith*. He had volunteered for China to get first-hand material for future strips.

Most of the Army pilots in the group had P-40 experience –William D. McGarry and Robert P. "Duke" Hedman came from the 1st Pursuit Group at Selfridge Field; Frank Schiel and George B. McMillan the bird. The three remaining pilots on board were Robert T. "R.T." Smith and Paul J. "Pea" Greene, the two Army flight instructors from Randolph Field, and Edward F. Rector. Most of the ground personnel in the group were crew chiefs: Jasper J. Harrington, John Cornelius, and Preston B. Paul from the 20th Pursuit Group at Hamilton Field, Harold L. Osborne, Leon P. Colquette, Estill E. Durbin and James P. Dyson. Also on board were two radiomen: Robert J. King from the 20th Pursuit Group at Hamilton Field, ex-Navy man Mickey Mihalko and armorer Keith B. Christensen, also from the Navy's VF-22 at San Diego Naval Air Station, California.[34]

The voyage was very pleasant and routine from San Francisco to Hawaii. Three days out of Honolulu, however, the ship's captain announced a detour through Brisbane, Australia, en route to Manila to avoid German surface raiders. The extended course added a week to their voyage. Time was beginning to wear on the men, as observed in a letter home from R.T. Smith:

> *"Things have been pleasant enough aboard, but we're getting bored as the devil with the same old routine day after day. The damn missionaries - "Holy Joes" to us - are driving us nuts with their constant gatherings around the piano and singing hymns by the hour. So we drive them nuts by playing hot swing records on the phonograph; Goodman, Dorsey, Artie Shaw, et al."* [35]

Not everyone was that annoyed by the missionaries. Tex Hill spent much of the voyage being "converted" by a very attractive and unattached missionary lady. According to R.T. Smith, "they seemed pretty chummy by the time they parted company in Singapore." [36]

Twelve days out of Honolulu, the *Bloemfontein* reached Brisbane, took on fuel, water and some fresh vegetables. They departed for Manila on August 11 and reached the Philippine capital on the 20th. After two days in Manila, they set a northerly course for Singapore. Then, with Singapore less than a day away, the "Bloom" turned around and headed south again – the captain was steering for the ship's home port of Batavia (now Djakarta), the capital of the Dutch East Indies. Another delay; they arrived in Batavia on the 26th and stayed four days. In this length of time, 37 days, the previous group on the *Jaegersfontein* had already reached Rangoon.

Two of the volunteers had had enough. When Ed Goyette called roll prior to leaving Batavia, Durbin and Dyson were missing. Some reported that they'd been complaining long and often about "the whole setup," so they were written off as jumping ship, giving them the dubious distinction as the first to leave the A.V.G. [37]

The twenty-three remaining volunteers arrived in Singapore on September 1 in the midst of a torrential downpour. British Customs were particularly inquisitive, wanting to know where each had come from, where they were going, how long they were going to be in Singapore, why were they going to Burma and what were they going to do there, etc. etc. The volunteers had the same doctored passports that the previous A.V.G. travelers had, and R.T. Smith was questioned about his occupation as a "plantation manager."

"And what kind of a plantation do you manage, Mr. Smith?" the customs official asked.

"Uh, rubber," Smith replied.

"Odd," the official said, "the last I heard, there were no rubber trees in Burma."

"Oh, well how about teak, then," Smith offered.

"That would seem a bit more likely," replied the official, a hint of a smile crossing his face.

The disaster wrought by the first contingent in July had apparently been forgiven - if not forgotten – for the *Bloemfontein* travelers were put up in the Raffles with no problem. They spent ten days in Singapore, shopping and sightseeing between rain showers, and were even treated to a memorable banquet hosted by the Chinese Consul General – memorable because of the effects that the mixture of generous amounts of "Johnny Walker" Scotch and shark's fin soup had on the Americans.[38]

Ed Goyette suffered the same problem as had his predecessors with regard to the transportation from Singapore to Rangoon; that seemed to be where the CAMCO travel plans broke down. He had been told in Batavia that the *Bloemfontein* wouldn't be going past Singapore, but that transportation to Rangoon was "in the process of being arranged." After nine days in Singapore it was indeed arranged. A small coastal steamer called the *Penang Trader* would take them on to Rangoon.

The accommodations didn't match those of the "Bloom," but the food was good. On the morning of the 15th, the steamer entered the mouth of the Rangoon River, an offshoot of the Irawaddy, and made its way up the wide channel, finally dropping anchor in the stream just off the Rangoon dock. After the obligatory customs routine, the men boarded two small motor launches to the docks, where they were met by Skip Adair and Bill Pawley. After a brief welcome by Adair, they were taken to the "Strand" for a few drinks and lunch, and at 4:00 PM were on their way to Toungoo on the now infamous "up-mail" train.

Following a five-and-one-half-hour train ride, they pulled in to the little station at Toungoo, where they were met by a brass band – three Burmese in ragged, red and black uniforms playing a cornet, a trombone, and a large bass drum – blasting away at their own rendition of *The Stars and Stripes Forever*. A half dozen of the earlier arrivals were on hand to greet the newcomers, and soon all were herded into the Studebaker station wagons and sedans brought in by the welcoming committee and driven out to Kyedaw. Half the pilots and ground crew were now in place. [39]

The new guys were given a day to rest up from their trip, which they spent in Toungoo buying the accepted dress of khaki shorts and shirts, mosquito boots or long socks, and pith helmets. Some got a little P-40 cockpit time when they got back to the field. The following day, September 17, they began their training program.

R.T. Smith, the former BT-9 instructor from Randolph Field, walked out to the flight line with Neil Martin, 3rd Squadron Operations Officer. The two climbed up on the wing of the big P-40 and Smith settled his six-foot-two-inch frame into the narrow cockpit. He had spent some time in the cockpit earlier going over the location of the gauges, knobs, levers, and switches, and he'd studied the pilot's manual thoroughly. Now he was ready to go. After a brief review of the procedures, Martin clapped him on the shoulder and said, "Have fun, R.T., there's nothing to it," and jumped down from the wing. As Smith described his first flight:

> *"The Allison, sometimes cantankerous to start, fired up immediately and soon I was taxiing out to the end of the narrow runway. There was nobody in the little wooden control tower, so after stopping just long enough to check the mags I closed the canopy and turned onto the runway, got lined up and firmly eased the throttle forward. That big prop began to bite into the humid air as eleven hundred horsepower thundered a message I'd never heard before. We began to accelerate rapidly, manifold pressure gauge showing 48 inches, 3,000 rpm on the tachometer, now holding lots of rudder to counteract the powerful torque that was trying to*

pull us off to the left, forward pressure on the stick to get the tail up. Now there was a clear view of the runway ahead and in a matter of seconds we broke free of the runway and started to climb. A quick upward tug on the landing-gear lever, and a moment later a glance at the position indicator on instrument panel showed that the wheels were retracting. More back pressure on the stick, nose pointed skyward, reach down quickly and crank a little nose-up elevator trim; now back off on the throttle and prop-pitch control until the proper manifold pressure and rpm for climb showed on the gauges. A quick glance at the other instruments; airspeed now up to 150, rate of climb indicator at about 3,000 feet per minute, altimeter needle winding steadily clockwise, oil and coolant temperatures in the green, and up ahead of that long nose a big patch of sky showing through broken clouds." [40]

After an hour and a quarter of putting the ship through a series of power-on and power-off stalls, slow-rolls, snap-rolls, loops, Immelmanns and spins, it was time to put the Tomahawk down on the ground. Smith said he felt "like a kid with a beautiful new toy."

He continued to describe the landing:

"The landing itself was uneventful, though the recommended procedure was unlike that for the BT-9 and most other airplanes of the time. Attempting a three-point landing with the P-40 was asking for trouble, as some of the ex-Navy pilots learned to their chagrin; with the tail down and nose high, engine throttled back, there was insufficient airflow to provide good rudder control, a sure invitation to a ground loop. The cure for the problem was to land in an attitude that more nearly resembled level flight, touching down with the main wheels only at about 90 mph, and holding the tail up with forward stick pressure until the speed slowed con-sider- ably, after which directional control could be maintained by judicial use of the brakes. Even so, there were to be ground loops galore in the next few weeks, even a few on take-off." [41]

Five days later, on September 22, the A.V.G. suffered their second fatality. Maax Hammer, who had come over with the *Bloemfontein* bunch only a week before, had taken a P-40 up on a routine flight. The weather was marginal, low clouds and thunderstorms in the area. Perhaps he was a little overconfident – after all, he was one of the A.V.G.'s more experienced P-40 pilots – but he decided to take the ship up anyway. During late afternoon, the R.A.F. office at Toungoo received a call from a village headman, a few miles south, that an aircraft had crashed during a thunderstorm and that the body of the pilot had been found in the wreckage. As darkness approached, a search party was organized. Equipped with flashlights and rubber boots to protect their legs from leeches and mosquitoes, the men found the wreckage about fifteen miles south of the field. The flashlight beams revealed Hammer's mangled body in the twisted remains of the cockpit. From the condition of the wreckage, it was obvious that recovery operations would have to wait until daylight. Two men volunteered to stay at the site, and the search party returned to the field.[42]

The next morning they brought the body out. At three in the afternoon Chaplain Frillman conducted his second funeral in two weeks, once again in the pilot's mess. They brought the tin-lined teak casket back to the British Cantonment Cemetery in Toungoo and Maax Hammer was laid to rest in Plot D-15 beside Jack Armstrong. [43]

No one knew for sure what had happened to Hammer's aircraft. The men speculated, however, that he had attempted to fly through a thunderstorm rather than around it. The cumulonimbus clouds associated with thunderstorms are known to contain violent up-drafts and down-drafts and turbulent air currents severe enough to toss even an aircraft as large as a P-40 around like a toy. It was surmised that as Hammer's plane was caught by these currents he became disoriented, lost control and spun in.

The day after Hammer's funeral, five pilots and one crew chief submitted their resignations. It can be argued which had the greater effect on their decisions – the heat, bugs, and generally poor living conditions, or the two recent fatalities. In any case, pilots E.J. Cook, John P. Kelleher, Maurice G. McGuire, Robert H. Power, and S.H. Wallace, along with crew chief Carl E. Schur, became the first to resign from the A.V.G. There would be more. [44]

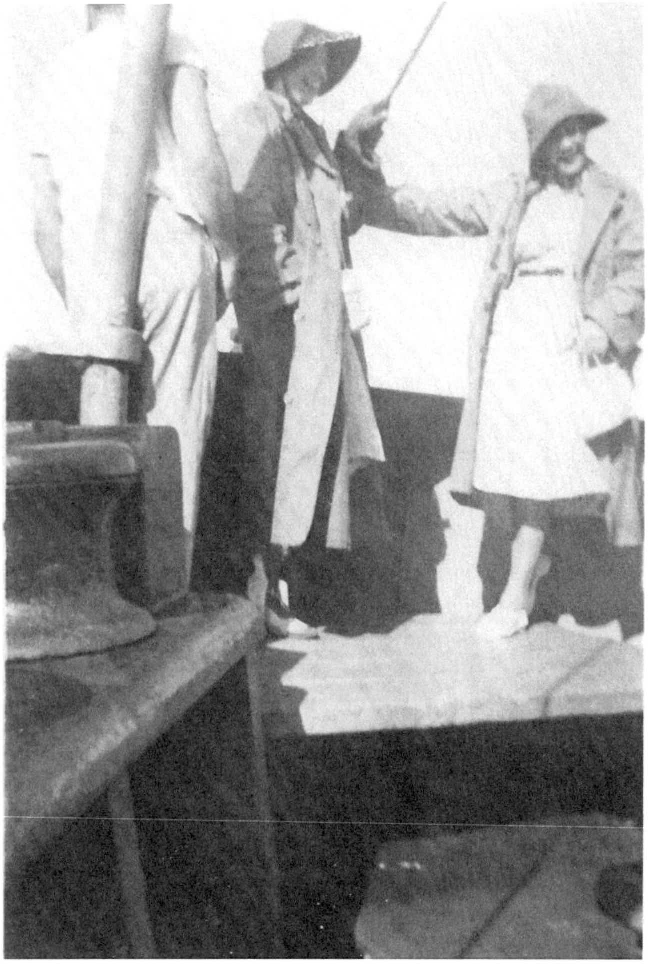

Above: Nurses Jo Stewart and Emma "Red" Foster aborad the Jagersfontein *during "King Neptune Watch" while crossing the Equator. (via John Williams)*

Above right: Interior of the pilots' barracks at Toungoo, "...complete with snakes, lizards, ants and sleeping accommodations." (via Lee Burgard)

Chennault and Harvey Greenlaw at Toungoo. (via John Williams)

Far left: John Dean Armstrong. (via George Paxton)

Left: Maax Curtis Hammer (via George Paxton)

From portraits by Raymond P.R. Neilson, commissioned by William D. Pawley for his 1945 booklet Americans Valiant and Glorious.

Second Squadron pall bearers carry the casket of John Armstrong to the St. Luke's Church of England cemetery in Toungoo. Left to right: Charlie Mott, Jack Newkirk, John Petach, Peter Wright and Noel Bacon. (via Tripp Alyn)

CHAPTER 6

More Accidents, Another Fatality and Resignations

The Java Pacific Line's *Zaandam* contingent (after transferring to the British freighter *Pancius* at Batavia) arrived in Rangoon on October 11 with Chuck Older, Frank Swartz, Ken Jernstedt, Tom Haywood, Bill Bartling and Johnny Farrell and thirty-two volunteer instructors for the CAF flying school. The *Klipfontein* was en route with twenty-eight more. Back in the United States, recruiting continued at a slow but steady pace. By October 18, 289 volunteers had been inducted into the group – 102 pilots and 187 ground personnel. Of that number, 217 had reported to Toungoo. The remainder were either en route or waiting for a scheduled departure.

By this time sixty-seven Tomahawks had been delivered to Toungoo.[1] Training accidents, however, had accounted for the loss of four; in addition to the three previous accidents, Frank Schiel lost a P-40 when he got into an inverted spin during a practice dogfight with Erik Shilling on October 7th.

During the mock combat, Shilling got on Schiel's tail. In an attempt to shake him, Schiel yanked on the stick, climbing at a steep angle into the sun with the hope of blinding Shilling if he followed. Shilling, however, leveled out and broke off the attack, before his aircraft reached a dangerous angle of attack.

Shilling was one of the A.V.G.'s most experienced fighter pilots, with considerable time in the Curtiss XP-36 and XP-37 – both predecessors of the P-40 – and knew of the tumble characteristics of the bird. If the aircraft stalled in an extremely nose-high attitude, it would begin a tail-slide as it stalled. Because of the long, heavy nose of the big fighter, it would then fall through with the nose making a complete circle as momentum carried it around. This would happen once or twice more, as the aircraft fell almost 10,000 feet before it would stabilize in a controllable dive. Since Schiel lost control of the Tomahawk at 8,000 feet, he didn't have a chance of recovering.

As Schiel's P-40 began to spin, Shilling followed it down, spiraling around the aircraft as it fell. Schiel finally bailed out at 3,000 feet. His aircraft impacted with a mighty explosion in a rice paddy as he floated down to land in a wide and deep *klong* (an irrigation tributary) that fed the rice paddies. He was picked up by a native who paddled out to him in a small boat.[2]

According to R.T. Smith:

> *"Inverted spins were not common, but when they did occur the P-40 tended to flatten out and flight controls were ineffective. Without a lot of altitude to play with, a pilot was wise to hit the silk before it was too late."* [3]

The Java Pacific Line's *Klipfontein* carried the fifth load of A.V.G. volunteers. Of the twenty-eight on board, ten were pilots and eighteen were ground crew. All of the pilots had signed up as a group from Naval Air Station San Diego. The former PBY pilots included Joe Rosbert, George "Pappy" Paxton, Tom Cole, Robert "Catfish" Raines, Hank Gilbert, Carl Brown, Ed Conant, Leo Houle, Morris Bohman, and Dick White. The ground crew, by contrast, were all ex-Army personnel. Joe Rosbert was selected to be in charge of the group.[4] Departing San Francisco on September 4th, the *Klipfontein* followed much the same route as did the previous liners – Honolulu, Batavia, Samarang, Singapore, and on to Rangoon.

The trip was uneventful, with the exception of some antics by two of the ground crew, Joe Ringey and George B. "Dutch" Jaeger. They provided enough diversion to keep Rosbert busy most of the trip. First they showed up late – and drunk – for departure roll call leaving Honolulu and they attempted to bring a water buffalo on board the ship at Batavia. When the ship pulled in to Singapore, the Raffles Hotel could accommodate only 16; the remainder would have to be put up at the Adelphi Hotel, about a half-mile away. Rosbert took advantage of the opportunity and put Ringey and Jaeger with the Adelphi crowd. The next few days were peaceful, then Rosbert received a message from the British manager of the Adelphi to see him immediately. Ringey and Jaeger were throwing a party, he learned, and were tossing empty beer and scotch bottles out of their window at taxicabs.

When Rosbert arrived at the Adelphi, the British manager was fit to be tied. He said that he had first sent a note to the room, requesting they be a bit quieter. When they ignored that, he called the local military police. The revelers responded to that action by inviting the two policemen in and having them join the party! As a final resort, the manager said, he went to the room himself and asked that they quieten down. His face and neck became redder and redder as he related the final indignity wrought upon his establishment. The ruffians grabbed him by the scruff of the neck and threw him bodily from the room! Rosbert promised to have all of them out of the hotel by the next morning.

Rosbert was as good as his word. The next morning he paid the bills at the Adelphi, then took his two miscreants to the dock and left them in the charge of the captain of the *Tarakan*, the Dutch freighter that was to take the volunteers on to Rangoon. They left Singapore on October 19th and headed up the west coast of Malaya. Ten days later, after stops at Port Swettenham, Malaya, and Penang, the *Tarakan* entered Rangoon Harbor. After a three-hour wait to clear customs, the volunteers were allowed to disembark and were met by Ed Pawley. By this time, the arrival procedure had settled in to a routine – get a bite to eat at the Strand Hotel and then board the up-mail Mandalay Express for Toungoo.

Joe Alsup, the last of Chennault's staff, joined the new arrivals at Rangoon and came up on the train with the group. Alsop, who had met Chennault in Washington, D.C. during the fall of 1940, had taken an interest in the A.V.G. during that first meeting. As a Washington insider, and familiar with the brass in the military command circles, he had wangled a reserve commission in the Navy and was subsequently sent to Bombay as an observer. Unhappy with the assignment, he discussed the problem with two close friends, Undersecretary of the Navy James E. Forrestal and Undersecretary of the Navy for Air, Robert Lovett. They suggested that he contact Claire Chennault, who was then back in Chungking. Alsop arranged for his itinerary to Bombay to take him through Chungking, where he met Chennault at one of Madame Chiang Kai-shek's garden parties. Alsop offered his services and Chennault, desperate for qualified staff help, said he'd be glad to have him work in administration and press relations if he could be released from the Navy.

Alsop's first attempt at resigning was angrily rejected, whereupon he sent a private telegram to Forrestal at his home telling him, in defect, that he had every intention of joining Chennault and that "they can come and arrest me if they choose." Forrestal arranged his release, and Alsop reported to Chennault at Kyedaw Aerodrome on October 29, 1941. As an accomplished journalist, Alsop's writing talents proved invaluable, as he turned out most of the correspondence from Chennault to Chungking, Loiwing, Rangoon, Singapore and Washington.

Morale at Toungoo was at a low point when the *Klipfontein* group arrived. Just four days earlier, Pete Atkinson, an experienced P-40 pilot from Martinsburg, West Virginia, had been killed while flight testing P-40 No. 1. At about 0900, as he nosed over into a power dive from 18,000 feet, the prop governor failed and the prop went out of control, rapidly accelerating the speed of the dive. The shriek of the runaway prop awakened R.T. Smith from his Saturday morning relaxation with, "the most God-awful sound I ever expect to hear," he wrote later. He described it as "an unearthly screaming roar, going up very suddenly to a high pitch and then a sound like an explosion." [6] Observers on the ground saw the tail separate at about 800 feet and then the rest of the aircraft disintegrated. Wreckage was spread over several acres.

The camp took Atkinson's death hard. A devout Catholic, he was popular with all, a "great guy," as one described him.[7] Olga Greenlaw arranged a Catholic burial for him, and the high mass ceremony was conducted by a bishop and twenty priests, with one-hundred choir boys in attendance. Atkinson's coffin then joined those of Armstrong and Hammer in Toungoo's St. Luke's Church of England cemetery, beside the graves of Hammer and Armstrong.

Chennault suspended flying training while his maintenance crews inspected the tail sections and wing roots of the P-40s. Ken Merritt had lost an elevator in flight earlier, but had managed to get the bird back on the ground without further mishap. As it turned out, the elevator fittings on both aircraft failed at the same point. No more flying was scheduled until Friday, October 31, when check-out of the new arrivals began.

Ed Conant was one of the first ex-PBY pilots up. As he came across the landing threshold, he leveled off twenty-five feet in the air, stalled out and dropped it in. The P-40 bounced as the landing gear took the force of the fall, then came down on one wing and the belly, and spun 180 degrees around before it slid to a grinding halt. Conant climbed out of the cockpit embarrassed but unhurt. The plane was a write-off.

Later that morning George McMillan had two feet of his right wing clipped off by Ed Leibolt's prop. Leibolt was parked ninety degrees to the runway, right on the edge. As McMillan hurtled past on his takeoff roll, his wingtip contacted Leibolt's idling prop. McMillan continued the takeoff roll, then came back around and landed without further incident.[8]

There were no mishaps the following day, but, as R.T. Smith recalled, there was "many a breath-taking moment during most of the landings." Monday, November 3, proved to be more exciting. Ed Conant (again), landing long, found he couldn't stop as he reached the end of the runway. He stood on the brake pedals and the P-40 stopped – a three-point landing, but on the main gear and nose, badly bending the prop.

Catfish Raines was next. As he began his takeoff roll a tire blew out. His P-40 veered off the runway and ran across a ditch, collapsing the landing gear. As the P-40 slid to a stop on its belly, the prop was bent back 90 degrees and one wing was bent.

Later in the day a crew chief taxied one plane into another, ruining both props. Finally, 1st Pursuit Squadron commander "Sandy" Sandell ground-looped on landing after returning from a cross-country to Lashio, proving that the Tomahawk could bite even experienced P-40 pilots. But five ships were put out of commission in one day! [9]

The next day only three P-40s were damaged. Henry Gilbert, newly-graduated from Navy flying school, ran his engine up to 3,500 rpm, well past tech order limits, and burned out the bearings. Bill Reed, another high-time P-40 pilot, landed too long and ran off the end of the runway. He hit a stump, nosed over, and bent the prop. A mechanic, bicycling while watching Reed's landing, pedaled into the tail of a parked Tomahawk, tearing off a piece of the aileron, putting the tenth plane out of commission. Erik Shilling, Group engineering officer, observed that, at the present rate of attrition, the A.V.G. would be out of airplanes in six weeks! [10]

Chennault was appalled at the abominable accident rate. The early morning ground school session on November 5th was devoted largely to flying safety and a long lecture on proper landing techniques. To emphasize his point, at Shilling's suggestion, Chennault announced that three-foot wide white lines was to be painted across the runway, 1,000 feet from each end, to mark one-third of the runway length. Henceforth, anyone touching down past the white line would be fined $50. [11]

Later in the day, poor Ed Conant washed out yet another aircraft, although (fortunately!) this time because of aircraft problems. As he came in to land, his left tire blew out, the landing gear collapsed, and the plane slid several hundred feet on its belly – another P-40 written off to the bone-yard, and Conant's third accident in four days of flying. [12]

To compound Chennault's problems, that same day two of the most recent arrivals – pilots Morris Bohman and Richard White – tendered their resignations, less than a week after they had reached Toungoo. The reasons they gave Chennault were several, but they boiled down to misrepresentation by the CAMCO recruiters. Although it had taken them six days to turn in their papers, they said they had decided on resigning within twenty-four hours of their arrival.

Chennault accepted their resignations, then sat down and drafted a scathing letter to Richard Aldworth, the retired Air Corps officer in charge of recruiting for the A.V.G.:

> *Dear Dick:*
>
> *Too many pilots recently sent out complain that combat conditions here were misrepresented to them by the members of the CAMCO staff with whom they discussed the Group. So many of them tell the same story, I can only conclude there is some truth in it. The chief misrepresentations named are four in number:*
>
> *(1) That combating unprotected bombers will be the Group's only task, and that no member of the Group will*

be asked to combat Japanese pursuit.
(2) That no member of the Group will be sent on an offensive mission.
(3) That no night flying will be required.
(4) That the Group's strategic mission is confined to protection of the Burma Road.

Every one of these statements is untrue, and when a new pilot learns they are untrue and complain to me that he has been deceived into volunteering, I have no recourse but to say I am sorry. Most pilots volunteering under a misapprehension of the Group's real nature are quite useless to me, and sending them to the Far East is a plain waste of their salaries and travel expenses. Captain Irvine and Captain Claiborne have been particularly mentioned as having painted a false picture, and I wish you would warn them and all other members of the CAMCO staff to discontinue the practice forthwith.

Furthermore, I wish you would require all members of the CAMCO staff to explain fully and frankly to each prospective employee the precise realistic probabilities of service in the Group, which are almost the exact contrary of the statements set forth above. I know this will make it much more difficult to fill the employment quotas; but the difficulty will be amply compensated for by the great saving of time, trouble and money wasted by the present system. I should much prefer to get fewer pilots, but all up to the job and ready for it, than to get the mixed lots I do now.

Two days later, he followed up with another letter to CAMCO headquarters in New York, citing specific examples of his complaints to Aldworth: [13]

...Typical of these problems is the case of Pilot Officer E.S. Conant, who reported at this station with nine other Navy pilots on October 29. Conant has the rating of a four-engine flying boat pilot. He informs me that for more than a year before your representatives accepted him for service with the A.V.G., he had never flown a land plane of any sort. He has no pursuit experience. I need hardly point out that a pilot so trained is hardly qualified for combat in Curtiss P-40s. I may add that the result of Pilot Conant's employment is precisely what might have been expected; he has cracked up three planes in the first week of flying. Furthermore, as he is a young fellow of some spirit, he still wishes to remain with the Group, and although he is obviously unsuited for the mission the A.V.G. must undertake, I can find no way of sending him home under the contract.

Two more of these ten Navy volunteers, Pilot officers Bowman and White, decided to go home within twenty-four hours after their arrival at this station. Their stated reason was that the conditions of service had been falsely described to them, but I am certain that they were also actuated by a sense of unfitness for the work. None of the remaining seven Navy pilots has the sort of experience we want, and although two or three of them may eventually prove to be useful members of the A.V.G., the process of adapting them to our purposes is exceedingly expensive....

Let me repeat, much money and much irreplaceable equipment has already been wasted, and the A.V.G.'s combat efficiency seriously lowered, by the employment policy that has been followed. I am aware that this policy makes it far easier to fill the employment quotas. But I prefer to have the employment quotas partly unfilled, than to receive pilots hired on the principle of "Come one, come all."

Chennault was filled with dismay as he reviewed the 102 application forms recording each pilot's flying experience. His original plans called for hiring only the most qualified pilots, between the ages of twenty-three and twenty-eight, with at least three years' experience in fighters. Only 12 met the criteria. His oldest pilot was Louis "Cokie" Hoffman, a 43-year-old former Navy enlisted pilot; the youngest was Henry Gilbert, just 21 and fresh out of Navy flying school at Pensacola. Twenty had flown P-40s previously, but most of those had less than two years of fighter time. The majority of the ex-Navy pilots had flown only multi-engine patrol bombers or single-engine dive bombers.[14]

Chennault was expecting another boatload of volunteers around the middle of November. From their records, they looked little more promising than the last bunch – few had any fighter time. However, in this group would be two ex-Marines that Richard Aldworth had recommended highly – Gregory Boyington and Curtis E. Smith. In a letter to Chennault on October 7, Aldworth had urged, "your particular attention to these pilots, especially Gregory Boyington... I investigated him thoroughly, and the more I got to know him, the better I liked him." He further said, "Curtis E. Smith's record speaks for itself. I feel you would be doing yourself a favor if you make them both Squadron Commanders." Boyington did fit the criteria – he was 28 and had six years of military flying behind him, mostly in fighters, Chennault noted from his records. He would wait and see.

The *Boschfontein* departed San Francisco on September 24, 1941. Curtis Smith and Boyington were on board, as expected. Smith, because of his age (35) and former rank of captain, had been selected to be in charge of the group en route. Two more ex-Marines, C.H. "Link" Laughlin, and Edmund F. Overend, were on board, as were eleven ex-Navy pilots – Robert F. Layher, Robert B. "Buster" Keeton, John S. Croft, John R. "Dick" Rossi, Lynn A. "Tex" Hurst, Freeman I. Ricketts, Frank W. Adkins, Ralph W. "Luke" Gunvordahl, Robert W. Prescott, Clifford G. Groh, Einar I. Mickelson, and Lewis S. Bishop. The Army pilots, in the minority on the voyage, included Charles R. Bond, Jr., George T. Burgard, and James D. Cross, all of whom had shared a last duty as ferry pilots for the fledgling U.S. Army Air Corps Ferrying Command, delivering Lockheed "Hudsons" from Burbank, California, to Montreal. Other ex-Army pilots on board were Donald R. Bernsdorf, Fred S. "Fearless Freddy" Hodges, and Robert C. "Moose" Moss.

The volunteers shared the *Boschfontein* with about sixty civilians, fifty-five of whom were missionaries en route to China. This soon proved to be a little awkward, since many of the volunteers, including Boyington, were traveling under cover passports labeling them as "clergy." The men never did admit their subterfuge, although Boyington had to decline an invitation once to give the Sunday service! [15] The voyage took forty-eight days, sailing by way of Hawaii, Australia, Java, and Singapore.

The long days of inactivity began to wear on the men, and tempers flared easily. Charlie Bond's diary entry for October 12th recorded:

> *"Everyone is becoming tense and irritable. Got into a hell of an argument with Jack Croft about something concerning the Neptune certificates. First time I was fighting mad on the trip, and I invited him on the deck. He wouldn't go. We're all drinking a lot."* [16]

The "Neptune certificate" mentioned by Bond was presented as a part of the Equator-crossing ritual. The *Boschfontein* group really made a grand ceremony of it. Starting at 1100, they held "King Neptune's Court" on the after deck. Lew Bishop was "Father Neptune" and Bond was his "Queen." Dick Rossi was the barber, and Luke Gunvordahl and Tex Hurst were the policemen. As Bond described the affair:

> *"Our faces were painted wildly with rouge and we dressed in shorts, for all of us expected to end up in the swimming pool. Dick had a large wooden straight razor and a pail of horrid-smelling goo that was shaving cream. It took several men to hold down the guy who got a shave, for the goo was made of flour, water, fruit juices, and fish soup which had been cooked days before and set out to spoil. A dish-washing mop was the shaving brush, and when the shaving cream was applied, a load of it always got in the poor guy's mouth. That, together with the fact that he was already skunk drunk, meant that he usually wound up tossing his cookies."* [17]

Before the celebration was over, fights had broken out between Bond and Croft (again!), Boyington and Keeton, and Gunvordahl and Prescott, during which "Zipper" Gunvordahl (a "very big fellow," according to Dick Rossi) knocked Prescott down. Laying there, a bit dazed, Prescott looked up at Gunvordahl and said, "Have you had enough?" Five days later, George Burgard and Frank Adkins tangled as a result of a bar argument. Curtis Smith stepped in to break it up and got a split lip for his trouble.[18]

When the group reached Singapore, they ran into Harvey Greenlaw and six A.V.G. members, who were in the city to pick up supplies for the home base in Toungoo. Greenlaw had acquired fifty P-40 tail-wheel assemblies from the R.A.F. which he placed aboard the *Boschfontein* for shipment to Rangoon. He spoke some words of encouragement to the group about the situation in Burma, but added that "there are some awful flyers there in Toungoo flying for Chennault," Bond wrote in his diary that night, "I don't know if he is speaking of their character or their flying ability. I will soon find out." [19]

Arriving at Rangoon pretty much on schedule on November 12, those on the *Boschfontein* turned out to be the final group of pilots to arrive in Burma. One more ship, the *Zaandam*, was en route, but it carried only six volunteers – all ground crew – and nine pilot instructors hired for the Chinese training school at Yunnanyi.

Even before the last contingent of pilots departed San Francisco, Chennault and the pro-China crowd were lobbying for a second volunteer group – this time a bomber force. In October President Roosevelt authorized 106 additional pilots and 181 ground crew personnel to be hired. Assistant Secretary of War for Air Lovett restricted the recruiters to thirty-seven bases, primarily those not visited the first time around.[20] Although many were recruited on this drive, none reached China before the Japanese attack on Pearl Harbor, and after that all en route volunteers were brought back into American military service (including Albert J. "Ajax" Baumler, who had scored 4½ victories during the Spanish Civil War) and who would return to China following the induction of the A.V.G. into the 23rd Fighter Group.

Meanwhile, back at Toungoo, the idea was developed of painting the snarling shark's mouth and baleful eye on the nose of the A.V.G.'s P-40s. Charlie Bond's diary records that on Saturday evening, November 15, 1941, about ten pilots piled into one of the group's Studebaker station wagons and went into Toungoo to the local Baptist mission, called the Bunker Memorial School and Church, for dinner.[21] Dr. Harry I. Marshall and the Reverend Chester L. "Roy" Klein, along with his wife, Alice, ran the mission. Since early August the Kleins and Dr. Marshall had extended the hospitality of the mission to the young Americans at Toungoo, and nearly every weekend found some of the A.V.G. at their dinner table enjoying an American-style, home-cooked meal. Following a dinner that included baked beans and home-made Boston brown bread, the group crowded into the living room for coffee and conversation. Among those there that night were Charlie Bond, Jim Cross, George Burgard, Gil Bright, Erik Shilling, Ken Merritt and Lacey Mangleburg. The men were seated around a rattan coffee table, on which were scattered several magazines and newspapers. As Charlie Bond recorded:

> *"I noticed a magazine section of a British newspaper on the table. I was attracted by the picture of an Australian P-40 on the front page that was exactly like ours, but with one exception - the lower nose section around the oil scoop inlets was brightly painted red, white, and green in the shape of a snarling tiger shark mouth wide open with teeth exposed."* [22]

"Gee, I'm going to paint my P-40 that way!" Bond exclaimed to the group. Heads nodded in agreement as each pilot envisioned what his own P-40 might look like with the snarling mouth and wicked eye painted on its cowling.

Reverend Klein reinforced the idea, commenting on sharks relating to Japanese culture. At one time he was based in Tokyo, he said, and told of the age-old Japanese fear of sharks:

> *"Japan depends on the sea as an important source of food," he continued," and pearl harvesting is a major industry as well. There have been many stories of fishermen being killed and pearl divers being destroyed by sharks, which has accentuated the myth of the invincibility of the shark and brought about a general fear of the creature."*

The more they talked, the more the group agreed that the shark mouth design was a natural choice for the A.V.G.'s P-40s. Before leaving, Charlie Bond asked Mrs. Klein if he could keep the magazine and took it back with him to the barracks that night.

The next day Bond, excited about the prospect of painting his airplane, bicycled into Toungoo to buy cans of red, white, green and black paint and brushes. He spent that day and much of the next outlining the markings from the magazine onto the nose of his P-40, No. 5, in chalk, then filling in with paint. [23]

Erik Shilling remembered the situation a bit differently. As he recalled it in his autobiography *Destiny*, self-published in 1963:

> *"After the meal, I picked up the rotogravure section of the* India Illustrated Weekly, *a Rangoon newspaper lying on the coffee table. Leafing through it, a picture of a German Messerschmitt 110 caught my eye… On second glance, I saw it had a shark's mouth painted on the nose of the fuselage. It didn't seem to fit and I immediately thought the P-40 would look better with the shark's mouth than the Messerschmitt. The P-40 already looked like a shark. I visualized that after it was painted, it would enhance its appearance making it look even more like a shark."* [24]

It would be a controversy between Bond and Shilling for the next 50 years, with each maintaining that he was the originator of bringing the shark's mouth design to the A.V.G. In all likelihood, both were right, with Bond taking a cue from the full-color cover of the *India Illustrated Weekly*, illustrating a Tomahawk of No. 112 Squadron, Royal Air Force, bearing the shark's-teeth marking on its cowl, while Shilling concentrated on an inside article with an illustration of the insignia of the Luftwaffe's Me-110s of the *Zerstörergeschwader* 76.

The same morning that Bond was busily painting his Tomahawk No. 5, farther down the flight line among the Third Squadron P-40s, Erik Shilling, accompanied by Ken Merritt and Lacy Mangleburg, chalked out the outline of a shark's mouth on the nose of his aircraft. Before going further, however, he asked Chennault for approval to use the design as a distinctive Third Squadron insignia. "No," said the Old Man, "but we can use it as a Group marking." At that, Shilling began painting in the outlined mouth and teeth. After finishing his aircraft, Shilling helped Mangleburg and then Merritt to paint theirs.[25]

The idea conceived by Bond and Shilling spread quickly throughout the Group and soon every Tomahawk at Toungoo bore the tiger shark markings that became the trademark the world over of the Flying Tigers.

The new guys began their checkout program on November 16. Ex-Navy pilots Ed Goyette and Noel Bacon had briefed them two days earlier on the flying characteristics of the P-40 and the local flight rules and regulations.[26] Most of the group were ex-Navy pilots with little or no experience in high-performance, single-engine fighters. Although R.T. Smith noted, "there were plenty of thrills again," during the landings, all went well, and no one ground-looped.

Only one accident marred the day's record. Jim Cross, flying No. 11 on his first flight, had an engine rod failure shortly after takeoff. As Cross described it:

> *"I saw some rice paddies dead ahead and went for them. By this time, fumes and smoke in the cockpit were getting so bad that my eyes were burning and tearing and I could hardly see. I got the airplane down near the rice, leveled, and waited until I could hear and feel the rice hitting the belly, pulled back and waited. She went on very nicely, slid through three dikes and on the fourth dike the tail went up to a 50 or 60 degree angle and fell back. I hit the quick disconnect on the harness and up I went. Bango! Back down. I tried again – same thing. Darned near knocked myself out on that one. I finally realized that when the plane tried to stand on its nose the canopy slid to the closed position. I finally got the canopy open, jumped out, ran 100 yards, then thought, 'Dumbo! There's nothing but more jungle out there.' So I returned to the plane, got the fire extinguisher and put the fire out."*

Cross was helped by orange-robed monks who had appeared at the scene. Outside of a bent prop and some minor skin damage to the underside of the fuselage, the plane wasn't badly damaged. Cross turned and headed back toward Kyedaw, four miles away. No one had seen him go down, and it had been two hours since he'd taken off. Those back at the base were beginning to become a little concerned. He soon trudged in, however, appearing at the door of the mess hall, a sheepish grin on his face – just in time for lunch.[27]

George Burgard checked out the following day. He was very comfortable with the P-40, as he wrote in his diary that night, "The P-40 is just another airplane – different in many respects, but all in all nothing to disturb a good flyer's equanimity." [28] His checkout came just in time, because the next day Freeman Ricketts came in with his gear partially lowered and tore up another airplane. Chennault promptly grounded the new pilots until further notice.[29]

The seventh and final contingent of volunteers arrived at the railroad station at Toungoo at 2130 on November 25. There were six ground personnel for the A.V.G., among whom were aircraft mechanics Irving J. Stolet and Joseph Peeden, and nine pilots who had been hired by CAMCO to instruct at the Chinese flying school at Yunnanyi. Four of them: E.W. "Bus" Loane, Lester J. Hall, Van Shapard, Jr., and Arnold W. "Red" Shamblin, would join the A.V.G. in the final weeks before the disbandment of the Group.

The following day Byron Glover brought in the 99th P-40, tail number P-8101. Now Chennault had 82 pilots and 62 aircraft in commission. Seventeen more Tomahawks were not flyable for various reasons, and 25 pilots were not considered combat ready. Indeed, four months later Chennault considered eighteen still not proficient enough for combat flying.

It was early in December when the squadrons adopted the nicknames and distinctive insignia for their units. During a bull session, First Pursuit Squadron pilots Dick Rossi, Bob Prescott, and Charlie Bond were tossing around ideas. In a flash of inspiration, Dick Rossi proposed that the "first pursuit" was Adam chasing Eve, and that the First Pursuit Squadron be called the "Adam and Eves" Squadron. Charlie Bond picked up on the idea and sketched an outline of an apple, within which was drawn two stick figures of a man chasing a woman.

Bond was as eager to decorate his ship No. 5 with the new insignia as he had been to paint shark's teeth on its nose, and was busy the next day with his paint and brush. Soon a big red apple adorned the side of the fuselage, just beneath the cockpit. Chennault, driving by in his Studebaker sedan, stopped to view Bond's handiwork.

"Charlie," he said, "it just won't work."

"Why?" asked Bond. "I think it's good."

"Because," replied Chennault, "it looks too much like the Japanese Rising Sun, Charlie. It just won't do." With that, Bond changed the color of the apple to green, since there were obviously green apples as well as red, and the Old Man was satisfied.[30]

The Second Squadron adopted the Chinese panda bear for their insignia. Squadron member Bert Christman, former professional cartoonist, volunteered to draw distinctive panda bears for each squadron pilot, reflecting a unique characteristic of each. For instance, he drew Tex Hill's panda wearing cowboy boots, six-shooter and hat, Moose Moss (from Doe Run, Georgia) was represented by a panda with a moonshine jug, while Gil Bright, who had so recently joined the "Caterpillar Club" after bailing out following his unfortunate midair with Jack Armstrong, was shown as a panda falling headfirst through the air reaching for his ripcord.

Oley Olson's Third Squadron chose "Hell's Angels" for their logo, borrowing the name from the early film epic of the same name about World War I flyers. Stan Regis, Third Squadron crew chief and a very talented amateur artist, painted a silhouette of a curvy, nude female in red with white wings on each squadron aircraft – each in a different pose.

Peter W. Atkinson. From a portrait by Raymond P.R. Neilson, commissioned by William D. Pawley for his 1945 booklet Americans Valiant and Glorious. *(via George Paxton)*

Paul Frillman performs burial service for Peter Atkinson, the third A.V.G. fatality. (Courtesy Atkinson family)

Initiation rites on the Boschfontein *during the King Neptune ceremony while crossing the equator. (via Charlie Bond)*

The "royalty" of the King Neptune ceremony included, from left to right: "Queen" Charlie Bond, "King" Lew Bishop (behind Bond), Bob Prescott and "Barker" Dick Rosssi. (via Charlie Bond)

Accidents...

Accidents...

And more accidents. (via P.J. Perry)

Curtiss Tomahawks of 112 Squadron, Royal Air Force. It was one of these, flown by an Australian pilot, that inspired Charlie Bond and led to the markings that were immortalized by the Flying Tigers.

Squadron artists painted their respective Tomahawks with their individual emblems. Left to right: 1st Pursuit Squadron: Adam & Eves, 2nd Pursuit Squadron: Panda Bears; 3rd Pursuit Squadron: Hell's Angels. (via Flying Tigers Association)

CHAPTER 7

North to Kunming

Throughout November and into December tension mounted. Following the oil embargo by the United States on Japan in the summer of 1941, it was evident that Japan would be forced to look elsewhere for the oil to fuel their expansionist drive.

As summer gave way to fall, Japan began to pull back her experienced divisions in China, replacing them with Chinese puppet troops. Military intelligence reported Japanese build-ups in Canton, Haiphong, Hainan Island and Formosa – all strategic staging areas. The new Japanese Zero was reported for the first time at Tan Son Nhut Aerodrome in Saigon, and Japanese aircraft in Indochina tripled during November. As Chennault observed, "It was no secret that the stage was being set for a drive to the Dutch East Indies as the minimum objective and an all-out offensive to drive the Western powers from the Pacific as the maximum possibility." [1]

From mid-November on, the camp buzzed with talk of moving up to Kunming. On November 21, Chennault called a morning meeting to brief the men on what to expect when they arrived in China – customs and conditions. He told the group they would be under the directives of Generalissimo Chiang Kai-shek and would effectively be in the Chinese Air Force. He went on to say that they would have a support force of nearly 1,000 Chinese. John Williams, an old acquaintance of Chennault's from his Maxwell Field days, was already in Kunming, overseeing the preparations for the group's arrival. Toward the end of his lecture, Chennault reiterated what he had said in countless chalk talks over the past 90 days: "If you use it right, the P-40 can whip the Zero. The Zero can't match you in a dive." And he warned the men again, "Never, never try to do dogfight with a Zero. Hit and run - hit and run! Dive and then use your speed to climb back to altitude and do it again! But stay together – always fight in groups of at least two." [2]

By November 25, rumors were flying that the Group would be leaving in a week or ten days. On the 29th the First Squadron took the opportunity to have a picnic before they left, however, taking a truck out into the mountains east of the city. Charlie Bond commented in his diary that "I was the only one that did not drink beer. Had I drunk it, I am sure that I would have had a merrier time. Everyone got tighter than a tick." On the way back, Pappy Boyington sighted a stray cow on the road, climbed out of the truck and wrestled it to the ground. That night, Bond wrote that, "In a bull session tonight I learned that we are leaving for Kunming the day after tomorrow." Gunnery training flights increased and the ground crews sweated to peak each Tomahawk in their charge.

Just before dawn on November 26, 3,300 miles to the northeast, a 33-vessel Japanese task force (six aircraft carriers with 414 aircraft aboard, two battleships, two heavy cruisers, nine destroyers, one light destroyer, eight tankers and five small, two-man submarines) under the command of Vice-Admiral Chuichi Nagumo slipped out of Hitokappu Bay, in the Kiril Islands of Japan, heading due east.[3] Such was the secrecy of the operation that no one at Pearl Harbor – much less the men at Toungoo – would have guessed that an attack on the United States fleet was underway.

Among the busiest of the ground crews during this period were the armorers. The P-40s had been delivered with only the World War I vintage ring-and-bead sights. After arriving at Toungoo, however, the aircraft were retrofitted with newly-purchased optical sights. It was the armorer's job to install and adjust the new sights, a difficult task under the best of conditions, much less in the heart of the Burmese jungle using primitive equipment. There were five or six armorers per squadron, and 3rd Squadron pilot R.T. Smith summed up the pilots' admiration for them all when he said (speaking of 3rd Squadron armorers Joe Poshefko, Charlie Baisden, Clarence Riffer, Keith Christensen and Paul Perry), "These five guys did the work of a dozen or more who would be found in any pursuit squadron back in the States." [4]

Sunday, November 30, was no rest day for the Group. The men were rousted out at 0600 with the word that all were to be ready to leave for Kunming on three-hour's notice – orders from the colonel.[5] Kunming had been heavily bombed on Saturday and needed the A.V.G. badly.

The remainder of the day was spent in completing final preparations on the aircraft. The First Squadron pilots, as relayed by George Burgard, "loaded all the guns ourselves and spent most of the day checking the airplanes," after which he took a late afternoon gunnery mission to check his guns. The Third Squadron got their last two P-40s bore-sighted and fired-in on the range. [6]

Monday, December 8 (on the western side of the dateline), began like any other day. The pilots were roused at 0530, had breakfast, then went to the flight line to wait for the day's flying schedule to be put up. Some lounged outside beneath the overcast sky, others sat around inside the teak and bamboo ready room.

About 0700, someone ran into the ready room shouting that the Japanese had bombed Honolulu. The men were stunned at first, and their first reaction was one of disbelief. They laughed and kidded about it – someone was trying to pull a joke.

Twenty minutes later, a jeep pulled up. The driver was the editor of the little Group newspaper that came out weekly. He thrust several mimeographed sheets of paper into one of the pilot's hands and sped away toward headquarters.

The pilot quickly scanned the top sheet then yelled, "Holy hell! We're in a war!" Everyone then crowded around, trying to get a look at one of the sheets. The headlines, all in capital letters, read:

> JAPS ATTACK PEARL HARBOR! OUR RADIOMEN ARE IN CONTACT WITH CORREGIDOR, HONOLULU AND OTHER PARTS OF THE WORLD - WILL KEEP YOU POSTED.

Robert "Buster" Keeton and a few others ran to the radio shack a short distance away to see if they could get a radio report. Singapore Radio was announcing that both Singapore and Manila had been attacked along with Honolulu.[7]

The news of the Sunday morning attack on Pearl Harbor first reached Toungoo by way of a Japanese-controlled radio station in China, and later confirmed by radio station KGEI in San Francisco.

Chennault's first action was to send a telegram to Madame Chiang Kai-shek, suggesting that the A.V.G. be moved to Kunming immediately, since "we are not prepared for combat operations here." [8]

While he waited for a reply from Chungking, Chennault moved quickly to put the A.V.G. on a war footing. Steel helmets, gas masks, and guns were issued to all ground personnel, and before noon he had issued Field Order No.2 announcing 24-hour alert status. He designated the Third Squadron to be the assault echelon, Second Squadron for support, with the First Squadron to be held in reserve. Aircraft on alert would be fully armed and loaded, and all personnel would carry side arms.[9]

Shortly after, Chennault added the last volunteer to the Group. An ex-pat German automobile mechanic who lived in the same housing complex in Kunming as Chennault, Gerhard Neuman, had impressed Chennault with his skills. Asked if he would join the A.V.G., he immediately accepted.

Many diaries of A.V.G. members recorded the day's events. All expressed outrage and the readiness to go to war. First Squadron pilot George Burgard wrote:

> *"Within an hour we had every ship in alert formation and all of us were waiting for a visit from the Japs. They didn't show up today, but we all expect a wave of them tomorrow. We have started on 24-hour duty. My plane is ready and so am I."* [10]

All were up at 0430 the following day. No raids had come during the night, but few had been able to get much rest. After breakfast at 0500, the pilots went to the flight line to relieve those who had pulled alert duty through the night. Morning fog still lay heavy on the field and the pilots lounged around the alert shack near their planes.

As the fog burned off, First Squadron pilots George Burgard, Jim Cross, Charlie Bond, Bob Little, Bob Neal, and Robert A. Smith started the engines of their P-40s and taxied into position for takeoff. They climbed out to 18,000 feet and headed east to patrol the Burma-Thailand border. After three hours of fruitless searching, flying up and down the border, they landed and resumed duty with the 12-plane ground reserve, while the Second Squadron took their place on airborne alert.

The procedure was to be repeated every two hours. At 1605 someone spotted a metallic glint high above the field. Bond, Cross and Little, along with a Second Squadron flight were scrambled to intercept the high-flying intruder, obviously a Japanese reconnaissance ship. The Japanese, however, had too much altitude and were too far ahead for the P-40s to intercept.[11]

Chennault, peering through his binoculars toward the east, ordered the ground alert pilots to their planes. One by one, Allison engines were started and the P-40s taxied into take-off position, each pilot tensely waiting for the first Japanese bomber to appear. After an hour of cockpit alert, the "all clear" siren was sounded and the pilots taxied their aircraft back to the dispersal areas. The cause for alarm, they found out, was an R.A.F. Tiger Moth that had been spotted, but not identified, in the distance.

Two more air raid warnings were sounded during the afternoon, but both turned out to be false alarms. Excitement was generated, adrenalin started flowing, engines were started – but no bombers appeared. Twilight faded into darkness as the day crew turned over their alert duties to the six night-flyers.

A few weeks earlier, Chennault, realizing the need for a photo-reconnaissance capability, put Group engineering officer Erik Shilling in charge of the project. Selecting P-40 No. 53, the ship he had flown earlier in the Second Squadron, he scrounged an aerial camera from the R.A.F. in Rangoon and had it installed in the Tomahawk's small baggage area behind the cockpit. A hole was cut in the bottom of the fuselage beneath the camera, and a trip-wire was jury-rigged to snap the shutter from the cockpit. To lighten the ship, the four wing guns were removed, the gun barrel apertures and other gaps were sealed with fabric, glued and covered with twelve coats of butyrate dope. Finally, a coat of wax was applied over the entire aircraft. The result was a photo-reconnaissance P-40 that was a good 20-mph faster than the stock Tomahawk and, according to Shilling, "better than 50 mph on the Zero."

On December 10, Chennault had had enough of waiting on the enemy, and ordered Shilling up to find the opposition. Shilling took off in the photographic ship, escorted by Second Squadron pilots Bert Christman and Ed Rector.[12] They flew to Tavoy, Burma, an R.A.F. base on the Gulf of Martaban, where they stopped just long enough to refuel. Off again, they set course for the Thai capital of Bangkok. Thailand had acquiesced to the Japanese in a bloodless surrender, and Don Muang Aerodrome, just outside of Bangkok, was reported to have become a staging base for the Japanese.

Shilling urged the light photo-recce P-40 up to 26,000 feet, while Rector and Christman flew below, on the lookout for intercepting Japanese fighters. As they climbed to altitude, Rector recalled that he "rode the wobble pump" the last 50 miles in and for over l0 miles out, to maintain pressure in his fuel tanks in the rarefied upper atmosphere.

They arrived over Bangkok shortly after noon. From the ground, the high-flying P-40s were mere specks in the sky, and the bright sun overhead further shielded them from sight. The trio weren't prepared for what they saw; on the tarmac of Don Muang set row upon row of Japanese fighters and bombers – nearly fifty of each. Further south in the harbor were over two-dozen ships, and military vehicles and equipment lined the dockside.

Shilling rolled steeply to insure that he was just above the target area, then made his photo run. Although Shilling's P-40 was unarmed, Christman and Recter had full loads of ammo. It was all they could do to keep from diving down to make a strafing pass on the field. However, Chennault had warned them, "Get your pictures and get home. Fly high. No fighting." Their mission accomplished, the three banked toward the west and headed for Toungoo.

When the film was developed and Chennault saw the results, he exploded. He told those around him, "If we had just a dozen bombers right now, we could wipe out the whole Japanese air offensive in Southeastern Asia in twenty minutes!"[13]

The night of December 10 was clear with a full moon – perfect for a night raid. The wail of the air raid siren jolted everyone awake at 0310. The six P-40s on alert were quickly airborne, taking off on the moon-lit runway.

George Burgard, Charlie Bond and others arrived at the field at 0325, jumped into their ships and started the engines. Since the alert flight was already up, however, their orders were to stand by until the bombers were sighted. All they saw were flashes of light in the distance to the south. The Japanese had come alright, but their night navigation was poor and they dropped their bombs on a field 35 miles away. [14]

An hour later the all-clear siren sounded, and those in the bomb shelters outside the barracks filed back inside, hopefully to resume their disturbed sleep. At 0500, Chennault called in the airborne alert flight. Although the moon was still up, the runway slashed out of the jungle was hardly visible. Lanterns were hurriedly placed along each side of the runway as the six P-40s circled overhead. No one had practiced night landings at Toungoo and this would be a real test.

David L. "Tex" Hill was the first to make a landing approach. As his ship came down final, those on the ground could tell he was too high. As he hurtled past the field's mid-point, he was still airborne. Then his wheels touched and he rolled down the runway... and off the end. The landing gear of No. 48 gave way as the P-40 plowed through vegetation off the end of the strip. When the dust finally cleared, Tex climbed out, unhurt, but his plane was a write-off.[15] The remaining five ships were ordered to continue circling, while a station wagon and one of the Studebaker sedans were brought into position on each side of the runway, their headlights aimed down the strip. With the aid of the makeshift runway lights, all five landed safely.

Group Captain E. R. Manning, senior Royal Air Force commander at Mingaladon Aerodrome near Rangoon, had continually urged Chiang Kai-shek to place the A.V.G. under his operational control and send them south to help defend the port of Rangoon. Chennault resisted the transfer vehemently. Manning had stubbornly refused to cooperate with Chennault with regard to providing the A.V.G. with an adequate air raid warning system, and Chennault had grave misgivings of his fighter tactics; above all, he didn't want to lose authority over his Group. [16]

On December 11, Chiang Kai-shek cabled Chennault, "Keep the A.V.G. at Toungoo for the time being, and cooperate with the British." Chennault worked out a compromise with Group Captain Manning, agreed to by Chiang, whereby one squadron would be sent to Rangoon, under operational control of the Royal Air Force. Command of the squadron in Rangoon was to be retained by Chennault, however, leaving the A.V.G. squadron the freedom to exercise their own tactics in support of R.A.F. missions. Chennault wrote later that, "Manning agreed to provide housing, transportation, food, and communications for the American squadron at Rangoon. This he failed to do." [17]

Chennault selected Arvid Olson's Third Squadron to be the first to go to Rangoon and ordered Olson to prepare to move. At 0930 on the 12th, eighteen shark-nosed P-40s took off from Kyedaw for the thirty-minute flight south to Mingaladon Aerodrome, ten miles north of Rangoon. There they joined No. 67 Squadron, Royal Air Force, who were equipped with sixteen Brewster Buffalo fighters, with another sixteen in reserve. Although an R.A.F. squadron, No. 67 was manned primarily by Australian and New Zealand pilots. Twelve Bristol Blenheim twin-engine bombers rounded out Rangoon's air force. [18]

Living conditions at Mingaladon were certainly an improvement over Toungoo – two men to a room instead of open-bay barracks living and good food in the mess hall. Olson's "Hell's Angels" had little time to luxuriate in their new surroundings, though. They were up at 0530 on the 13th, with sixteen aircraft on alert. There were two air raid alarms that first day and all sixteen P-40s, accompanied by the R.A.F. Buffalos, took off to intercept. After patrolling for an hour and a half, however, they returned to base without sighting a single bomber. The next several days were passed in much the same way – air raid warning, race to the aircraft, start engines, wait, switch off, wait, then the "all clear." Another false alarm.

Back at Toungoo, the First and Second Squadrons were preparing for the anticipated move north. On December 15 Chiang Kai-shek ordered Chennault to leave one squadron in Rangoon but to move the rest to Kunming.[19] Two days later, an olive-drab four-engine bomber with the U.S. star on the wings and fuselage landed at Mingaladon. As the B-24's big props wind-milled to a stop, Lieutenant General George Brett, the U.S. commander for Southeast Asia, and his staff stepped out of the big bomber. He had come in from his headquarters in India for discussions with the R.A.F. and Chennault in Rangoon. Chennault flew down from Toungoo that night in a Blenheim bomber provided by the R.A.F.[20] Before Chennault left for Rangoon, he laid the plans for the First and Second Squadron's move to Kunming.

On the 17th, a C.N.A.C. transport landed at Toungoo to pick up the first load of people and equipment. Irving P. Gove, crew chief for George Burgard, and six other mechanics were aboard the transport that afternoon as it climbed out for Kunming.[21] The following morning Chennault returned in an R.A.F. C-47, accompanied by Lieutenant General Brett and staff. After a short tour of Kyedaw, the general and his entourage boarded the C-47 and were off again. At 1100 Chennault, now dressed in the uniform of a Chinese Air Force colonel, called his squadron leaders together and told them to prepare for immediate departure for Kunming.

The first of the P-40s took off at 1230. Just as George Burgard's aircraft broke ground, however, his engine cut out and he settled back onto the runway, running off the end and into the jungle. The plane was a washout, but Burgard wasn't hurt. Next, Freeman Ricketts taxied into the back of Chennault's staff car, cutting the car in half and ruining the prop. Then, Second Pursuit Squadron pilot Noel Bacon taxied back in with a sticking brake. Burgard and Ricketts were given additional airplanes, Bacon had his brake fixed and they were finally off, with Bacon leading. Thirty minutes later the three appeared back over the field. Ricketts peeled off and landed as the other two turned north again. As he rolled his canopy back, his crew chief shouted up, "What's wrong?" Ricketts replied, "Oil temperature's off the peg," as he stop-cocked the throttle and switched the engine off.

Half an hour later, two more P-40s appeared from the north. Lynn A. Hurst's engine was cutting out and Frank Swartz was escorting him back. All-in-all, a very disappointing departure for the A.V.G.

Later that afternoon a C.N.A.C. DC-3 came in for Chennault and pilots Don Knapp, Bob Prescott and Dick Rossi, along with a group of crew chiefs and radio men. Loaded as it was with men and equipment bound for Kunming, it used most of the runway on takeoff and finally staggered into the air as the end of the strip approached, its wheels almost grazing the tops of the jungle vegetation. [22]

The two-and-one-half-hour flight to Kunming was a challenge within itself for the P-40 pilots. It was 670 miles as the crow flies over rugged, mountainous terrain. There were no maps available – the pilots were given only a small, mimeographed sheet showing the basic route and mileage. [23] As George Burgard recalled, "It was the roughest country I've ever flown over. Not a place to land, even in a parachute, for over 500 miles." [24]

Newkirk's Second Squadron climbed out to 15,000 feet, while the First Squadron continued on to 21,000. There were sixteen ships in each formation – the one take-off accident and three aborts left each squadron two shy of their full strength of eighteen.

As the P-40s leveled off at their assigned altitude, they throttled back to 190 mph indicated airspeed to conserve their precious fuel. The P-40B carried only 134 gallons of fuel, which allowed a range of 730 miles at normal cruise settings.[25]

Although the First Squadron employed "weavers" – two aircraft flying a criss-cross pattern above the formation as lookouts for Japanese attackers – Newkirk opted not to use the procedure since the two weavers, using a higher manifold pressure, would be dangerously low on fuel when they arrived.

All had been warned to watch for Japanese interceptors and each pilot's head was on a swivel during the entire flight. Surely the Japanese wouldn't pass up the opportunity to destroy the bulk of the A.V.G. on their way to Kunming. The nearly 700-mile flight passed uneventfully, however. The monotony was broken only once, when John Dean's oxygen tube came loose, causing him to pass out.

His plane began a slow descending turn as his hand relaxed its grip on the stick. He passed through several thousand feet before he came to, recognized his problem and climbed back to rejoin the formation.[26]

After two-and-one-half hours in the air, they caught sight of their destination through a break in the clouds. The field lay just northwest of a huge lake and southeast of the city of Kunming. With a signal from their flight leaders, the P-40s dropped back into single file and began their descent through the clouds into Kunming.[27]

As they flew over the city, smoke was rising in several columns among the buildings. The Japanese raid that was reported just before they took off had been successful. As they approached the field, they could see bomb craters from previous Japanese raids.

One after the other, thirty-four planes touched down on the 5,500 feet long, compacted earth runway at the Wu Chia Ba aerodrome, and were directed to individual parking areas. Throttles to idle-cutoff, ignition switches off, and each pilot slid back his canopy, taking in the crisp, cool Kunming air. At 6,500 feet altitude and 670 miles further northeast than Toungoo, the difference in climate was felt immediately. As First Squadron pilot Joe Rosbert later recalled:

> *"My plane was almost immediately swarming with Chinese clad in quilted, indigo blue outfits. They were all smiling broadly enough to crack their faces. It was the first time they had ever seen such a large number of planes land at their airfield. When they measured the amount of gas remaining, it showed only five gallons, just enough to wet the bottom of the tank! Anyway, we had arrived in China, at last!"* [28]

The pilots piled into waiting station wagons and were off on the bumpy three-mile ride to their quarters. Hostel No. 1, as the building was called, was a converted dormitory of Southwest University, a former girls' school outside of Kunming. It would be the home of the Second and Third Squadron pilots and most of the staff. Hostel No. 2, consisting of low, one-story adobe brick buildings, was situated near the airfield, and would be the new quarters of the First Squadron pilots and the ground personnel.

"What a difference," each pilot thought as they were shown to their rooms. Each had an individual room – no more open-bay barracks living – furnished with a comfortable bed, chair, table, a chest of drawers, and a desk. A small charcoal heater in the center of the room helped to break the December chill. Best of all, there was hot water available for shaving and bathing. Although some still complained about their facilities, George Burgard expressed the feelings of most:

> *"With every passing hour I find it harder and harder to believe that Kunming is such a tremendous improvement over Burma. This is really great. The quarters are splendid, soft cots, furniture, showers, good food, a nice bar, excellent reception at the hands of the Chinese, and a good field to operate out of for a change. We have been issued good equipment, jackets and the like, and things are really nice."* [29]

Crew chiefs labored throughout the rest of the day and into the night refueling, rearming and correcting mechanical problems several pilots reported upon landing. By dawn on the 19th, thirty-four aircraft were on alert and combat ready.

The 19th passed quietly. Pilots were up at 0430 and out at the flight line before dawn. Breakfast, to the delight of all, was real bacon and eggs. Even the food was good!

Around the field there were lights flashing in the pre-dawn darkness as crew chiefs warmed up the engines of the P-40s. To guard against a surprise attack by the Japanese, Chennault had ordered a four-ship patrol to be airborne at dawn.[30] After two hours of boring holes in the sky, the four landed.

Later in the day, Chennault sent George Burgard, Dick Rossi, and Jack Croft on a reconnaissance mission to survey two nearby fields at Yangling and Chanli. Both were found to be suitable for P-40s.[31] Charlie Bond went in the back seat of a Chinese-piloted BT-13 to check out Mengtzi, a field about 125 miles southeast of Kunming. He brought back word that the field was unusable – deep furrows were dug across the entire area.[32] Two air raid warnings during the day got everyone's adrenalin pumping, but both turned out to be false alarms. By late afternoon they had given up on the Japanese.

Remains of Tex Hill's No. 48 after his December 10, 1941 night landing. (via Lee Burgard)

David L. "Tex" Hill was unhurt (but very embarrassed!) as he climbed out of the wreckage of No. 48. (via Bob Andrade)

The bar in Hostel #1 was a popular watering hole. Left to right: Joseph Pietsker, Wayne Ricks, "Fuzzy," the Chinese bartender, Joe Sweeney, Frank Van Timmeron, Leo Schramm and Sharon McHenry. (via E.J. Harris)

Hostel #1, home of the 2nd and 3rd Squadrons in Kunming. (via P.J. Perry)

Harvey Greenlaw and Chennault enjoy a leisurely smoke at Wu Chia Ba aerodrome in Kunming as John Williams looks on. (via Flying Tigers Association)

A few of Chennault's Group Staff in Kunming, early 1942. Left to right; Bill Wyke (Assistant Adjutant), John Williams (Communications), C.E. Smith (Adjutant), C.B. Adair (Supply), Tom Gentry (Chief Flight Surgeon), Chennault, E.W. Bruce (Dentist), Harvey Greenlaw (Executive Officer), Ed Conant (Transportation) and Lewis Richards (Assistant Flight Surgeon). (via Flying Tigers Association)

CHAPTER 8

"Next Time, Get 'em All"
– *Claire L. Chennault*

The Japanese airfield at Hanoi, 300 miles to the east of Kunming, hummed with activity in the early morning chill of December 20. Ten Kawasaki Ki-48, Type 99, twin-engine bombers, Allied code-name "Lily," of the 82nd *Chutai* (Squadron), 34th *Sentai* (Group), were being loaded with bombs as their four-man crews attended pre-dawn briefings. At approximately 0900, squadron commander Capt. Tatsjiro Fujii pushed the throttles of his Ki-48 forward. As commander of the 82nd *Chutai*, he was leading the flight this morning. His heavily-laden bomber moved slowly at first, then picked up speed. As he climbed out he circled to allow the other nine bombers in his *chutai* to form up. With Fujii's Lily in the lead, Corporal Kinosuke Yamaguchi moved in on his right while Corporal Isamu Sgt. Isamu Tsuji took his left wing, and Sgt. Shoigorou Abe moved into the slot. The second element of three, behind and to the right, was led by Lt. Shigeru Funamoto, with Sgts. Masayoshi Kameyama and Sabarou Yamamoto on the right and left wings. As the last element in the standard V-of-Vs formation, led by Lt. Gouichi Suzuki with Sgts. Ikuji Keshikawa and Kukuo Fujimoto, moved into formation, Capt. Fujii turned to a westerly heading and began a cruise-climb to 20,000 feet in heavy cloud.[1]

Colonel Chennault lit another cigarette as he sat at his desk in his headquarters building adjacent to the operations shack. He had sent up a pre-dawn patrol at 0600, but they had landed after at 0730 with nothing to report. He glanced at the clock on the wall... 0945. "Will this be another quiet day?" he wondered. "Where are the Japanese?" Suddenly his telephone rang. It was Colonel Wang Shu Ming himself, commander of the Chinese Fifth Air Force and Chinese chief of staff for the A.V.G. His report to Chennault: "Ten Japanese bombers crossed the Yunnan border at Laokay heading northwest." This was no false alarm! The alarm bell sent pilots racing for their P-40s.

Chennault had briefed his battle plan to all the pilots earlier: A four-plane element of Newkirk's Second Squadron would make the interception, while a second four led by Jim Howard would orbit over Kunming in a defensive patrol. Sandell was ordered to remain west of Kunming with the sixteen ships of the First Squadron as a reserve force.[2] As the position reports filtered in, the points on the plotting board in the operations shack drew a clear course line of the bombers path. They were headed straight for Kunming. When the plot showed the Japanese force to be about fifty miles east, Chennault stepped outside and fired a yellow flare, signaling his pilots to start engines.

The cool morning air was blasted by the sound of twenty-three Allison engines coughing to life. Barely had the engines settled down to a roaring growl before a red flare arced up over the runway. Take off!

One-by-one, the shark-nosed fighters eased out of their parking spots and began rolling down the taxiway. Jack Newkirk was the first to take the runway, followed by his flight members Bert Christman and Gil Bright.

Newkirk's left hand moved the throttle forward smoothly as he added right rudder to counteract the torque of the 1,040-horsepower engine. The fighters accelerated quickly in the cold air and soon the first flight of three was climbing out to the east.

They were followed by Jim Howard's four-ship flight. As they climbed to orbit over Kunming, Sandy Sandell's First Squadron P-40s took the runway. Among those in Sandell's fourteen aircraft that morning were Charlie Bond, Einar Mickelson, Bob Neal, Fritz Wolf, Don Bernsdorf, Bob Little, Bill Bartling, Jim Cross, Joe Rosbert, Ed Liebolt, Carl Brown, and Louis "Cokey" Hoffman. Climbing out to the east, they broke up into three elements – Sandell took five as the assault element, Neal plus three formed the support echelon, and Little took Bond, Rosbert, and Liebolt as a reserve echelon.

Tex Hill was on alert duty with Newkirk's flight early that morning. When the anticipated Japanese raid hadn't appeared by mid-morning, he left the alert tent to take a short coffee break. Hearing the sound of engines, Tex ran out of the building, coffee cup in hand. His plane was already taxiing out, a stand-by pilot in the cockpit. He could only watch helplessly as the P-40s climbed out.

Ed Rector, who was to have had a day off because his P-40 had been scheduled for periodic maintenance that morning, saw the aircraft taking off and ran toward the flight line. The crew chief still had the cowling off of No. 58 when he arrived at his airplane. "Get the cowling back on," Rector yelled as he jumped in the cockpit and buckled his parachute harness. He took the runway as the last of Howard's flight lifted off. Making a "rolling" take-off check, he was soon airborne, throttle to the firewall in an attempt to catch the others. As he climbed out, Rector recalled thinking to himself, "This is it. Finally, combat."[3]

By this time, Chennault, Greenlaw, and Chennault's interpreter, P.Y. Shu, had moved from the operations hut to a grassy hill overlooking the field. There they entered a heavily-timbered, pyramid-shaped structure covered with earth that served as a combat operations shelter. Inside was a radio, telephone and plotting board. As Chennault flipped the switch to the radio receiver, they were met only with crackling static. Then suddenly:

"There they are!"

"No, no. They can't be Japs!"

"Look at those red balls."

"Let's get 'em."

And then silence.[4]

To the east of Kunming, high above at 22,000 feet, Newkirk's flight of three had sighted the ten Lily twin-engine bombers below, silhouetted against the 3,000 feet overcast. The three could hardly believe their eyes – there they were, ten enemy bombers sailing along in tight formation. This was what they'd been training for so long! After a moment of hesitation, not unlike a novice hunter with "buck fever," Newkirk gave the signal to attack, as he rolled to the left and dove on the Japanese formation.

Newkirk and his pilots weren't the only ones who couldn't believe their eyes – no one was more surprised than Capt. Fujii in the lead bomber. The Japanese had enjoyed complete air superiority over China since 1939. This mission to Kunming was to have been just another leisurely flight – a "milk run," as the Americans would say. Then suddenly, from out of nowhere, came these three fighters, gun muzzles flashing.

The fighters dived past the front of the formation, continuing on down into the clouds below. Fujii, unnerved by the unexpected attack, signaled his formation to jettison their bombs and began a left turn back toward the east.

By the time Newkirk and his flight pulled out of their dive and began their climb back to the bombers' altitude, they had lost sight of the formation. As three pairs of eyes searched the skies above, one pilot spotted the bombers retreating to the east, just above the cloud deck. Newkirk and his flight gave chase, but the five-mile lead the bombers had proved too much, and he reluctantly signaled to break it off. His guns had jammed and his two-bit radio was cutting in and out. "What a disappointing first combat," Newkirk thought as the three Tomahawks turned back toward Kunming.

Back on the ground, Chennault was monitoring the sporadic radio calls of Newkirk's flight and trying to second-guess the Japanese bomber commander. Picking up his radio mike, he directed Sandell, orbiting at 22,000 feet west of Kunming, on a vector toward Iliang, about 30 miles to the southeast.[5]

Charlie Bond described the action from that point:

> *"We leveled off at twenty thousand feet in an area southeast of Kunming, where we sighted Japanese raiders - ten bombers in tight formation... We were practically full throttle at twenty thousand, indicating 225 mph, and it still took us a good eight to ten minutes to overtake them and get into attacking position."*[6]

At approximately 1030, Sandell attacked first with his flight of six, followed by Bob Neal's reserve quartet. Bob

Little's four P-40s stayed high to cover the attack from any possible fighter escort accompanying the bombers.

Weeks of training went by the board - Chennault's admonition to "fight in pairs" was quickly forgotten in the heat of the A.V.G.'s first battle. As Bob Neal recalled:

> *"We went in in rat-race formation, everybody chasing the tail of the plane ahead. We opened fire and the bombers seemed to fall into pieces. I saw pieces of engine cowling fall off into space. Glass from the gun turrets flew in all directions. Engines smoked and caught fire. Tails just crumpled and fell off. It was the queerest thing I ever saw. Then the air was so full of P-40s dashing all over the place that I worried more about colliding with a P-40 than about the Japs."* [7]

All the action was too much for Bob Little and his support echelon. With Charlie Bond on his wing and Joe Rosbert flying wing for Ed Liebolt, Little took his four into the battle. Charlie Bond continues:

> *"Bob Little... dove down into the lower clouds to come up from below and the right quarter, while I climbed and veered off to the right to attack from above and off their right beam. I charged all my guns and turned on my gun sight and gun switch. The Japs were in a tight formation. As we closed in, they lowered their 'dustbin' rear-fighting guns to defend against us. As a result they slowed down."* [8]

Bond rolled to the left and dived. As the nearest bomber filled his gun sight, he squeezed the trigger. Nothing! Looking down he realized that in his excitement he had turned the gun switch off. He broke down and away, then climbed for a re-attack. This time all guns worked and he saw tracers entering the bomber's fuselage. After his second pass, he recalled seeing only seven bombers.

Cokey Hoffman, the 42-year-old "old man" of the A.V.G., pulled in behind one bomber, raking the rear gunner's compartment with 50-caliber slugs. Lowering his flaps to slow his speed, he rode the tail of the stricken bomber, pouring lead into the fuselage until it dived earthward out of control. [9]

Following his attack on the rear of the formation, Sandell came out of his 400 mph dive and swung around for a front quarter attack. He scored hits on a bomber's right wing, then saw his tracers march up to the wing root. Pieces of engine cowling flew off and heavy black smoke began to billow from the right engine. A moment later it spun out of formation, exploding in a fireball as it struck the ground.[10] Two more bombers fell to Fritz Wolf's guns. As he described the action:

> *"I attacked the outside bomber in a V. Diving down below him, I came up underneath, guns ready for the minute I could get in range. At 500 yards I let go with a quick burst from all my guns. I could see my bullets rip into the rear gunner. My plane bore in closer. At 100 yards I let go with a long burst that tore into the bomber's gas tanks and engine. A wing folded and the motor tore loose. Then the bomber exploded. I yanked back on the stick to get out of the way and went upstairs.*
>
> *"Then I went after the inside man, I came out of a dive and pulled up level with the bomber just behind his tail. I could see the rear gunner blazing away at me, but none of his bullets were hitting my plane. At 50 yards I let go with a long burst, concentrating on one motor. The same thing happened and I got No. 2. The bomber burned and then blew up."* [11]

During the mêlée, Ed Rector finally caught up with the battle. He had missed Newkirk's attack, but arrived right in the middle of the First Squadron action. Recording his battle in his subsequent combat report, he wrote:

> *"The bombers were flying in a beautiful V-formation and I picked out the last plane on the left side of the formation and started a curve of pursuit from a high-perch position. I had heard about target fixation and*

it happened to me during that first sortie... I was closing in on the bomber, firing all the while... It was at the last second, no, the last split second that I shoved the stick forward and went under the bomber. I don't know how close I came, but I can still see every detail. The camouflage paint, the rivets on the wings, and the tail gunner slumped over his gun... As I climbed up for another pass, I looked down and saw that the bomber was still in formation, but streaming flames and smoke. He slowly dropped his nose and finally went straight in." [12]

Throughout the attacks, the Japanese bombers maintained a tight formation – as one bomber went down another slid into its place. As they sped toward the east, some with engines trailing plumes of smoke, those "dustbin" gunners who survived kept up their defensive fire.[13]

Finally, aware of their rapidly decreasing fuel state, Sandell called off the attack and ordered his squadron to return to base. The 10-minute battle and pursuit of the bombers had taken them over two-hundred miles to the east of Kunming and they would need every drop of fuel left to get back safely.

Back at Kunming, Chennault was pacing nervously. It had been a long while since he had heard any radio calls. Jim Howard's flight had landed much earlier. As Howard climbed down from the cockpit, he reported that they hadn't seen a thing.

Newkirk, Christman and Bright landed next. Newkirk sheepishly admitted that they had made only one pass before his guns jammed and his radio failed, and they gave up trying to catch the fleeing bombers.

At long last, those waiting on the ground heard the sound of aircraft engines in the distance and soon they could see the P-40s – first a flight of four, then a pair, another pair, then a couple of single ships. Nearly half an hour separated the first from the last, but as the ground crews counted each approach they were relieved that all fourteen had returned. The first ones back made a low pass over the field, some rolling slowly in the traditional "victory roll." Then they pulled up into the landing pattern, dropped their gear, and one by one touched down and taxied in.

The big three-bladed prop on the lead aircraft had barely swung to a stop before the ground crew swarmed around the plane. "Did you see any Japs? How many did you get? Was anybody hurt?" The scene was repeated at each P-40 as fourteen weary but very excited pilots climbed out of their aircraft.

They soon gathered around the Old Man, each one chattering like a schoolboy after a winning baseball game. "Hold it, hold it – take it easy," Chennault told the excited group of pilots. "You did a good job, but not good enough. Next time, get 'em all," he said, a trace of a smile playing around the hard line of his mouth.[14] Turning on his heel, he said, "Now, let's go in and talk about it," and led the sweat-soaked pilots into the operations building. For the next hour – lunch be hanged – he went over the entire battle in as much detail as each pilot could remember. He pointed out their mistakes and once more advised them how to "get 'em all next time." Then he let them break for lunch.

The official score for the day, after each pilot gave his combat report, was four confirmed destroyed, which, by mutual consent, were equally shared by the 15 pilots. They knew that they had damaged most of the others severely, but held their claims on the conservative side.

As the A.V.G. celebrated their first victories, Capt. Fujii was leading his surviving bombers back to Hanoi. The entire three-plane element led by Lt. Funamoto had been destroyed. Virtually all of those remaining had sustained battle damage of varying degrees. Most were streaming smoke as they made their landing approach at Hanoi. One crash-landed on the runway, another landed with only one main gear down and careened off the edge of the runway in a cloud of dust. The remaining five landed without further incident.[15]

Until the A.V.G. arrived at Kunming, the Japanese had bombed the city with impunity. It had served almost as a training ground for their more inexperienced crews. Never again, however, while the A.V.G. was based there, did the Japanese attempt another bombing raid on Kunming.

Following the debriefing, some went to the mess hall for lunch, while others made their way directly to the bar in Hostel Two to celebrate the day's activities. Their exuberance was dampened by the loss of Ed Rector, however. Had he been shot down by a lucky tail gunner? Had he run out of gas and crash-landed somewhere? No one knew, of course. They only knew that he was their only casualty of the day.

It would be a week before the mystery was solved. After multiple passes at the bombers, Rector's guns overheated and jammed. By this time, Sandell's flight was already on the way back home. With only primitive maps of the area, Rector found himself "temporarily disoriented," as lost fighter pilots prefer to say. Pulling off to the side of the retreating bombers, he paralleled their course, then did a 180, judging that heading would lead him back to Kunming. He maintained his course until his fuel low-level light began to blink red.

Descending below the cloud deck in search of an auxiliary field, Rector found himself flying down an ever-narrowing canyon, and, gritting his teeth and praying a lot, he climbed back through the overcast expecting to impact a mountain any moment. Luckily he broke through into bright sunlight, but with mountain ridges protruding through the clouds on both sides. With the red fuel light glowing steadily now, he racked the P-40 around and headed for a Chinese village he'd just flown over. Picking an area that had been freshly plowed, he brought the Tomahawk around, lowered the flaps but not the gear, cut the switch and flared for a reasonably soft belly-landing.

As he climbed out of the cockpit, he could see literally hundreds of Chinese running toward him from the village above. After an initial tense moment, he identified himself as an American flyer and was taken back to the village. That night the Chinese treated him to a lavish dinner and the next day helped him to strip the plane of guns, ammunition, and anything else worth salvaging. With additional help from the Chinese peasants, Rector made the 90 miles back to base inside of a week.[16]

Later that afternoon, Harvey Greenlaw interrupted the revelers to announce that they should straighten themselves up a bit to receive an adoring throng. Joe Rosbert recalled the event:

> *"Soon we heard a band coming towards the field and a long procession appeared at the entrance. Led by the Mayor were hundreds of people, each one carrying something. He made a speech while we all stood in line listening. Although we could not understand, I could tell what he was saying by the thankful looks on the faces of those grateful people. Little girls with pretty faces crowned by neatly cut bangs, stepped up and placed long pieces of purple silk around our necks and bouquets of flowers in our hands. We were like a bunch of embarrassed school boys. I could feel my face flush, along with the rest. What we had done seemed small to us, but to them it meant a feeling of security in place of one of fear. The Chinese and American anthems were sung and gifts, which included native food, were distributed."* [17]

That night Chennault, his staff, and all the pilots were invited to a celebration banquet in Kunming hosted by General C.C. Wang, Chinese commander of the Kunming area. The amount of food was exceeded only by the toasts to the "brave American airmen." The men were grateful for having survived the air battle – but some wondered whether they would survive the celebration!

Back in the United Sates, the story of the A.V.G.'s first combat was picked up by the *New York Times* (on page 27) two days later. It was covered more extensively by Henry Luce, *Time* magazine editor for the Far East in the December 29 issue of the magazine in an article entitled, "Blood for the Tigers:" [18]

> *"Last week, ten Japanese bombers came winging their carefree way up into Yunnan, heading directly for Kunming, the terminus of the Burma Road. Thirty miles south of Kunming, the Flying Tigers swooped, let the Japanese have it. Of the ten bombers... four plummeted to earth in flames. The rest turned tail and fled. Tiger casualties: none."*

Flying Tigers! And so was born the nickname for the A.V.G. that resonated in the imagination of readers worldwide, and for which the American Volunteer Group would forever be known.

The Legend Begins. *Oil on canvas by Roy Grinell. The painting depicts Vice Squadron Leader Robert H. Neale in a diving attack on a Japanese Kawasaki Ki-48 during the Japanese raid on Kunming, December 20, 1941. Neale would score only a "probable" on that date, but would go o to become the A.V.G.'s top ace with 13 confirmed aerial victories. (Courtesy Roy Grinnell Studio)*

Right: Robert H. Neale, Vice Squadron Leader, 1st Squadron. (via J.J. Harrington)

Below: Diagram of the ten-plane formation of the Japanese 82nd Chutai/*34th* Sentai *that attacked Kunming on December 20, 1942. (Drawn by 82nd* Chutai *pilot, Gouichi Suzuki, who flew in No. 6 position on the raid, and translated by his grandson, Fumio Watanabe)*

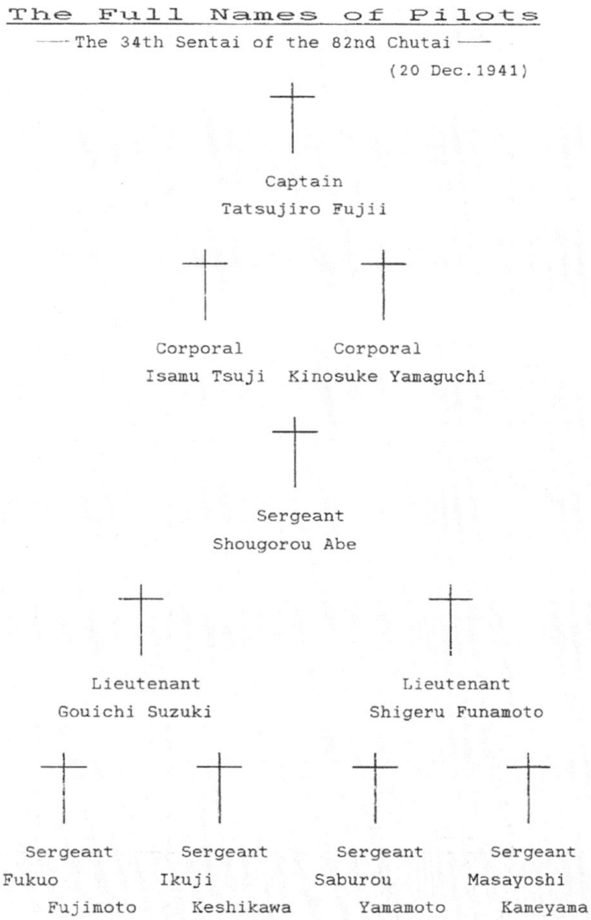

Kawasaki Ki-48 Army Type 99 twin-engined light bomber. Wingspan: 57' 3"; Length: 41' 4"; Height: 12' 6"; Maximum Speed: 298 mph at 11,485 feet; Range: 1,230 miles; Crew: Pilot; Bombardier/Gunner; Radio Operator/Gunner; Navigator/Gunner. Bomb Load: Eight 33lb or six 110lb bombs.

CHAPTER 9

Christmas Over Rangoon

The Third Pursuit Squadron, "Hell's Angels," learned about the Kunming air battle by radio on the night of the 20th. The men were "mad as hell" that the First Squadron got into action before they did.[1] Olson's Rangoon contingent had had a week of alert duty with the British – seven days of tense, yet monotonous, waiting punctuated only by air raid warnings on the 13th and 19th that turned out to be false alarms.

Things were so quiet that Chennault considered pulling the Third Squadron out of Rangoon and sending them on up to Kunming.[2] When that word was circulated, it raised spirits all around. When it didn't happen, the men settled back into nervous tedium. R.T. Smith's diary reflected the thoughts of most:

> *"I wish somebody would make up their mind what the hell they want to do with us and then do it. This is the damnedest war I ever saw."[3] When word came that Sandell's pilots had engaged the enemy and knocked down four, their anger and frustration at not being in on it is understandable.*

Olson's men didn't have much longer to wait for a piece of the action, however. Sunday, December 21, passed quietly, although normal alert duty was carried out. The following day a flight of four took off to intercept a reported intruder 30 miles east of the field, but returned with no sighting.[4]

That afternoon, a radiogram came in from Kunming ordering the movement of the three Curtiss-Wright CW-21 "Demons" from Rangoon to Kunming. CAMCO had brought the prototype interceptors to Mingaladon earlier for flight testing by Byron Glover, preparatory to building more at the CAMCO factory at Loiwing, China. When Erik Shilling first saw the little wasp-waist interceptor with the 1,000-hp radial engine, he asked CAMCO manager Walter Pentecost for permission to try it out. Impressed with the CW-21's 5,200-feet-per-minute rate of climb (more than twice that of the Tomahawk), he later approached Chennault to add the little fighter to the A.V.G. inventory. Initially cool to the idea, Chennault then decided to have the three flown to Kunming for further evaluation. Oley Olsen hastily drew up "SQUADRON ORDERS NUMBER 26," relieving Shilling, Lacey Mangleburg and Ken Merrit from the Third Squadron and assigning them to the Group Headquarters section, with instructions to "…proceed to Point 'A' (Toungoo) for minor maintenance and proceed to Point 'X' (Kunming) as soon as practicable…" [5]

"As soon as practicable" meant right now, and the three prepared for a late-evening takeoff. Arriving at Toungoo just before sunset, they put on an impromptu airshow for the benefit of the assembled ground crew, demonstrating the capabilities of the CW-21.

The next morning, the 23rd, it was finally the "Hell's Angels" turn. About 1000, the alert telephone installed in Olson's office rang. As Olson answered, he heard the cryptic orders, "Clear the field." No other explanation – just, "Clear the field." Guessing that a Japanese attack was imminent, Olson hurried to the flight line and ordered his pilots to "scramble." [6]

Very shortly, ten of the fourteen available P-40s were moving down the taxiway. Two pilots, Robert P. "Duke" Hedman and Tommy Haywood, were experiencing engine trouble and would take off late. Across the field, ten R.A.F. Buffalos were already taxiing into position and taking off, some singly, some in pairs.

Charles H. "Chuck" Older and Ed Overend had the day off and were just about to start into Rangoon for some shopping and sightseeing. As the sound of aircraft engines reached them – both A.V.G. and R.A.F. fighters taking off – the two stopped. Hopping a ride to the field, they found two P-40s were left, both unattended. Sirens were wailing and the mechanics were running to the shelters.

As Older sprinted for the nearest plane, he called back to Overend, "I'm going to take a ship up. You can either follow me or run for cover!"

"Fine," Overend replied, "but where are we going?" As he wriggled into the parachute harness, Older called over his shoulder, "I don't know – let's just get up!"[7] Five minutes later they were airborne, Older in front with Overend a mile back in trail.

Meanwhile, the first aircraft off were climbing through 8,000 feet several miles to the east. The first instructions from R.A.F. Fighter Control came crackling through their headsets – "Enemy approaching from the east!." That was all – no information as to the size of the force, altitude, heading, or whether they were bombers or fighters.

What they didn't know, neither the pilots in the P-40s and Buffalos, nor the R.A.F. Fighter Control, was that their combined force of twenty-four fighters would be pitted against a Japanese air armada of ninety bombers and fighters – thirty-three Mitsubishi Ki-21s out of Bangkok – eighteen from the 98th *Sentai* and fifteen from the 62nd *Sentai* – and twenty-seven more from the 60th *Sentai* based at Phnom Penh, Cambodia. They were escorted by thirty fighters – fixed-gear Nakajima Ki-27 Nates and the more modern Nakajima Ki-43 Oscars from the 77th *Sentai* based at Tak, 200 miles due east of Rangoon and just across the Thai border. The Japanese aircraft had taken off just after dawn from their respective fields in Thailand and Cambodia, rendezvousing near the Burmese border. There they split into two waves – one to attack the docks at Rangoon, the other to hit the base at Mingaladon.

George B. McMillan, Third Squadron Vice-Squadron Leader and one of the first to get airborne, circled the aerodrome once, waiting for his flight to join up. As R.T. Smith, P.J. Greene and Hank Gilbert slid into position on his right and left wing, he received the warning call from Fighter Control. Fire-walling the throttles, he began a climbing turn to the east.

After takeoff, Older and Overend climbed out at full throttle. As they passed through 8,000 feet, they saw McMillan's flight of four in the distance, a thousand feet higher. Eventually reaching the formation, they joined up as the last element.[8]

Fifteen miles east of Mingaladoon, McMillan caught sight of the Japanese bombers. There were 27 in a tight V-of-Vs formation, sailing along at 12,000 feet over the Gulf of Martaban. Above and about a mile behind were thirty fighters. Further to the south he could see another large formation of enemy bombers.[9]

The bombers were still 3,000 feet above the P-40s when McMillan rocked his wings violently, signaling his flight to attack. McMillan and Older continued head-on toward the formation, while Smith, Greene, and Gilbert swung around in their climb for high side passes from the rear. As McMillan opened fire from the front quarter, Older banked sharply directly beneath the formation. As he described the action:

> *I began climbing up underneath them, keeping almost directly below them as I climbed. When I reached a point about a thousand feet below and directly underneath the enemy formation, I leveled off and picked up some additional speed. With this extra speed, I was able to begin making attacks by pulling up rather sharply, closing in from below, and firing at close range. I attacked the leader of the 27-plane formation first. As I came up under the leader, I saw the bomb-bay doors open on all the bombers and knew we were getting close to Mingalon Airport. I aimed at one of the planes on the left side of the formation and after my first attack, smoke began pouring from the port engine. I rolled out to the side below the formation and came back for another attack. I aimed again at the same plane and closed to about 75 yards. I gave it a long burst and the bomber suddenly nosed down out of the formation with smoke streaming behind. I saw it roll over into almost a vertical dive and disappear below. I continued making attacks from below, this time aiming at the leader of the formation. I closed in from below to about 50 yards, and gave the leader a long burst into the bottom center of the fuselage. I saw the bomber explode from below, and flame and smoke seemed to pour out from the bottom of the fuselage. I saw debris falling from this plane immediately after the explosion, and the bomber nosed straight down with flames and smoke pouring from it."*[10]

Older took several hits from the bombers' guns, but none hit a vital spot. He climbed above the formation for a front quarter attack, but he found that he was finally out of ammunition. With two bombers to his credit, he turned for home.[11]

Meanwhile, R.T. Smith, P.J. Greene and Hank Gilbert had begun chewing up the formation's left flank. After several passes, one Ki-21 separated from the formation. Smith was right behind, whipsawing the hapless bomber in multiple passes. As he closed in to 25 yards the bomber blew up, a piece of the plane striking his right wing.

Suddenly, the attackers became the attacked as several radial-engine fighters flashed down on the Tomahawks. Three headed for Smith, who racked his P-40 around to meet them head-on, getting off a quick burst at the leader before diving away.[12]

P.J. Greene met the fighters in a climbing turn, firing at the lead ship from 300 yards. The Japanese fighter fell away, diving toward the ground. Another half-rolled directly in front of him as he squeezed off a burst at point blank range. Pieces flew off the enemy plane as it dived away. Greene didn't have time to watch where the damaged fighter fell, for at that moment .303-caliber bullets from several Japanese guns smashed into his ship, severing control lines and knocking great chunks out of his wings. As the stricken Tomahawk began to go into a spin, Greene slid back his canopy, snatched at his lap belt, and leaped over the side. A rush of cold air hit him in the face as he dropped away from his spinning plane. His first impulse was to pull the ripcord – and he did. Only after the chute popped and blossomed above him did he realize that he had bailed out at 8,000 feet, right in the midst of a swarm of Japanese fighters. His worst fears were realized as he watched a fighter turn into him. One Japanese fired a long burst, but miraculously missed him. A second fighter then took his turn, cutting his engine to circle the helpless pilot, firing bursts all the way down to 2,000 feet. After what seemed like an eternity, a rice paddy rose up to meet him and he was safely down. Although his chute had several holes in it, he hadn't been scratched.[13]

Hank Gilbert wasn't so lucky. Four attacking Nakajimas hit his P-40 before he had time to react. Rolling inverted, with flames streaming from both wings, he dived straight into the ground four miles from the field. The A.V.G. had lost their youngest pilot.

Throughout the attacks, the Japanese bomber pilots maintained admirable flight discipline. Just as noted during the December 20 raid on Kunming, as one bomber left the formation, another would close in to take its place. Now they were on their bombing run. As they approached Mingaladon Aerodrome bomb bays opened and each bomber dropped its eight-ton load.

The bombs fell with devastating accuracy. The operations office took a direct hit, as did one of the R.A.F. hangars, destroying several aircraft inside. A gas dump went up in flames, and an anti-aircraft battery manned by Burmese soldiers was wiped out by a direct hit. The runways were pockmarked with bomb craters and the hospital and barracks buildings were damaged by shrapnel. Several Japanese fighters, following on the heels of the bombers, dived down to make strafing passes on the burning aerodrome. Although no A.V.G. ground personnel were injured, ten R.A.F. troops and thirty Burmese soldiers were killed.

Two miles above, R.T. Smith was diving away from the fireball of the Japanese fighter that had blown up scant yards in front of him. Using the energy his plane built up in the dive, he zoom-climbed above the retreating bombers and attacked again. He spotted a straggler that had dropped back from the formation being attacked by an R.A.F. Buffalo and a P-40 (which he learned later was George McMillan's Tomahawk). He joined the trio, and made multiple passes at the bomber. Soon its nose dropped and it went down at a 30-degree angle, smoke pouring from its engines. Smith broke off the attack then and turned toward Rangoon, where another air battle was raging.[14]

Meanwhile, Ed Overend, separated from the flight when the battle was joined, was waging a one-man battle. As he flew ahead of the bomber formation, the bombers began a turn to the northeast, continuing on to a southerly course out to sea. He made three rear-quarter attacks with no apparent effect, then caught a straggler about 200 yards away. After several passes, the bomber's left engine caught fire and it fell off into a seventy-degree dive. Overend followed it down to 4,000 feet when he was jumped by three Japanese fighters.

Pushing his stick forward, he dived away until the airspeed indicator reached 400. Pulling out of the dive fifty feet above the ground, he looked back for the pursuing fighters, but he was alone once again. The bomber crashed northeast of Syrium, a village on the Irawaddy River southeast of Rangoon.[15]

While McMillan and company were engaged over Mingaladon, Parker Dupouy's flight of six was taking on the second wave of bombers attacking Rangoon. Dupouy had initially been given orders to orbit at 20,000 feet over Syrium. He led the first element of three P-40s with Ralph Gunvordahl and Bill Reed on his wing. Neil Martin led the second element, made up of Ken Jernstedt and Bob Brouk.[16]

Two Third Squadron pilots took off just ahead of the bombs. Both Tommy Haywood and Robert P. "Duke" Hedman got off late because of engine trouble. Haywood climbed out to 20,000 feet on an easterly heading. Spotting the bomber formation 5,000 feet below, he did a quick wing-over and dived to the attack. Picking the tail-end bomber on the right flank of the "V" formation, he opened fire at 500 yards and kept the trigger depressed until he was within 75 yards. The bomber rolled out of the formation, spinning down out of control.[17]

Haywood continued his dive through the formation before pulling back to climb for a re-attack. At that moment he caught sight of a Japanese fighter and bent the P-40 around. The Japanese pilot did the same and the two fighters raced toward one another at a 500-mph closure rate. Haywood got off a good head-on burst before each turned away. Then began a curving tail chase as each tried to get lead on the other. After two complete 360s, it was obvious to Haywood that the Japanese fighter was turning tighter and would soon have the advantage. He felt his plane shake as bullets thudded into the wings and fuselage of the P-40. "The Old Man was right," he thought, as he rolled away and dived, throttle to the firewall. "You can't turn with one of these Jap fighters." Craning his neck to see if he was being followed, he caught a glimpse of his adversary, plus two of his pals who had joined him. The heavier P-40 soon outdistanced the three Japanese fighters, however, and there were no aircraft to be seen when Haywood leveled out. His battle was over, though – one of the ten rounds that pierced his airplane had severed his left aileron cable and he was lucky to be able to bring his crippled ship back to the field.[18]

At 1035, Dupouy sighted the Japanese formation approaching Rangoon from the south. They intercepted the bombers ten miles south of the city at 18,000 feet. Their first pass was a well-coordinated head-on attack, diving slightly below the bombers to hit the Sallys from beneath. From that point on, it was "every man for himself," though, as the whole flight of P-40s broke up to make individual attacks. Dupouy's flight – just as did McMillan's group over Mingaladon and the First Squadron over Kunming three days earlier – promptly forgot about the teamwork that Chennault had tried to hammer home in weeks of "chalk talk" blackboard sessions back at Toungoo.

Dupouy later reported:

> *"I broke away down and to the right of the bomber formation and made a series of attacks from the rear quarter below on the rear bomber. While doing this I saw two other bombers leave the formation flaming. One caught fire in the center of the fuselage immediately after a P-40 had attacked it, and the other had its right engine on fire. I was making an approach directly below the leader when I saw the leader's bombs drop; most of the other bombers dropped at the same time. I saw many of them hit in the river south of Rangoon. The bombers turned 180 degrees and started towards the bay in a slight dive. When I ran out of ammunition they were at 12,000 feet headed due south and indicating about 230 m.p.h."* [19]

Ralph Gunvordahl broke away after the first pass, swung back around and began a climb from beneath the formation. At 300 yards range, he pressed the trigger on the stick's handgrip. Lead from four .30-caliber and two .50-caliber guns spewed forth as he closed in to 100 yards. Flames erupted from the bomber's right engine as it began to drop away from the formation.

Gunvordahl continued his attack on the bombers as they turned south over the water. Joined by P-40 No.99, he they made several passes on a straggler, finally setting it ablaze. It wasn't until he ran out of ammunition that he turned for home.[20]

Bill Reed lost contact with Dupuoy and Gunvordahl after the first front-quarter pass, but pulled back up and began raking the bomber formation's flank from 200 yards. Swinging around to the left of the "V" formation, he came in on a beam attack, setting one bomber's left engine afire. The damaged bomber dropped back 200 yards. Reed made one more pass before turning his attention to the main formation.[21]

After sighting the Japanese bombers, Neil Martin broke away from McMillan's element and led his two wingmen above and to the left of the bombers. With a waggle of his wings to signal the attack, he dove down on a left beam attack. The top turret gunners tracked the P-40s as they approached, opening fire at 700 yards. It seemed that all the fire converged on Martin's lead plane, for the Tomahawk shuddered visibly as bullets from several bombers tore into the P-40. His stricken plane dropped away, trailing a plume of fire and smoke all the way to the Gulf of Martaban.[22]

With their element leader gone, Jernstedt and Brouk broke up to make individual attacks. As the formation turned west after salvoing its bombs over Rangoon, Jernstedt moved in on the rear quarter. His fire took effect this time, as the bomber began to fall back. A slight kick of the rudder brought his aim on the next bomber in the "V." Jernstedt squeezed off a short burst which must have hit a vital spot, because it immediately burst into flames and fell off in a blazing dive.[23]

By this time Duke Hedman had reached the battle. Climbing above the formation, he made a series of overhead passes with little visible effect. He then pulled back and hit the bombers from the rear, firing first on the last bomber in the string and then moving up one-by-one to the leader, giving each a short burst.[24]

The Japanese bombs hit Rangoon with devastating effect. Two large department stores were destroyed, as was an oil refinery, and the docks and railway station were badly damaged. As the bombers retreated, the fighters dived down to strafe the crowds in the streets, killing hundreds.[25]

Tomahawks and Buffalos continued to harass the Japanese as they flew out over the Gulf of Martaban, turning back only when out of ammunition or low on fuel.

Once back on the ground, they tallied the results of the 45-minute battle. The A.V.G. pilots claimed 15 destroyed. This was pared down later to 11 destroyed, five probably destroyed, and ten damaged when R.A.F. confirmations came in. The Brewster pilots accounted for seven. The A.V.G. had lost two pilots and three P-40s in the air battle, plus two badly damaged in the strafing attack on the field. Their ground personnel, protected by slit trenches, suffered no injuries. The R.A.F., by comparison, lost five planes and pilots in the air, and six more on the ground. Forty were killed during the bombing and later strafing attacks. Over 1,000 civilians died in the streets of Rangoon.

Meanwhile, at Toungoo, the three CW-21 ferry pilots, learning of the raid on Rangoon, hastened their departure from their former training base – Toungoo could be next! Taking off at 1400, Shilling led the flight, since he was the only one who had flown the Rangoon-Kunming route before. Navigation would be difficult at best, since they carried only a primitive mimeographed copy of a map.

The leg to Lashio, where they refueled, was uneventful, except for an occasional back-firing of Shilling's engine; they were able to use the Irawaddy River as a navigational aid. About 1600 they took off from Lashio for the final leg to Kunming. After an hour-and-a-half, Shilling's engine began to run rough again, backfiring constantly and gradually losing power. Selecting what appeared to be a clearing alongside a mountain, he put the ship down for a forced landing, crashing through trees that severed the left wing and broke the fuselage just behind the cockpit.

Merritt and Mangleburg circled the area in the darkening twilight, uncertain what course to fly. Finally, as both realized that it would be futile to try to find their way to Kunming in the dark, each attempted an emergency landing. Ken Merritt found a dry river bed about 10 miles north, circled once and came in gear-up. Although badly shaken up as the CW-21 bumped and slid down the riverbed, he was able to climb out in one piece.

Mangleburg circled for a few minutes more, then attempted several approaches before making a final approach to land alongside a stream with gear down. He misjudged his height in the darkness and his prop tips hit the water in the stream. Yanking back on the stick, he tried to climb out. His landing gear clipped the edge of a stair-stepped rice paddy, however, and the ship flipped on its back. Ammunition from all four guns exploded on impact, and the wreckage was immediately engulfed in flames.

Meanwhile, Shilling had climbed from his cockpit to find himself surrounded by hill tribesmen, none of whom spoke a word of English. Armed with ancient flintlock rifles and long knives, they refused to let the "foreign devil" venture more than a few feet from his cockpit. To insure that their strange visitor wouldn't wander off, they soon put up a bamboo fence surrounding the aircraft.

Shilling spent a restless night in the cockpit of his plane while a number of tribesmen stood guard beyond the fence. The next morning, they gave him a bit of rice and began to cover the airplane with branches and leaves.

Then Shilling remembered his portable wind-up phonograph and stack of records that he'd put beneath his seat. He'd bought it in Rangoon, along with a stack of jazz records, to take to Kunming. Reaching beneath the seat, he was relieved to find that it hadn't been badly damaged. He put on one of the three unbroken records, wound it up and introduced the astounded tribesmen to the world of jazz.

At the end of the record, they indicated by sign language that they wanted more. Shilling, figuring he had a good thing going, countered with gestures that he wanted more food. For the rest of the day, Shilling ate his fill as he wore the needles down on the record player.

That evening, just before dark, a runner appeared with a message that Chinese soldiers would be coming for the American flyer. He learned later that he had crashed just 15 miles from a spotter's post and that the incident had been reported by radio to Kunming almost as soon as he went down.

The following morning a contingent of Chinese soldiers arrived to escort him to their post. Within two more days both he and Ken Merritt were back in Kunming.

At Rangoon, there was no way of knowing whether the morning bombing raid would be followed up by a second attack, so Olson launched a patrol as soon as the planes could be refueled and rearmed. After two hours of searching the skies to the east and south, the P-40s landed. There was no further sign of the enemy.

Fire continued to rage in Rangoon throughout the day, and a pall of smoke hung over the city. As the fires diminished, British workmen and volunteers cleaned up the rubble and tended to the wounded, but were powerless to stem the mass exodus of Burmese and Indian laborers to the countryside. This had been Rangoon's first taste of war and the native workers wanted no more of it. Consequently, the city ground to a halt within days, resulting in shortages of fuel, food, basic services and other essentials. The British commander imposed martial law to prevent wide-scale looting and rioting.[26]

The following day was spent on tense alert, waiting for the "other shoe to fall." McMillan, Haywood, Smith, and a few others spent five hours in the air on combat air patrol, but nothing was sighted. A report of nine enemy ships headed for Rangoon proved false.[27]

That afternoon, Japanese-controlled Radio Bangkok, broadcasting in English, announced the destruction of Rangoon's air defenses and warned that December 25th would bring a special Christmas present – poison gas and paratroops.[28]

Several of the A.V.G. went into town that evening to shake off some of the tension. They found it to be like a ghost town, with only two or three bars still open. Fortunately, they met a group of seamen from the freighter *Tulsa*, in port there, who invited the pilots on board ship, where they had dinner and shared American cigarettes.[29]

Olson was left with only thirteen flyable P-40s following the air battle of the 23rd. Nevertheless, he maintained dawn-to-dusk patrols on the 24th. R.T. Smith recorded in his diary:

> *"Spent 5 hours on patrol today with Mac & Haywood and others. Nothing sighted. About 1300, while most were still in the mess hall, the air raid alarm sounded. Everyone poured out of the building and ran for the slit trenches. After a few minutes, the all clear sounded. Another false alarm. Some heard that three enemy planes had been sighted, but no one could confirm."* [30]

Christmas Day, dawned clear, hot and muggy in Rangoon. In the topsy-turvy world of the Southern Hemisphere, it would be another hot, humid day. The Burmese workers had fled the area following the Japanese bombing raid on the 23rd and no fresh food had been delivered to Mingaladon Aerodrome in two days. The men enjoyed a breakfast of stale bread and canned beer.[31]

Thirteen pilots were on alert that morning, ten of whom had been in on the air battle of the 23rd – Vice Squadron Leader George McMillan, R.T. Smith, Duke Hedman, Chuck Older, Tommy Haywood, Ed Overend, Curtis E. Smith,

Parker Dupouy, Ken Jernstedt, Lew Bishop, Bob Brouk, Freddy Hodges and Bill Reed.[32]

At 1130 the air raid alarm sounded. Olson ordered all thirteen planes off. McMillan led a flight of seven, while Dupouy led the remaining flight of six. Sixteen Buffalos of the R.A.F.'s No. 67 Squadron took to the air at the same time.

As they climbed out, McMillan received orders over the R.T. to climb to 15,000 feet and orbit over Syriam, 10 miles southeast of Mingaladon and just across the river from Rangoon. Dupouy and his flight of six were directed to patrol over Rangoon.[33]

Stung by their losses on the 23rd, the Japanese planned their Christmas Day attack to be the knockout blow for the Rangoon defenders. This time they sent 98 bombers and 57 escorting fighters – 71 twin-engine Mitsubishi Ki-21 Sally heavy bombers from the 12th, 60th and 62nd *Sentais*, escorted by 25 Nakajima Ki-43 Oscars of the veteran 64th *Sentai* – all from Bangkok – and 27 Mitsubishi Ki-30 Ann light bombers from the 31st *Sentai* out of Lopbure, Thailand, with a 32-plane escort of Nakajima Ki-27 Nates from the 77th *Sentai* at Tak.[34]

The 64th, bloodied in three years of fighting in China and Burma, was led by Major Tateo Kato, with nine victories at the time. Among his pilots that morning was another future fighter ace, Yohei Hinoki. The Japanese force would hit Rangoon and Mingaladon in three waves.

After circling Rangoon for half an hour, Dupouy spotted the bombers – a formation of 30-36 twin-engine Sallys at 20,000 feet just west of Mingaladon heading east. Behind and 2,000 feet above was a loose swarm of fighters – perhaps 10 to 15 Oscars.

Pushing the throttle to the firewall, Dupouy began the climb to intercept. Stretched out in loose formation around him were Freddy Hodges, Bill Reed and Ken Jernstedt. Lew Bishop and Bob Brouk, the other members of the flight, had separated from the others shortly after takeoff. Unnoticed by the Hell's Angels pilots in their determination to reach the bombers to the northwest, a second formation of Sallys, Anns, and Nates was bearing down on Rangoon from the east.

As they approached the Sallys, at least nine defending Oscars attacked the P-40s from above and behind. Flashing past the Tomahawks, they turned to engage the shark-nosed fighters. Dupouy took one Oscar head-on, while his wingman, Freddy Hodges, turned with one of the Oscars, putting a burst of .30- and .50-caliber slugs into his engine at 50 yards range. As Dupouy described the encounter:

> *"... The first fighter I engaged I met in a head on attack and I fired directly at his nose from close range. We passed very close and in the next 360... I fired a burst directly at the nose of his ship and he rolled over on his back and dove straight to the ground, apparently out of control. I saw him hit in the mud flats southeast of Thongwa. This fighter had square wingtips and an inline engine. I believe it was a ME 109. R.A.F. Flying Officer Bingham Wallace saw this fighter and also believed it to be a ME 109."*[35]

As another Oscar made a head-on pass on Bill Reed's ship, Reed held the trigger down for a long burst, watching his tracers disappear into the Japanese fighter. The Oscar dove away out of control.[36]

Ken Jernstedt, the fourth member of the flight, was experiencing engine trouble and couldn't stay with the others as they began their climb to intercept the formation. As he watched the Japanese fighters descend on the other P-40s, an Oscar disengaged himself from the formation and dove toward him. Jernstedt shoved the stick forward in a long dive and soon outdistanced his pursuer. At 4,000 feet his engine began to run smoother and he climbed back to the east to rejoin the battle. He didn't have to search long – he soon sighted a bomber formation below and off his port wing, with several Nakajima fighters to the rear. He turned to engage the last fighter in the string, opening up with all his guns. The Oscar faltered, did a half roll, and went into a right spiral with smoke streaming from his engine. Jernstedt followed him briefly, but was forced to break off when the other Oscars turned in his direction turned in his direction. Discretion being the better part of valor, he firewalled his throttle and turned back toward Rangoon.[37]

Meanwhile, an Oscar had tacked itself onto Parker Dupouy's tail, but he soon shook it by diving away. Zooming up once again, he chased the bombers ten miles out into the Gulf. Then, spotting another formation of bombers coming from Rangoon at 20,000 feet, he climbed to intercept. Coming in on a head-on attack, he got off a quick burst on the lead bomber.

As he broke away, he ran into three Oscars, but again dived out of their range. Bill Reed then appeared and joined on his wing. Low on fuel, the two headed for home. Suddenly they saw three Oscars five thousand feet below on an easterly heading. A quick wing-over and the two P-40s were on a pursuit curve diving to intercept the Japanese fighters. Low fuel or not, this was too good to pass up. Dupouy and Reed opened fire simultaneously at 30 yards range. The Japanese never knew what hit them. The fighter on the right dove toward the ocean, his engine flaming. Pulling up in a sharp climbing turn toward the Oscar on the left, Dupouy misjudged his overtake and his windscreen was suddenly filled with the Japanese fighter. Before he could react, Dupouy's right wing sliced into the Oscar's left wing root. As the two separated, Dupouy's P-40 was minus four feet of its right wing, and the Oscar, flown by 64th *Sentai* pilot, Lt. Hiroshi Okuyama, spun out into the water. Escorted by Reed, Dupouy managed to nurse his damaged ship back to Mingaladon. [38]

At 1210 McMillan sighted the enemy – twenty bombers in a tight V-of-Vs formation, accompanied by about 15 fighters a mile or so in trail. They were passing over Mingaladon five thousand feet above and heading east. Having already dropped their bombs on Rangoon, they had turned over Mingaladon and were now streaking for home.

McMillan and his wingman, R.T. Smith, reached the Japanese formation over Thongwa, 20 miles east of Rangoon. Climbing toward the left flank, McMillan fired at the leader but saw his tracers fall behind the bomber. Closing in on the formation, he brought his sights to bear on a second bomber midway back in the stream. Another burst from his four .30s and twin .50s, and the bomber exploded. Continuing down the left flank, he shot both engines out of another and watched it fall off in a steep dive toward the water below. A cannon shell from the stricken bomber tore through McMillan's right wing, but didn't do serious damage.

Spotting a "straggler" below, McMillan dived to pick him off. The enemy gunner began firing at the P-40 at about 700 yards, but McMillan bored on in to within 100 yards, firing all the way. Suddenly he felt his aircraft shudder as it took a hit in the engine, but at the same instant saw the bomber explode. As he dived away, he watched the engine temperature gauge begin to rise rapidly. His cooling system had been shot out and Glycol coolant was streaming out of the ragged holes in the cowling. McMillan left the battle to make a wheels-up crash-landing in a rice paddy eight miles east of Thongwa.[39]

R.T. Smith hit the last bomber on the left flank of the "V," knocking out its left engine. As it dove out of formation, smoke pouring from its engine, Smith swiveled his head to see two fighters closing in from the rear. Banking hard left, he came around to meet the nearest fighter head on. He began firing at 400 yards, continuing until the fighter passed below him with not more than ten feet of clearance.

Smith racked the P-40 into another tight left turn, looking for his opponent. He saw him below in a gentle right turn, smoking badly. Suddenly the aircraft burst into flames and dove into the Gulf of Martaban.

Turning back toward the bombers, Smith hit the left flank again. After four or five ineffective passes, his bullets found their mark on the right engine of the last bomber in the "V." As the Sally began to trail smoke, he made two more passes at it before it dropped out of formation and, now smoking badly, dove off toward the Gulf.[40]

Duke Hedman made his first pass with McMillan and R.T. Smith. The three broke up for individual attacks and Hedman's next attack, again from the low rear quarter on the formation left flank, brought him into 50 yards before his target exploded. Half-rolling away and down, he came back up again from the rear, selecting the last bomber. He continued his fire until flames began to appear from the belly of the bomber. He watched it nose over into a 45-degree dive.

Rolling away from his second victory, Hedman saw a "V" of three bombers turn away at a 45-degree angle to the main formation. Turning into the three from the right, he moved into close range again and put a short burst directly into the bomber's right engine. As the bomber did a slow half roll to the right, smoke pouring from its engine, Hedman moved on up to the lead ship. Holding his fire once again until 100 yards, he gave the bomber a long burst. As if the pilot had fallen against the control wheel, the bomber pitched down into a steep dive, emitting black, oily smoke as it fell. Suddenly, Hedman's canopy was shattered by a burst from the remaining bomber's gunner, with one of the bullets lodging to the left of his headrest. Six inches closer and he would have been a statistic.

Hedman banked away sharply to see a Nakajima Oscar to his right at 500 yards, turning to dive on another P-40 200 yards below. He turned inside the Oscar and opened fire at 300 yards. The Japanese fighter immediately burst into flames and went straight down. The P-40 that was the intended target was flown by Curtis Smith. Smith continued his dive out of the battle area and flew back to Mingaladon, where he landed with nearly a full load of ammunition and an undamaged airplane. Chennault removed him from flying status and appointed him Group Adjutant.

As Hedman dived away from his fifth victory, he turned to see another Oscar on his tail, not 200 yards away, gun muzzles flashing. He half-rolled and split-essed from 20,000 down to 5,000 feet. When he recovered, he was alone and flying in and out of dense cloud. He flew on instruments from the Gulf to the satellite field near Pegu, 35 miles northeast of Rangoon. By the time he brought his crippled ship in, his damaged engine was turning up only 1500 RPM and his bullet-riddled fuel tanks were nearly dry.[41]

While McMillan, R.T. Smith, and Hedman were working over the left flank, Older, Haywood and Overend were attacking the right. Older led a two-plane section with Haywood on his right wing, and they stayed glued together throughout most of the battle – just like Chennault had stressed. Attacking from below and to the right of the outboard "V," Older and Haywood opened fire at 200 yards and fired short bursts until the bomber was less than 250 feet away. They rolled away for a re-attack and closed in again to 75 yards. As Older reported later:

> *"...smoke was streaming from several bombers on the right flank. One of them burst into flames, nosed down to a steep dive, and disappeared below. Another bomber pulled out of the formation with smoke trailing behind it and started down. I followed it down and saw it roll over on its back into a vertical dive and disintegrate in the air. I continued making attacks from below on the bombers and saw two more go down in flames after close attacks, in which Haywood and I fired together, almost side by side."*

As he continued chasing the bombers out over the Gulf, he noticed Haywood was no longer with him. Looking back, he saw a Nakajima Oscar 500 feet above, heading in the opposite direction. Turning tightly, Older brought his P-40 beneath the tail of the unsuspecting Japanese fighter and opened fire at 50 yards. The Oscar's rudder fell away as Older's bullets smashed into the fuselage and black smoke began pouring from the stricken fighter. Starting a downward spiral, the Oscar's wings came off as Older followed it down to watch it crash into the Gulf of Martaban.[42]

Overend fell into trail behind Older and Haywood as the three made their first attack. Selecting the wingman of the leader of the last section in the "V," he led his target by half a plane length and began firing at 200 yards. The bombers left wing caught fire and it rolled slowly to the left and spiraled out of the formation. As he pulled away for a re-attack, it seemed that the sky was raining bombers. Overend saw a wing falling past him, tumbling end over end, and "two more bombers going down, one blazing underneath." On his fourth pass, he damaged a second bomber and watched it glide steeply toward the ground. His P-40 had taken a hit, however, and he began losing power rapidly as his manifold pressure fell to five inches. His right wing dropped and he found that his stick was jammed. By holding full left rudder and rolling in full left trim tab, he managed to bring the wing up. He eased his plane down to a crash-landing in a rice paddy eight miles east and south of Kayan.[43]

Bob Brouk and Lew Bishop had separated from the four others in Dupouy's flight shortly after takeoff. Seeing nothing of the enemy after thirty minutes of patrolling, they returned to base. On final approach, Brouk received orders from the tower not to land and to climb back to 15,000 feet. As the two P-40s climbed out, they saw one of the retreating bomber formations about 20 miles south of Rangoon.

The two pilots separated, each to make individual attacks. As Brouk closed in on the bomber formation, he spotted an escorting Oscar a mile or so behind the bombers. Diving behind and below the Japanese fighter, he pulled up into a zoom climb and opened fire at 200 yards. After a three-second burst, the Oscar rolled over and spun down out of control. As four more Oscars appeared, Brouk dove away.[44]

While Brouk was engaging the fighters, Bishop concentrated on the bombers, raking the formation with multiple passes. He damaged one, observing smoke coming out of the fuselage, before he, too, was chased off by a pursuing Oscar.[45]

Haywood, finally separated from Older, formed on Smith's wing. After making separate attacks on the retreating bombers, they finally gave up the chase and turned north for Mingaladon.

Ten of the thirteen P-40s made it back to Mingaladon, landing between 1300 and 1330. George McMillan was down in a rice paddy 25 miles southeast of the base near Thongwa, and Ed Overend crash-landed 30 miles east, near Kayan. Duke Hedman recovered his damaged fighter at the satellite field near Pegu, bringing his ship in for a smooth landing despite having to fly with a smashed canopy. Parker Dupouy made it back to Mingaladon, wheeling his Tomahawk onto the runway at 142 mph to compensate for the loss of four feet of his right wing and half his right aileron. With the exception of 15 holes in R.T. Smith's P-40 No. 77, the remainder of the Tomahawks suffered no damage. The R.A.F., however, didn't fare so well – they lost nine of the 16 Buffalos they sent up, along with six pilots. The Brewster Buffalo, while matching the Japanese Oscar in armament and top speed, was no match for the nimble, little Japanese fighter's maneuverability.

On the ground, damage was heavy. The bomb hits were concentrated on the southeast corner of the field, ironically the area that the ground personnel had gathered for shelter. The men's barracks were heavily damaged and operations clerk L.P. Clothier and armorer P.J. Perry were slightly wounded by flying shrapnel. Rangoon was also plastered once again. What services remained after the raid of the 23rd were brought to a halt and thousands more of the native residents streamed out of the city into the hills and countryside.[46]

The Japanese had paid a high price for the damage they inflicted on the ground, however. The A.V.G., after all claims were sifted through, were credited with the destruction of 24 aircraft – 15 bombers and nine fighters. The A.V.G. had their first "aces" (although it's impossible to say who of the two was the first). Chuck Older shot down two bombers and a fighter to add to his two bombers on the 23rd, and Duke Hedman made "ace-in-a-day" by bringing down four bombers and a fighter. Other credited victories included three bombers for McMillan, two bombers and a fighter for R.T. Smith, two bombers each for Overend and Haywood, two fighters each for Dupouy and Reed, and a fighter each for Brouk and Jernstedt. Added to the seven credited to the R.A.F.'s No. 67 Squadron, the Japanese lost a conservative 31 aircraft, twenty percent of their attacking force – and that figure didn't take into consideration the multitude of "probables" and "damaged" claims, many of which crashed in the Gulf of Martaban and were therefore unconfirmable.

Back on the ground, once the tension of combat was relieved, the pilots excitedly swapped combat stories, their weaving hands becoming Tomahawks and Oscars or Sally's in retelling the action. In R.T. Smith's words:

> *"Our morale was higher than a kite now, and it was evident that suddenly the 'Hell's Angels' Squadron had come of age; in a matter of forty-eight hours we had progressed from an unknown quantity to a battle-tested unit, a dozen or so who could now lay legitimate claim to being fighter pilots."*[47]

That afternoon the Third Squadron was treated to a Christmas feast. CAMCO chief Bill Pawley was at Mingaladon during the air battle, and observed the interweaving contrails from a grove of banyan trees near the field. Before the last P-40 landed, he ordered a panel truck of food and liquor brought out from the CAMCO supply store in downtown Rangoon.

In the sweltering Burma heat, the men sat down to makeshift tables under the trees and enjoyed chicken, ham, fresh vegetables, cakes and pies – and all the beer and Scotch they wanted to wash it down with.

Although they didn't make it back in time for the feast, both George McMillan and Ed Overend were brought back that evening, an hour or two apart. They were a little worse for the wear, but in good spirits. Their safe return was the "icing on the cake" for the day's activities, and the men celebrated far into the night.[48]

Curtiss-Wright CW-21 "Demon" interceptor sent to China for evaluation in 1941. If successful, 22 were to be assembled at Loiwing by CAMCO. It had a top speed of 314 mph and service ceiling of 34,000 feet.

Erik Shilling (shown here in the cockpit of his Tomahawk) led the ill-fated ferry trip of the three CW-21s on 22 December, 1941. (via George Paxton)

Ken Merrit survived his crash-landing with minor injuries. (via Chuck Baisden)

Lacy Mangleburg, portrayed by artist Raymond P.R. Neilson for Bill Pawley's booklet "Americans Valiant and Glorious." (via George Paxton)

Paul Frillman performed the burial service for Lacy Mangleburg on December 23, 1941. (via Bob Andrade)

Christmas Over Rangoon. *Oil on canvas by Roy Grinell. The painting depicts Charles Older diving away after an overhead attack on December 25, 1941. Older would become one of the first two A.V.G. aces (along with "Duke" Hedman), adding three victories to the two on December 23. (Courtesy Roy Grinnell Studio)*

Parker Dupouy stands beside his P-40 No. 72 after landing from the Christmas Day air battle minus 4 feet of his right wing following a head-on collision with a Japanese "Oscar." (via C. Baisden)

Four of the Christmas Day defenders in front of R.T. Smith's P-40 No. 77. Left to right: Ken Jernstedt, Tom Haywood, Chuck Older and R.T. Smith, (via R.T. Smith)

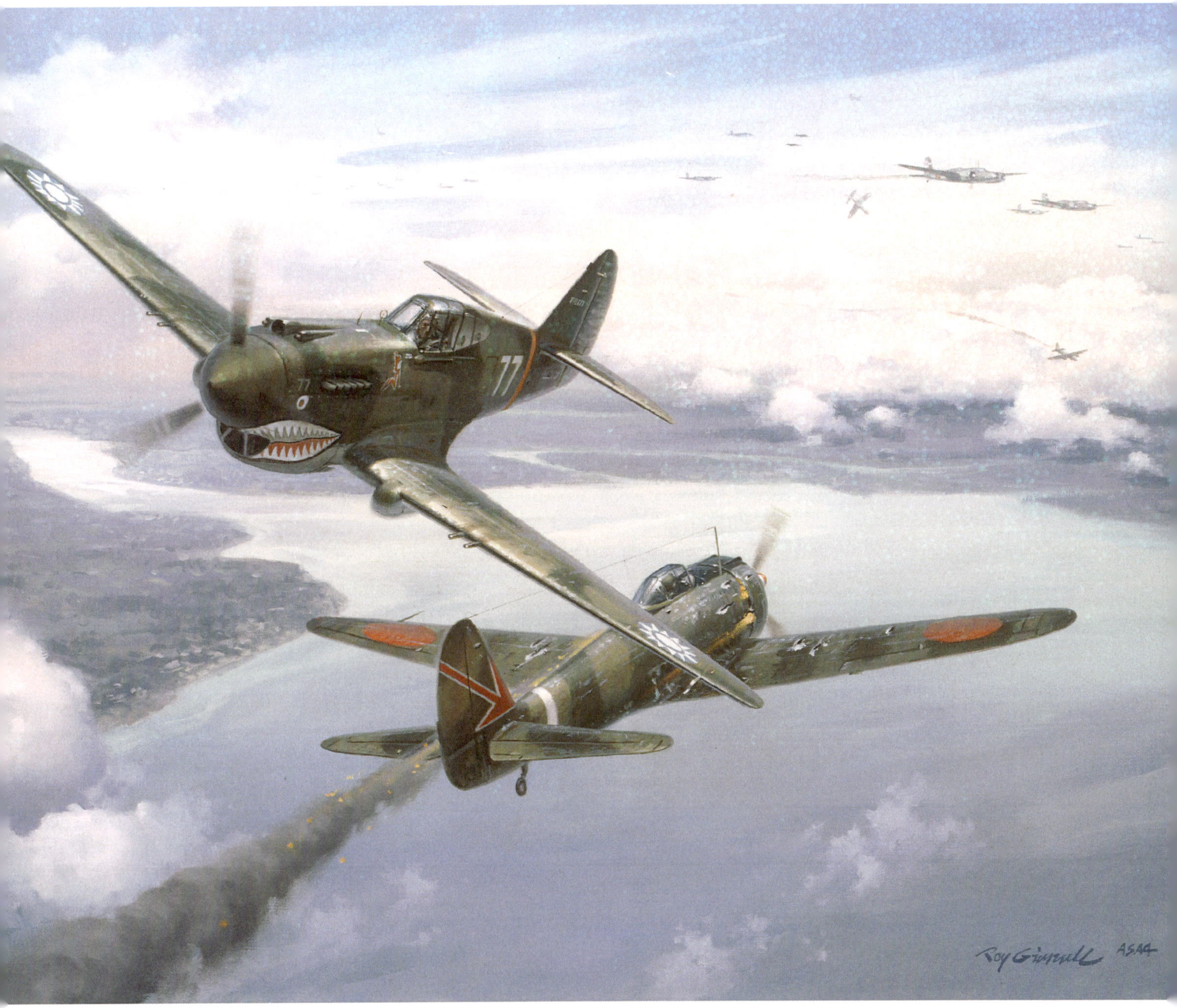

Greetings from the A.V.G. *Oil on canvas by Roy Grinell. Painting shows R.T. Smith meeting a Nakajima Ki-43 "Oscar" in a head-on attack on December 25, 1941. Smith would be credited with three victories on that Christmas Day. (Courtesy Roy Grinnell Studio)*

CHAPTER 10

The Pandas Get Their Turn

Alert was resumed on December 26 and Olson sent a radiogram to Chennault in Kunming, via Goyette at Toungoo, reporting the previous day's battle:

> ALL PILOTS RETURNED AFTER BATTLE ON 25 DEC STOP SHOT DOWN 10 FIGHTERS 9 BOMBERS BRITISH ACCOUNTED FOR 6 STOP LIKE SHOOTING DUCKS STOP WOULD PUT ENTIRE JAP FORCE OUT OF COMMISSION WITH GROUP HERE STOP HAVE ELEVEN PLANES LEFT STOP SENDING PILOT GREENE TO KUNMING VIA LOCKHEED STOP

As regular as clockwork, the air raid warning sounded at 1000. Due to the combat damage of the day before, however, only seven of the surviving 11 Tomahawks and about an equal number of R.A.F. Buffalos got airborne. They climbed out to 22,000 feet and patrolled for two hours, but saw nothing.

That afternoon at 1500, a lone Japanese plane dropped leaflets over Rangoon warning the population of an imminent invasion by parachute troops. Although many suspected the warning was a Japanese trick to keep them on edge, a full alert was maintained throughout the clear, moonlit night. Of course, no troops appeared.

Another 1000 air raid alarm on the 27th got everyone's adrenalin pumping, but it was cancelled by the time the pilots reached their ships. Later, a radiogram from Chennault arrived which raised everyone's spirits:

> CONGRATULATIONS AGAIN UPON RETURN OF TWO MISSING PILOTS FROM YOUR MAGNIFICENT VICTORY XMAS STOP AM SENDING STRONG REINFORCEMENTS SOON AS TRANSPORT PLANE AVAILABLE ADVISE MANNING STOP DO YOU WISH PULL OUT YOUR UNIT FOR REST HERE AFTER ARRIVAL NEW UNIT STOP IF SO WOULD LIKE TWO OR THREE VOLUNTEER PILOTS AND CREW TO REMAIN WITH NEW UNIT FEW DAYS AT LEAST UNQUOTE BELT 22 THOUSAND ROUNDS 7.92 AMMUNITION THERE AND REQUISITION 75 THOUSAND MORE IF AVAILABLE FROM MAYMO.

Officer Harvey Greenlaw followed with a second wire, spelling out the details:

> SECOND SQUADRON WILL RELIEVE YOU TOMORROW EXCEPT FOR FOUR VOLUNTEERS FROM BOTH PILOTS AND CREW CHIEFS STOP C.N.A.C. LEAVING HERE APPROXIMATELY SIXTEEN THIRTY TODAY SEND MEN AND PILOTS THAT CAN BE SPARED BACK ON TRANSPORT STOP YOU WILL FLY AIRPLANES NEEDING REPAIR TO LOIWING ON BEING RELIEVED PERSONNEL RETURNING KUNMING FOR REST STOP REGARDS AND CONGRATULATIONS UNQUOTE PUT COL CHENNAULT'S YELLOW TRUNK AND MY LEATHER TRUNK ON C.N.A.C. BEST REGARDS[1]

Ed Rector, Peter Wright and Lynn A. "Tex" Hurst were the first of the Second Squadron replacements to arrive, flying in on the C.N.A.C. DC-3 on the morning of the 28th. Squadron Leader Jack Newkirk, commander of the "Panda Bears," and Noel Bacon were the next to arrive, landing their P-40s on Mingaladon at dusk the following day.

Newkirk had departed cold and cloudy Kunming at 0930 that morning, leading eighteen Second Squadron P-40s.[2] One turned back for engine trouble, but the remaining seventeen flew on to He Ho where they refueled before going on to Toungoo. Newkirk and Bacon arrived at Toungoo at 1530, refueled and took off for Rangoon. The remainder of the squadron reached Toungoo at 1700, too late in the day to continue on.

Several crew chiefs and four pilots from the Third Squadron also arrived by train from Mingaladon to catch the C.N.A.C. plane to Kunming. Curtis Smith, one of the four pilots, was obviously still smarting from being relieved from combat flying. He began drinking that evening and got into an argument with medical orderly Charlie Cribbs. Smith finally slugged him and then pulled out a knife. Two others rushed in and grabbed Smith before the situation got out of hand. As he calmed down, Smith made a half-hearted apology and said that he was "on edge, and didn't know what I was doing." [3]

The fifteen remaining Second Squadron ships touched down at Mingaladon at 1030 on the 30th. Within half-an-hour, ten Third Squadron pilots, some in ships barely flyable, had departed for Toungoo, en route to Kunming. They found their old base at Kyedaw to be almost a ghost town by this time, with only a few pilots still undergoing P-40 transition training and a handful of mechanics to repair the damaged planes. Ed Goyette, another older pilot excused from combat flying by Chennault, was left in command of Kyedaw – by this time known as "Point A." [4]

Olson, Reed, Greene, Adkins, and Laughlin decided to remain at Mingaladon for three or four days until their ships could be flown out, and George McMillan was still there because of engine trouble experienced on takeoff the day before. McMillan and Adkins, however, left the next day for Toungoo to fly two P-40s out to Kunming.[5]

Olson worked with Jack Newkirk getting the Second Squadron settled in throughout the day, then turned in early to get some rest before departing for Kunming via Calcutta on a C.N.A.C. DC-3 at 0400 on New Year's Day. Newkirk's "Panda Bears" spent a quiet New Year's Eve. The fireworks would come later.

After two days, the "Panda Bears" were settled in. The tension lay heavy on the men, just as it had when the Third Squadron arrived at Mingaladon three weeks earlier. They waited in the alert shacks, in the barracks, in lounge chairs on the flight line. Not even a false alarm broke the silence.

Anxious to get his men into combat, Newkirk planned for a raid on January 3 on the Japanese-held aerodrome at Tak, 170 miles to the east, just across the border into Thailand. Intelligence indicated that a large force of Japanese fighters and bombers were based there. Indeed, although Newkirk couldn't have known at the time, 31st *Sentai* bombers had staged out of Tak for the Rangoon raid of the 23rd, and fighters from the 77th *Sentai*, also based at Tak, participated in both of the December attacks on Rangoon.

Newkirk selected Jim Howard, Bert Christman, and Tex Hill to accompany him. Taking off shortly after dawn, they climbed out to 10,000 feet. Christman developed engine trouble, however, and had to turn back, leaving the flight of three to proceed to the Japanese field.[6]

As they approached Tak from the west, Newkirk could see Raheng Aerodrome, just outside of the town, its single runway showing up as a brown slash across the green jungle. Rows of aircraft lined the field. Dipping his wings, Newkirk signaled his wingman to the attack, and peeled off in a curving dive. Jim Howard was next, followed by Tex Hill.

As the three screamed down toward the field at 250 miles per hour, Newkirk was the first to spot a pair of Nakajima Ki-27s circling the field at 2,000 feet. He leveled out and bored in on the nearest Nate, coming in from the stern. Suddenly aware of the attack, the Japanese jinked right and then back left. Newkirk's bullets had found their mark, however, because the stricken fighter began trailing smoke as it continued its roll to the left. Seconds later it was inverted in a dive that carried it to the jungle below.

While Newkirk was busy with the fighters, Jim Howard and Tex Hill continued their dive toward the field, leveling out fifty feet above the trees, bordering the runway. With Hill a quarter mile in trail, Howard lined up on a row of parked aircraft and began firing. They could see figures running wildly about as they flashed past. In less than ten seconds they had traveled the length of the field and Howard began his pull-up.

Suddenly, a Japanese fighter dived down "out of nowhere" between Howard and Hill. Ignoring Hill, the Japanese attached himself to Howard's P-40 and began firing. As Hill described it:

> *"This guy was on Jim Howard's tail, just really eatin' him up. And so I pulled in right behind him. I didn't even look. We had a ring and bead gunsight on the airplane, but I didn't even look through that. I was just looking through the windscreen there at the traces and just kinda hosin' him down – just kept pulling the nose right around on him till he just blew up."* [7]

While Hill was occupied with the fighter on Howard's tail, another Japanese fighter dived on Hill's P-40 and scored hits on both wings, but miraculously missed the cockpit and engine. Hill broke left and up, only to find two more fighters diving head on into him. Again, four .30-caliber guns converged on the light Japanese fighters and it blew up. Hill's aircraft took additional hits in the head-on encounter, however, and a 7.7 mm round holed his propeller, throwing it out of balance. "The dang thing nearly jumped out of the airplane," Hill said later.[8]

As the trio climbed out to the west, with a 50-mphr speed advantage over the Nates they easily outdistancing their Japanese pursuers. They only had a chance to make one pass on the field, but were still able to destroy four aircraft on the ground and four in the air. Unfortunately, Newkirk was otherwise occupied and Howard was oblivious to his attacker, so intent was he on his strafing run. Therefore, no one besides Hill witnessed the destruction of the fighter he shot off Howard's tail and he received confirmation for only one aircraft destroyed.

On the way back to Rangoon, Hill found an rpm setting that minimized the vibration set up by his holed prop, and the trio landed at Mingaladon with no further complications. Back on the ground, however, Hill walked over to Howard's aircraft and, grinning, said, "Good thing I shot that Nip off your tail, Jim."

"What do you mean?" replied Howard. "There wasn't anyone on my tail."

"Just take a look at your ship," smiled Hill, pointing to the several bullet holes punched into the tail and rear fuselage of Howard's P-40.[9]

That night Newkirk and his men celebrated at Rangoon's Silver Grill Restaurant. George L. "Pappy" Paxton was one of the last to get back to the base, falling into bed about 0400. He was to be on "standby" only, and could rest all day.

Newkirk and company had stirred up a hornet's nest with their raid on Raheng aerodrome. The next morning the Japanese returned in force to pay their respects – 31 Ki-27 Nates from Raheng headed west, bent on vengeance.

With only an hour's sleep, Paxton was jarred out of his slumber by an air raid alarm. Suddenly wide-eyed and awake, he pulled on his flight suit and boots and rushed outside to jump into a waiting Studebaker station wagon for the one-hour ride to the satellite field where several of the P-40s had been dispersed. Bombs had been dropped on the field, but fortunately they were duds and no aircraft were damaged. At first light, Paxton and the other pilots climbed into their cockpits, started their engines and one-by-one took off for the return flight to Mingladon.

As Paxton's P-40 touched down, it swerved to the right onto the grass. The aircraft's nose straightened out as he "gunned" the engine with a quick burst of the throttle. Suddenly, however, he saw Hank Geselbracht's aircraft taxiing across his nose, not fifty yards away! As Paxton braced for the collision, Geselbracht's P-40 cleared his right wingtip – barely. Continuing to taxi, Paxton rolled across a shell hole and his tail wheel folded. As he started to turn, his left main gear folded, leaving the P-40 on its belly. Paxton climbed out of the cockpit, walked to the alert tent and laid down to continue his rest that had so rudely disturbed four hours earlier.

At 1130 another air raid alarm sounded. Someone rushed in to tell Paxton that one of the alert pilots was missing and that he'd have to fly. He and five others – Hank Geselbracht, Ken Merritt, Bert Christman, Gil Bright, and Frank Swartz – ran to their P-40s and twenty minutes later they were off the ground. They split into two sections of three, Geselbracht leading one section with Paxton and Merritt, while Christman led the other section.

At first it looked like a false alarm. They had circled the field at 21,000 feet for 45 minutes with no sign of the enemy. Geselbracht's section spiraled down to 11,000 feet. Christman's flight followed, a few hundred feet higher.

Suddenly the sky was full of silver fighters. As Merritt broke to the left, Paxton reached for the gun switch. As he did so, his cockpit seemed to disintegrate – bullets penetrated both sides of the cockpit and canopy and smashed into his shoulders, side and leg. It was like an eternity as the deadly pounding continued.

As if in slow motion, the P-40 seemed to hang in the air and every movement Paxton tried to make seemed to take forever. The smell of oily smoke burned his nostrils as he fell off in a spin. He considered jumping, then decided to stick with the plane. He stopped the spin, but was still in a dive. His outside vision was obscured by an oil-covered canopy; his instruments, also covered with oil, were useless.

As Paxton started his pullout, questions rushed through his mind – was he high enough, was the fighter still on his tail, would the old '40 hold together? He eased back on the stick and the riddled Tomahawk leveled off at 2,000 feet. He tested the controls. They were OK. The engine still responded, although it was smoking. His shoulder, side and leg hurt like hell, but he fought to keep from passing out.

Just then, pieces of the shattered plexiglas canopy began to fall away and he could see out again. Paxton looked down to find that he was just north of the field – one turn and he'd be lined up with the runway. He pushed the gear handle down and the indicators showed the gear down and locked. He lowered the flaps as he turned on final.

As the wheels touched the ground, Paxton let out a huge sigh of relief. Then, as the plane rolled out, it veered to the right. He kicked left rudder, overcorrected, and the rudder locked. As the plane swerved to the left, Paxton braced himself for the crash. At that moment the gear folded and the P-40 slid to a stop in a cloud of dust.

Cutting the ignition switch, Paxton reached for the canopy handle. It wouldn't budge. He tried the emergency release. It wouldn't work, either. As a last resort, he tried to get his good leg up to kick out the glass – no luck. Paxton sunk back in the seat, exhausted. After what seemed like hours, he saw a car racing toward him across the field. Two men jumped out, forced his canopy back, and gently lifted him out of the cockpit.[10]

Paxton's No. 41 was a complete write-off. Tex Hill later walked around the crippled bird, amazed that Paxton had survived the attack at all. Although Paxton's diary records 61 holes in his P-40, Hill said, "I stopped counting at 63."

Meanwhile, two more P-40s had been shot down in the attack that Ken Merritt estimated took no more than five minutes.[11] Christman's aircraft was riddled by 7.7 mm bullets as the Nates dove on the six P-40s. He dove away, but his engine ran only five minutes before it froze and he stepped over the side in a "nylon letdown". Everything had happened so fast that he scarcely realized that he had suffered two bullet creases along his jaw and that another Japanese bullet had come close enough to tear out his left earphone from his leather helmet.

Gil Bright, who had survived the mid-air that took Jack Armstrong's life back at Toungoo in October, also took a hail of bullets in the attack. He managed, however, to guide his ship down to a wheels-up landing in a rice paddy. As soon as the riddled Tomahawk slid to a bumping halt, he jumped from the cockpit and ran from the plane, scant seconds before it caught fire and blew up. He suffered only minor facial burns, but later joined Pappy Paxton and Bert Christman in the hospital.

It had been a clear victory for the Japanese. Ken Merritt managed to shoot down the aircraft that hit Paxton, but the score for the day was Japanese 3, A.V.G. 1. It would be the only time the Japanese would enjoy such a kill/loss ratio with the A.V.G.

That night the Japanese launched the beginning of a series of night raids that would last for the next five days. At 0230 the air raid alarm sounded. Thirty minutes later, 27 bombers converged in Mingaladon from three different directions.12 Three P-40s, flown by Charlie Mott, Tom Jones and Pete Wright, rose to meet the night raiders but didn't make contact. The bombers didn't have much better luck – their bombs fell in a scattered pattern in the vicinity of the field and only succeeded in destroying General Wavell's Douglas DC-2.[13]

The night attacks continued. With each air raid alarm ground crew and pilots were roused from their sleep to pile into cars and trucks to drive to dispersal areas, where they'd try to get some rest. At midnight on January 6 the Japanese returned with five bombers that bombed and strafed the field, slightly damaging three Tomahawks.

The A.V.G. pilots were becoming more angry and frustrated with each raid . Although the alert flight rose to meet the invaders, they hadn't made contact with the Japanese once.

Analyzing the situation, three of Newkirk's pilots – Jim Howard, Peter Wright and Gil Bright – after observing the Japanese on two consecutive nights, devised a plan to combat the irritating "bed-check Charlies." They noted that the Japanese flew in a loose formation of five-to-seven planes as they approached the target. Just before they reached the field, they turned on their running lights and closed up the formation into a tight V. "We felt if we could get one of our Curtis P-40s into the air, it would be quite easy to bag a bomber or two," Peter Wright recalled, "especially if they kept their running lights on." [14]

The problem, however, was how to recover the interceptor – Mingaladon had absolutely no facilities for night flying. This was soon resolved with the help of the R.A.F., who provided eight kerosene lanterns, six white and two red. The plan was to have one of the ground crew, when the pilot was in the landing pattern, drive out onto the runway and place the lanterns at predetermined positions. In this manner, an amount of light for landing would be provided for a minimum period of time. "We decided to use only one plane at a time to eliminate the danger of collision at night," Wright continued. "Besides, the small force of enemy bombers was just the right size for one plane to handle." [15]

Wright's night "in the barrel" came on January 7. Earlier in the evening he had taken his mechanic, John B. Hauser, aside to brief him on the procedure for the night's operation. When the alarm sounded, Hauser would drive his jeep to the far end of the runway and turn around, shining his lights down the centerline of the runway. Using the jeep's headlights as an aiming point, Wright would make his takeoff. When he was ready to land, he would flash his running lights as he circled the field at 2,000 feet. At that instant, Hauser was to position four of the white lanterns on the runway's left side and the remaining two on the right. The two red lanterns would be placed on the approach end of the runway.

With everything in readiness, Wright turned in for a little shuteye in the alert tent while Hauser started and ran up the Tomahawk's engine every hour.

About 0245, Wright was awakened by the telephone in the alert tent. Operations reported "heavy engine noises 60 miles to the east." Squadron Leader Jack Newkirk and Ken Merritt were waiting outside in Newkirk's Chevrolet sedan as Wright rushed out of the tent, still zipping up his coveralls. He jumped into the sedan and the three sped off toward his parked Tomahawk, No. 44. With the lights of Hauser's jeep shining from the far end of the runway, he lifted off at 0310.

As he circled the field, Wright received conflicting reports from the ground controller. First the enemy was coming from the east... then the north. Suddenly he saw yellow flashes to the north just as the voice on the radio reported bombs falling. Damn! The Japanese had obviously missed their target and he had missed the opportunity for an interception. (The Japanese bombers, six single-engine, fixed-gear Mitsubishi Ki-30 "Anns" from the 31st *Sentai* at Phitsanulok, Thailand, had indeed become lost and dropped their 661 pounds of bombs on rice fields several miles to the north of Mingaladon.)

Disgusted at his lost opportunity, Wright descended to 2,000 feet, pushed his canopy back, flashed his running lights per the prearranged procedures, and prepared to land. Below, standing beside his sedan parked on the edge of the runway, Jack Newkirk was monitoring the approach. Ken Merritt, however, had succumbed to the warm Rangoon night and was asleep in the back seat of the car.

With gear down, Wright turned from base onto final and pushed the flap-actuating button on top of the stick to lower his flaps. At that moment a rubber gasket in the flap-actuating valve handle blew and a spray of hydraulic fluid spurted out, instantly drenching his face and chest. "It felt as though someone was standing about two feet in front of me spraying me with a garden hose," Wright said later.

Descending through 300 feet, with his airspeed still at 120 mph, Wright wiped frantically at his oil-covered goggles. Unable to clear the oily film, he pushed his goggles up and leaned his head out the left side of the cockpit into the slipstream in an attempt to see the fast-approaching runway. Describing the incident, Wright said:

> *"I caught a momentary glimpse of the two red lanterns disappearing beneath the leading edge of my left wing. I knew I was over the edge of the runway - about fifty feet up and settling fast. I waited a split second and then pulled back the stick. I hit the ground in a left skid, and the landing gear gave way with a lurch. For an instant I thought I would be okay, as the plane skidded along on its belly. Suddenly there was a terrific crash and I violently cracked my head against the windshield. The plane spun around and stopped with a jar."* [16]

As he sat dazed in the cockpit, Wright was barely aware of hands reaching in to pull him from the cockpit. He gradually came to, wiping the hydraulic fluid and blood from his eyes. Only then did he discover, illuminated in the headlights of approaching vehicles, the solid object that had caused his abrupt stop. It was Newkirk's sedan, now sheared in two by the Tomahawk's prop. Ken Merritt, or what was left of him, was still in the back seat.

The next morning Newkirk took the offensive again with a two-prong attack on Japanese-held Thai aerodromes. Charlie Mott, Percy Bartelt, Moose Moss and Gil Bright took off at 1100 to raid Mae Soht, while Newkirk himself led a flight of seven Tomahawks later that afternoon back to Tak.

Diving from out of the sun, Mott led his flight in a 400-mph pass down the airstrip. Thirty- and fifty-caliber bullets smashed into a row of six Japanese aircraft parked to the side of the runway. The strafing passes ended with a gut-wrenching pull-up at the end of the field. As they made a climbing left turn, they looked back to see several burning aircraft in the wake of their attack. Rolling out parallel to the runway, Mott brought his flight around for a second pass. The Japanese were caught by surprise the first time, but they were ready for the diving P-40s the second time around. Again, sixteen .30-caliber and eight .50-caliber guns blasted the parked Japanese planes, but as the Tomahawks flashed past this time, they could see the winking muzzle flashes of a score of anti-aircraft guns.

As they pulled up from the second pass, Charlie Mott's P-40 separated from the four, smoke pouring out of the engine. He managed to coax a shallow climb out of his damaged fighter, but he knew it was hopeless. He slid the canopy back, unbuckled his seat belt, and stepped over the side. His chute popped almost immediately and got one swing before he crashed through the upper branches of a tall tree. With his chute snagged, he found himself swinging twenty feet in the air. Wriggling out of his harness, he dropped to the ground, breaking his pelvis and some ribs, with a compound fracture of his left arm and a badly broken foot. As he lay on the ground, he was soon surrounded by a Japanese search party. Charlie Mott had earned the dubious honor of becoming the A.V.G.'s first P.O.W. (In 1945, Mott would escape from a Japanese labor camp and join a guerilla force before making his way to India.)[17]

Later in the day Newkirk and his flight of seven, along with six R.A.F. Buffalos, paid a return visit to Tak. As Jack Newkirk described it:

> *"The formation left the base at 1635 and flew in loose line astern with weaver at an altitude of 10,000 feet. While crossing the border we were able to take advantage of a cloud, in order to remain out of sight of Mesoht. The first section dove straight into Tak aerodrome out of the sun. The second dove from the southwest, and the third from the northwest, in that order. On the first dive, I was unable to distinguish my target, except the operations building. However, on reaching the field, I saw four enemy aircraft, a few trucks and several ground personnel. On the next dive I attacked an I-97 fighter which was parked by the operations building. It was silver-colored and had branches on it, and when I looked back it was in flames. On the same dive I fired on a truck which was diving across the field. The truck swerved and ran into the burning plane. On the next dive I fired on a plane across the field from the building, several pieces fell out of it and it collapsed on the ground."* [18]

The A.V.G. claimed eight destroyed at Mae Soht – Bright got one, Moss and Bartelt two each, and Mott was credited with three with three more destroyed at Tak – for the loss of one P-40 and pilot.

Newkirk kept the pressure on the Japanese, sending his pilots on strafing missions from Mesarieng in the North to Taway in the south. On January 9 he led another raid on Tak, with Bright, Bob Layher, and Noel Bacon. The pace was beginning to tell on both the pilots and the planes, however, and, following another afternoon strafing mission on the 10th, he sent a report to Chennault, which read in part:

> ...JANUARY 10TH NINE P-40S AND SIX R.A.F. BUFFALOS RAIDED ENEMY BASE AND DESTROYED TWENTY FOUR AIRPLANES THREE TRUCKS AND STRAFED TROOPS STOP FOUR AM ENEMY MISSED TARGET ANTI AIRCRAFT EFFECTIVE WHEN USED STOP

CONFINING SQUADRON TO DEFENSIVE OPERATIONS STOP AVAILABILITY LESS EACH DAY ESTIMATE EFFECTIVENESS HERE FIVE MORE DAYS PROVIDED EQUIPMENT ARRIVES STOP UNIT DESIRES TO STAY AND FIGHT END.

In response to Newkirk's radiogram, on January 12 Chennault ordered eight P-40s from "Sandy" Sandell's First Squadron ("Adam and Eve") to reinforce the Rangoon defenders. Second Squadron Leader Robert H. Neale was napping at 0930 when Sandell came in to wake him up to tell him that he was to take two four-plane flights to Rangoon around noon.[19] With only three hours to prepare, Neale rushed around getting planes, pilots and equipment ready to go. Neale selected Billy McGarry, Bob Little, Pappy Boyington, John Croft, Dick Rossi, Bill Bartling, and Frank Schiel to go with him.[20] They were airborne at 1330, but because of high headwinds they had to go into Lashio to refuel. After a quick refueling, they were off again. After thirty minutes, however, Neale developed engine trouble and turned back for Lashio, his flight returning with him. The eight remained overnight in the R.A.F. barracks at Lashio while the mechanics repaired Neale's engine. They took off at 1100 on the 13th and arrived at Mingaladon two-and-one-half hours later without further incident.[21]

By the time the contingent from the First Squadron arrived, several British residents in Rangoon, in a demonstration of gratitude to the A.V.G., had opened their homes to the Americans, and many pilots were being assigned to various British homes. For the first time in the thirty days since the A.V.G. had arrived in Rangoon, the pilots were able to leave their tents and enjoy a relatively restful night.

Bob Neale was met by Jack Newkirk as he climbed down from the cockpit of his P-40, No. 7.

"Good to have you with us, Bob," Newkirk said as he clapped him on the shoulder.

"Well," replied Neale, "we couldn't let you guys get all the glory."

"The Brits have opened their homes to us," Newkirk continued. "You and your men will be staying downtown at various homes. You'll be staying with a Mr. Scott." Bob Little joined Neale at the Scott residence that night, after they had dispersed all their aircraft. Neale recorded in his diary that night that he "...had a lovely dinner and Little and I went to bed very comfortable although I expect our sleep will be disturbed by the Jap bombs." [22]

He was right. At 0445 the air raid alarm sounded. Little got up and went to the air raid shelter, but Neale chose to stay in bed and went back to sleep. He was awakened forty-five minutes later by the dull explosions of bombs falling in the distance. Neale looked out the window to see about ten bombers circling the area, again with their running lights on. He later learned that no appreciable damage had been caused, and two of the bombers fell to British antiaircraft fire. But again – no fighters were able to make contact in the dark. It was the 12th consecutive night of raids on Rangoon.

First Blood. *Oil on canvas by Roy Grinell. Painting shows David L. "Tex" Hill climbing out of his strafing run on the Japanese held Raheng Aerodrome outside of Tak, Thailand on January 3, 1942. It was indeed Tex's first aerial combat and first victory. (Courtesy Roy Grinnell Studio)*

George L. "Pappy" Paxton. (via Lee Burgard)

Paxton's bullet-riddled Tomahawk No. 41, after his air battle on January 4, 1942. (via George L. Paxton)

CHAPTER 11

The A.V.G. Strikes Back

While the Japanese concentrated their attacks on Rangoon – the Burmese capital had been bombed nightly for the past 11 days – the Third Squadron "Hells Angels" and the remainder of Sandell's First Squadron were occupying themselves with card playing, bird hunting, test flights and occasional reconnaissance missions. The Japanese hadn't returned to Kunming since their abortive bombing raid on December 20; and, quite frankly, the men were getting bored. Charlie Bond expressed the feelings of most when he wrote on January 12:

> *"Same old routine! I wonder where the hell the Japs are? The guys in Rangoon have started going on strafing missions in Thailand. Do I envy them. This sitting on my fanny is getting me down. I wish we had some action. Perhaps this is what caused all the morale problems lately, and it is getting to me."* [1]

The "morale problems" Bond referred to included many things. Squadron Leader Sandell wasn't well-liked by his men. Some termed him a "martinet" and thought him too preoccupied with military procedure. Most in the A.V.G. had become accustomed to a loose casualness on the ground and rebelled at regulation Army discipline.

Rumors were also flying about the A.V.G. being inducted en masse into the Army Air Forces, now that America was in the war. With over half the A.V.G. pilots ex-Navy types, this rumor didn't sit well at all. Many felt the "rumor" was near to being fact when they were required to turn in a written report of their military backgrounds on January 4. Three days later they learned that Chennault had been recalled into the U.S. Army as a brigadier general, and had gone to Chungking to confer with Generalissimo Chiang Kai-shek. Most by this time were certain the meeting was to discuss the status of the A.V.G.[2]

By January 13, several pilots at Kunming had decided to resign and go back home. Chennault warned that all who left at this time would receive a dishonorable discharge and would have to pay their own way back home. This was viewed by many as a hollow threat, since they were technically civilian employees of CAMCO and, as such couldn't be"dishonorably discharged." And the $200 boat fare from Calcutta to the United States didn't faze a pilot who was making at least $600 a month.

Two of the ground personnel, administrative clerk Larry C. Moore and radioman Kenneth C. Sanger, were dishonorably discharged on the 13th. Two romantically involved men were not publicly accepted in 1941. They would return to the United States to write articles about the A.V.G. and served as "technical advisors" for the Republic Picture's movie *Flying Tigers*. Two days later Third Squadron pilot Ralph N. Gunvardahl resigned.[3] Three more pilots – Ronald R. Knapp, Donald R. Bernsdorf, and J. Haule – resigned on the 16th, and Einar "Mickey" Mickelson and John Farrell asked to be relieved from flying combat and were reassigned to fly the Group's Beechcraft.[4]

On the 17th, the Panda Bears had 28 P-40s back in commission, and Neale was ordered to send two pilots back to Kunming. With no volunteers, he had his pilots draw straws. Pappy Boyington and John Kroft would be on their way back to Kunming.

Earlier that morning, Chennault ordered "A" Flight of the Third Squadron to make a reconnaissance flight over the Japanese airfield at Hagiang, 100 miles northwest of Hanoi. As Vice Squadron Leader George McMillan and his five flight members were running up the engines on their P-40s, Harvey Greenlaw rushed out in a jeep to tell McMillan that the warning net had reported aircraft crossing the Indochina border in the vicinity of Mengtzi, 124 miles south of Kunming. They were to scrub the reconnaissance mission and take off immediately to intercept.

The six lifted off at 1005 and began their climb to the south. Almost immediately, Kunming radio ordered three to intercept and three to return to base. Only two received the message, however, and McMillan, Chuck Older, Tommy Haywood and Erik Shilling proceeded on.[5]

At 1017, leveling off at 20,000 feet, the flight spotted three Mitsubishi Ki-21 "Sallys" cruising 4,000 feet below, headed northwest. Part of the 14th *Sentai* based at Hanoi, their target for the morning was Kunming. [6]

Unseen by the Japanese, the Tomahawks peeled off in elements of two. Diving behind and below the Japanese formation, two P-40s climbed back up to attack from the right rear quarter, while the other two attacked from the left.[7]

After two passes by the P-40s, the Sallys wheeled around to the right to make a run for home. As the A.V.G. pilots continued their attacks, all three bombers began trailing smoke.

Chuck Older opened up on the lead ship, which instantly caught fire and was last seen diving in flames into the overcast below. The P-40s chased the remaining two Mitsubishi's south until they too disappeared in the overcast at 8.000 feet, both smoking badly.

The Japanese wouldn't have gotten as far, had it not been for faulty armament on all four P-40s. McMillan, Haywood, and Shilling all wound up with only one .30-caliber operating, and Older's .50-caliber gun jammed several times.

Despite their problems, Chinese intelligence confirmed all three destroyed. Older received credit for one-and-one-half, splitting a credit with Haywood, and McMillan was credited with the third.

After nearly a two-week lull, things began to stir once again in Rangoon. On January 12, the Imperial Japanese 15th Army, under the command of Lieutenant General Shojiro Iida, had begun to move from western Thailand into Burma, with the initial objective of cutting the supply link between Rangoon and Kunming.[8]

Tavoy, a Burmese city on the coast of the Andaman Sea, was one of the first to come under attack. On January 19, the R.A.F. commander at Rangoon ordered six of his Blenheim twin-engine bombers to proceed to Tavoy to evacuate the beleaguered British troops defending the city.

At 1015 the Blenheims took off, escorted by Second Squadron pilot Frank Lawlor with Dick Rossi from the Third Squadron on his wing. Flying straight east across the Gulf of Martaban, they turned south at the coast toward Tavoy.[9]

As they reached the airfield at Tavoy, they were attacked by two Ki-27 Nates, diving out of the sun, who were momentarily joined by four other Nates. After fifteen minutes of dogfighting neither side had scored any damaging hits and the Allied aircraft, running low on fuel headed north for Moulmein, 100 miles southeast of Rangoon across the Gulf of Martaban, where they were instructed to refuel before returning to Rangoon.

Panda Bear Vice Squadron Leader Jim Howard, with Johnny Petach, Bob Layher and Percy Bartelt, were at that moment climbing out toward Moulmein to provide top cover for the earlier flight's refueling. As they approached the base, they climbed further to 15,000 feet, where they loitered for forty minutes, flying top cover for Lawlor and Rossi as they refueled for their return to Mingaladon.

At 1140, they took up a heading for Mae Soht, a Japanese-held community 60 miles east of Moulmein, just across the border into Thailand. They overflew the aerodrome outside the town, then turned back west to make their attack. Howard and Petach peeled off to make the initial strafing run while Bartelt and Layher flew top cover at 9,000 feet.[10]

As the two flashed across the airfield, they were surprised to see it practically deserted. Howard then saw a bomber at the edge of the field and opened fire. It turned out to be a wrecked hulk from a previous raid.

As they pulled up from the strafing run, Bartelt gave Layher permission to join in the fun and he dived on the field, falling in trail with Howard and Petach as they began their second pass. By this time, however, the field had come alive and the three Tomahawks received heavy ground fire as their machine guns raked the outlying buildings.[11]

Looking back on the field as he climbed out, Howard spotted a Kawasaki Ki-32 light bomber taking off. The three wheeled around and each made a pass on the hapless enemy aircraft. Incredibly, although each scored some hits, none hit a vital spot and the bomber continued around to land. He overshot the field, however, and circled once again. The Japanese pilot lined up for the runway with the three Tomahawks in hot pursuit. The bomber was trailing black smoke as its wheels touched down. It rolled for almost 100 yards as the smoke turned to flame, yawed violently to the left and exploded.

Howard signaled his two wingmen to join up and climbed out for Rangoon, each sharing a third credit for the destruction of the bomber.[12]

The next day Jack Newkirk led a six-plane escort mission back to Mae Soht. Flying cover for six R.A.F. Blenheims, Newkirk, Bob Neale, W.C. "Moose" Moss, Bill Bartling, Hank Geselbracht and Bert Christman were forced to a lower-than-normal cruising altitude to keep the bombers in sight. Three miles southeast of the enemy aerodrome, they were jumped by eight Ki-27 Nate fighters making a high-side attack.

All hell broke loose as P-40s broke right and left toward the 7.7mm machine gun fire from the stubby little fixed-gear fighters. In the next few minutes of combat, Newkirk shot down two of the Japanese attackers, and Moss got a third before bailing out of his battered Tomahawk as flames licked back over his cockpit.

The A.V.G. escort mission was successful; the Blenheim's plastered the Mae Sot aerodrome. But they limped back to Mingaladon minus one aircraft and pilot – and Bert Christman landed with a punctured tire and twenty-five holes in his aircraft. Moose Moss, after bailing out at 1,500 feet, landed safely seven miles inside Burma. Picked up by Burmese tribesmen, he traveled 14 1/2 hours by bullock cart, four miles by boat and 14 miles by car to reach Moulmein. From there, he caught a ride in an R.A.F. Blenheim back to Mingaladon.[13]

After nearly three weeks of A.V.G. raids on Japanese airfields, the Japanese struck back on January 23. Three days earlier the Japanese Army had penetrated the Thai border and were moving westward into southern Burma toward Moulmein. Their plan was to sever the supply lines between the harbor at Rangoon and Kunming, thereby denying China seaborne resupply.[14] The Imperial Japanese Army Air Force had concentrated 180 bombers and 120 fighters for an all-out assault against the combined Allied air defenders at Rangoon.

It was 0810 when Jack Newkirk and seven more P-40s lifted off the field to escort six Blenheim bombers on another raid on Tak. At about the same time, 25 Nakajima Ki-27 Nates from the 50th *Sentai* were taking off from their base at Nakhon Sawon. They were the first wave, planned to soften up the Allied defenses; they would be followed by a wave of bombers to finish the destruction.[15]

Newkirk and his flight encountered a solid undercast as they flew east toward Tak and turned back at the border. While Bob Neale, Bill Bartling, and Bob Little orbited as top cover, Newkirk took the remaining flight into the satellite field at Pegu for refueling.

Newkirk's flight was still being refueled when the air raid alarm at Mingaladon sounded at 1010. Tex Hill and Frank Lawlor, lounging in the alert shack, ran for the two available P-40s.[16]

The Japanese fighters didn't see the two lone defenders until they sailed through their formation, diving from above. They each got a Nate on their first pass. With Lawlor on his wing, Hill turned behind a pair of Japanese fighters and shot down another. Lawlor's target turned as he was lining up his guns and he missed. Boring in closer, he was rewarded with pieces of the Japanese plane flying off before it went down in flames.

Meanwhile, Bob Neale and his flight element received word of the raid as they circled over Pegu. Ten minutes later they reached the swirling combat. By this time three Hurricanes and two Buffalos had joined in as well, cutting the odds to two-to-one. Bill Bartling managed to knock down a Nate, but then took a hit in his engine. He turned toward Pegu and barely made it in to a crash landing on the field.[17]

Bob Neale pursued the retreating Japanese for 50 miles out into the Gulf of Martaban. Firing on a "tail end Charlie," he was rewarded with smoke and then flames as the Nate fell toward the Gulf. It would go down as another unconfirmed claim, however, since there were no Allied witnesses and the wreckage was lost (as many would be) beneath the water in the Gulf.[18]

No sooner had Neale landed when a second air raid sounded at 1230. Ten took off to intercept – Jack Newkirk, Ed Cole, Bob Neale, Ed Rector, Frank Lawlor, Percy Bartelt, Gil Bright, Bert Christman, Johnny Petach, and Frank Swartz.[19] Noel Bacon had to wait for a new starter to be installed in No. 59, but managed to get off ten minutes after the others.[20]

Leveling off at 10,000 feet, they spotted the enemy formation – twelve Ki-21 Type 97 Sallys in a V-of-Vs formation escorted by 25 Ki-27 Nates flying in pairs above and behind the bombers. The bombers were from the 31st *Sentai* out of Phitsanulok and the fighters were from the 77th *Sentai* based at Tak. As Jack Newkirk reported later:

"We singled out the bombers and made an attack on them from three directions – dead ahead and the port and starboard beams. After several repetitions of this, the left hand plane which I was following fell out of the formation about 75feet but joined up again. Then the formation turned; one plane gave several large puffs of smoke and flame and finally went down near the satellite field. After diving 3,000 feet the fighters broke off and climbed back up; one of them lagging behind. I caught him just past satellite and after one burst his wing came off and he went down." [21]

Moments later, Newkirk was attacked from above by several Nates. Diving away, his engine began to cut out and lost power. As he made his approach to Mingaladon, he found that the flaps on his bullet-riddled P-40 wouldn't come down. Screaming down the runway in a no-flap landing attempt, he overshot the runway and nosed over past the over-run on the north end. Dragging his chute out of the cockpit of the still smoking Tomahawk, he asked for another aircraft as soon as the ground personnel reached him.

Meanwhile, the battle was still raging 10,000 feet above. In as many passes, Percy Bartelt flamed three of the Mitsubishi bombers, Gil Bright downed one another Johnny Petach got another.

Frank Lawlor devised an effective tactic during the battle; he would dive on a Japanese fighter, pull up into a zoom climb into the clouds just above the formation, then do it again. He knocked down two more Nates with this technique, which brought his score to four for the day.[22]

Bob Neale made a head-on pass on the bomber formation, then dived below. As he chandelled up for a rear attack, the tail gunner on his intended victim sprayed him with his single 7.7 mm gun. His left wing badly damaged and his engine hit as well, Neale left the battle and made a forced landing at the satellite field at Pegu.

Ed Rector and Bert Christman met the Japanese formation forty miles northeast of Rangoon. They overflew the bombers at 15,000 feet, then Rector, with a waggle of his wings signaling his wingman to attack, peeled off toward the rear of the formation. Screaming down at 400 miles per hour, Rector noticed too late that he was pulling too much lead to fire effectively and slid across the top of the formation.

As he pulled up he craned his neck around to see Christman coming in on his own attack. But just behind Christman was a flight of escorting Nates, boring in for the kill. Rector bent the P-40 around, flying head-on into the flock of Nakajima fighters.

Christman, his attention still riveted on the bombers, was hit full force by the diving fighters. With his engine holed and propeller windmilling, he slid his canopy back and stood up in the cockpit, preparing to bail out. At that moment a random bullet struck him at the base of his skull and his lifeless body plunged from the burning Tomahawk to the ground 12,000 feet below. His chute never opened.[23]

Noel Bacon, finally arriving on the scene, sailed into the bombers. As he wrote in his combat report:

"I came in on the bombers' beam and fell in behind them just as they dropped their explosives. The bombs hit rice paddies seven miles northeast of Rangoon. I dodged the bombs they dropped and, still climbing, found four Jap fighters over me. Down they came and I got one head-on. He spun in aflame... I dove out of the fighting maneuver and came back up, firing on the bomber formation, bow to stern. The other three jumped me again and the sky looked like it was full of them. Got another one head-on and dove out with one on my tail. He put 12 slugs in me: one across the top of the engine, one through the cockpit at my right heel, and the rest in my gas tank and wings." [24]

In 15 minutes it was all over. The Japanese attack had been turned back, their bombs dropping harmlessly in rice paddies northeast of Rangoon. Although initial estimates following the heat of battle reported as many as 32 Japanese aircraft destroyed during the two raids, final confirmation figures reflected 21 destroyed – 11 fighters and 10 bombers – with another six listed as "probables." The R.A.F. accounted for four of the bombers for the loss of one Buffalo. The A.V.G. got the rest, but lost Bert Christman and four P-40s.

The Japanese returned to Mingaladon for another try on the 24th. Seven Ki-21 Sallys of the 14th *Sentai* took off from Bangkok and rendezvoused with 20 Nates, again from the 77th *Sentai* at Lampan.[25]

The alarm sounded at 1030 and seven P-40s got airborne – Tom Cole, Tex Hill, Jim Howard, Ed Rector, Percy Bartelt, Frank Schiel, and Raymond Hastey.[26] Bob Neale broke a tail wheel taxiing out and had to delay for repairs. As the Tomahawk climbed out, they were joined by two Hurricanes and four Buffalos.

Leveling off at 18,000 feet, they soon spotted the seven bombers, 15 miles east of the field, 4,000 feet below. As the American and British planes attacked, the Sallys opened their bomb bays and dropped a salvo of bombs, which fell well short of the airfield.

Ed Rector, Raymond Hastey and one of the Buffalo pilots each downed a bomber on the first pass. By this time, though, the escorting Nates had joined the battle and the air was soon filled with swarming fighters and burning bombers. After a second salvo of bombs, which hit on the southwest side of the field, the four surviving bombers, still maintaining tight formation, turned slowly back to the east, as the Tomahawks, Buffalos and Hurricanes continued making their attacks.

Bob Neale, arriving in the area late, saw the four bombers 1,000 feet below him. "Four Buffalos and myself went after the bombers," he recalled. "I closed on one to about 100 feet from the rear and blew up his starboard engine."

As Neale's first victim was falling, Tex Hill came in on a beam attack from the right, aiming for the bomber to the far right. Moments later the Sally disappeared in a ball of flame, reduced to burning pieces slowly drifting to the earth below. Neale continued:

> *"Only two left by now so I took after the leader. Although I was getting a hell of a lot of return fire, I didn't seem to be getting hit, so I closed to fifty feet and gave him a burst. The whole damn plane blew up. Part of it came back and tore up my right wing and aileron."*[27]

With that, Neale dove out of the battle and nursed his plane toward home, praying that his damaged aileron wouldn't tear loose. He was within 20 miles from Rangoon and flying at 1,000 feet when he was jumped by two Nates. Pushing the stick and throttle forward, he dove to treetop level, streaking for home. One Japanese shell had pierced his cockpit and smashed his microphone. "Too damn close," he commented later in the understatement of the year.

While Neale, Hill and the four R.A.F. Buffalos were polishing off the last of the bombers, the rest of the A.V.G. flight were engaged with the fighters. Percy Bartelt downed two Nates, bringing his two-day total to five. Tom Cole, Frank Schiel, Jim Howard, Ed Rector and Tex Hill each destroyed a fighter. The R.A.F. Hurricanes shot down three.

The A.V.G. and R.A.F. had routed the Japanese for the second day in a row. Seven bombers and ten fighters had been destroyed – this time with no losses. And the A.V.G. had two new aces, Tex Hill and Percy Bartelt.

The alarm on the field sounded again at 1300, almost a replay of the day before. Pilots rushed to what flyable aircraft remained and climbed out to meet the enemy.[28] While the Tomahawks and Buffalos patrolled at 20,000 feet, searching for the reported bombers, 20 Nates of the 50th *Sentai* from Nakon Sawon were streaking in at low altitude from the west for a strafing run on the airfield.

As the Nates approached the field, an R.A.F. Hurricane was landing in the opposite direction. The quick-thinking British pilot pushed the throttle to the firewall, sucked up his gear and flew head-on into the string of Japanese fighters. Surprised by the bold attack, the Nates broke right and left. They never regrouped, but disappeared rapidly to the east.

At 1800, the A.V.G. held the funeral for Bert Christman in the cemetery of the Church of Edward the Martyr. Chaplain Paul Frillman flew in to conduct the service, and most of the R.A.F., plus a large number of the English residents remaining in Rangoon, came out to pay their last respects for the popular American pilot. Ed Rector, Tex Hill, Johnny Petach, Hank Geselbracht, and Noel Bacon served as pallbearers, as the casket was lowered into plot C-93, beside the graves of Hank Gilbert and Ken Merritt. An R.A.F. rifle detachment fired volleys over the grave as a Lockheed 14 roared overhead.29 That evening the Japanese-operated Bangkok radio broadcast in English that seven Japanese bombers had been "lost in a severe electrical storm near Rangoon."[30]

Earlier that day in Kunming, Sandy Sandell selected 12 more of his pilots to add to the First Squadron complement in Rangoon. As before, the men drew cards for the "pleasure." [31] They would go in two flights of six, Sandell briefed. He would lead the first flight, with Boyington leading the second shortly after. It would be a two-hop flight, with a refueling stop at Lashio. They would plan to arrive at Mingaladon Aerodrome at dusk.

Flight details out of the way, Sandell continued to say that the Old Man had told him that the A.V.G. was definitely going to be inducted into the U.S. Army Air Corps. Some would be offered regular commissions, he said, and rank upon induction would be based largely on Chennault's recommendations. There was nothing to worry about right away, he added, because the CAMCO contracts would be honored for the remainder of the full year.[32]

Sandell led the first flight of six off Kunming's runway at 0955 on January 25, followed by Boyington's six 40 minutes later. Despite strong headwinds (again!), they made it in to Lashio, and had lunch at the C.N.A.C. Club while their Tomahawks were being refueled. Proceeding on to Mingaladon, the last of the twelve landed as scheduled just at dusk. An otherwise uneventful flight was spoiled by Charlie Bond blowing his right tire on landing. Fighting to maintain control, he skidded to a stop fifty feet off the runway with no more damage to No. 5 than a shredded tire.[33]

The Panda Bears were glad to see their additional reinforcements. As they swapped war stories at the Silver Grill that evening, the new arrivals also brought word of the A.V.G. being taken over by the U.S. Army. That news was about as welcome as a skunk at a picnic, since most of Newkirk's squadron was made up of Navy flyers, and few relished the thought of serving in the Army.

The "Adam & Eves," anxious to get into combat, didn't have long to wait. The morning after their arrival they were scrambled off the field on four alerts, each time climbing to 20,000 feet and patrolling an area east of the field.

Charlie Bond was amazed at the lack of ground control during an alert. As he observed:

> *"The takeoff procedure is a hell of a mess. The R.A.F. has three or four Brewster Buffalos and two or three Hurricane fighters, and we have twenty P-40s on the field. The wind direction does not matter when the scramble order sounds. The Brewsters dart off down one runway, and the Hurricanes take off in the opposite direction a few seconds later. In the meantime, our First Squadron is getting into the air on another runway which intersects the R.A.F. runway in the center of the field. Then came the blue green flights down the same runway but in the opposite direction."* [34]

After three false alarms, the Japanese finally came – 23 Nates from the 50th *Sentai* at Nakon Phanom. The fourth alarm sounded at 1045 and 11 P-40s got airborne. Jack Newkirk took the lead with his flight of four, with Prescott, Hoffman, and Boyington, followed by First Squadron Flight Leader Al Probst with Bond and Bright. Shortly after takeoff, Newkirk's radio began cutting out and he turned the lead over to Probst.[35] Five minutes after the first seven were off, Bob Neale scrambled with Billy McGarry and one other Second Squadron pilot.

As the Tomahawks were climbing through 8,000 feet, fighter control radioed "Bandits ten miles southeast, very high, approaching base." 8,000 feet above were 23 Ki-27 Nates from the 50th *Sentai* at Nakan Sawan, circling in a squirrel cage formation that Bob Neale said "looked like a bunch of buzzards."

The P-40s were still well below, climbing to the attack when they were spotted by the Japanese. With a quick wing-over, the Nates dived to meet the Tomahawks. Caught at a disadvantage, Probst and his flight dived away, taking advantage of the P-40s superior diving ability. From then on, it was every man for himself. Probst zoom- climbed back up and managed short bursts on three Nates as they came head-on before diving away again.[36]

As Bob Neale came out of his initial dive, he found himself alone for a minute. Climbing back up, he circled a couple of times and then spotted several Nates about 2,000 feet below heading east back to the Thai border. Unseen, he slid in behind "tail-end Charlie" and opened fire at 100 yards. The Nate's engine caught fire almost immediately and Neale watched it fall until it plunged into the Bay of Martaban.[37]

Second Squadron pilot Gil Bright got another Japanese fighter as they streaked for home. He fired a long burst from dead astern, then climbed above for another pass. It wasn't necessary, because the fighter was going down trailing smoke and flames.[38]

The P-40s pursued the Japanese formation until they were 120 miles east of Rangoon. Despite multiple passes, only First Squadron pilot Billy McGarry managed to bring down another Japanese fighter, his first victory claim.

Ten P-40s landed safely at Mingaladon after the air battle. Only one had failed to return. Louis "Cokey" Hoffman was last seen diving away after the first engagement. The third A.V.G. pilot killed in combat, Hoffman was the oldest pilot in the Group, outside of Chennault himself. At 42, married with children, he had almost as much flying time as the Old Man, and had been an enlisted pilot in the Navy since the early '20s. Ironically, the A.V.G. had now lost their youngest and their oldest. Later in the day, they recovered his mangled body from the wreckage of his plane found two miles from the field.[39]

That night a small flight of Japanese bombers returned to Mingaladon but did little damage. Charlie Bond and George Burgard, billeted in town with a Danish oil man, Mr. Jensen, watched the show in comfort. As Bond recalled:

> *"George and I had scotch and sodas with Mr. Jensen out on his front lawn - relaxation and pleasant camaraderie. The air-raid siren gave us about five minutes of warning. The Japs were trying to bomb the field to wipe out our aircraft. We watched all of this and even saw the reflections of the bursting bombs against the low-hanging clouds. Later came the rumbling sound of the bombs. What a show."* [40]

The following day Jim Howard took ten P-40s back to Tak for a strafing raid. As five flew top cover at 16,000 feet, the other five swooped down on the Japanese aerodrome. Anti- aircraft fire opened up on the Tomahawks, but no one was hit. But neither did the A.V.G. score any hits – the field was empty.[41]

The moon was full that night and the Japanese returned, right on schedule, this time, however, those on the ground were treated to the sight of tracers arcing into the three-bomber formation. An R.A.F. Hurricane night fighter had gotten airborne and scored a hit. The bomber immediately burst into flames and dived into the ground, exploding in a fireball that could be seen for miles.

They buried Cokey Hoffman at 0830 the next morning, January 28, in plot C-97 at the Church of Edward the Martyr. The ceremony was delayed for several minutes while the grave was lengthened. The Burmese gravediggers had miscalculated the length of the tall American pilot's coffin. Finally, covered by an American flag, the coffin was lowered into the grave.[42]

Sixteen P-40s, nine from the First Squadron and seven from the Second, were airborne at 1150 in anticipation of another Japanese morning raid – the Japanese were nothing if not methodical. Patrolling at 18,000 feet, twenty miles east of Rangoon, they spotted a large formation of Japanese fighters at their same altitude coming head on. Ten Ki-27s, again from the 50th *Sentai*, were joined this time by 23 Nates from Lampan. First Squadron pilot Frank Schiel, one of the first to spot the Japanese fighters, fired a burst to attract the attention of the rest of his flight, then turned to make a head-on attack.

The battle raged from 20 miles east to 20 miles west of the field, as 49 fighters swarmed around one another like a gigantic beehive. Squadron Leader Sandy Sandell reported:

> *"I was never out of sight of enemy aircraft. On my first attack, one fighter caught fire and dived toward the ground from 17,000 feet, disappearing from sight still in a dive. After two more unsuccessful head-on runs, I hit one more ship that similarly went down. About 20 minutes later, I made a head-on run to very close range and could see parts of the plane coming off."* [43]

Shortly after, Sandell's attention was diverted by a sudden rise in his coolant temperature gauge, and the cockpit soon began to fill with smoke. Diving for the field, he made a dead-stick landing. He kept the P-40 on the runway despite a tire punctured by Japanese bullets, and jumped out as soon as his aircraft rolled to a bumpy stop. He was no sooner clear of the cockpit when he looked up to see a Japanese fighter slowly circling the field. Suddenly, in a deliberate kamikaze attack, the Nate dived at the damaged P-40. A wingtip barely clipped the tail of Sandell's Tomahawk before the Japanese fighter exploded in a fireball scant yards beyond.

Frank Schiel, also of the First Squadron, chased four of the fighters as they retreated toward the east, finally shooting one down about 30 miles east of Rangoon.[44] Besides the two confirmed and one probable that Sandell got, Johnny Petach, Dick Rossi, and Bill Bartling got one each, for a day's total of six confirmed and one probable.

Charlie Bond was up early the next morning, January 29, along with several other pilots scheduled to drive to "Johnny Walker," the dispersal airstrip north of Rangoon, to fly their aircraft back to Mingaladon. After a half-hour of driving past countless rice paddies on the narrow, dusty jungle road, the headlights on the station wagon illuminated the long, pointed snouts of the Tomahawks parked beneath the jungle foliage.[45]

The field was deserted; the crew chiefs would be waiting for the planes back at Mingaladon. The pilots climbed out of the station wagon in the hazy pre-light of dawn, and soon the air was filled with the roar of Pratt & Whitney R-1830 engines warming up. One-by-one, they taxied out to the narrow dirt strip that passed for a runway and took off. Within five minutes they were circling for landing at Mingaladon.

Their aircraft in place, the A.V.G. pilots assumed the morning alert. The first siren came at 1100. Fifteen P-40s, two Buffalos and two Hurricanes got airborne and climbed to 22,000 feet. When 40 minutes of patrolling brought no results, they came back in.[46]

The next alarm sounded at 1515. This time the defenders climbed out to 19,000 feet and circled southeast of Rangoon. Soon they saw them – two formations of Ki-27 Nates, one high and one low – some 20 aircraft in all. Charlie Bond spotted one group, "a swarming beehive," as he described it, at ten o'clock low. As the P-40s dove to the attack, Bond described his heart-thumping excitement:

> *"I was breathing hard as I charged my guns. Gun sight switch on, I was almost vertical in a dive. The Jap fighter I picked out began turning tightly in a horizontal plane. I missed him by a mile."* [47]

Squadron Leader "Sandy" Sandell pursued one Nate through the lower cloud layer. Firing all the way down, he saw him crash in the jungle below. Pulling back up in a zoom climb, Sandell fired on a second Nate with a beam deflection shot. As the Japanese fighter passed in front of him, he banked tightly to pull in behind. Within seconds, the Ki-27 was in flames and disappeared through the clouds below. After a third attack, Sandell watched his victim's prop wind down to a stop before it, too, fell smoking into the clouds.[48]

Meanwhile, Squadron Leader Jack Newkirk and his Second Squadron Panda Bears, sighting a Japanese formation above to the east, turned west and climbed to 20,000 feet. The Japanese broke right and left as the P-40s attacked. Newkirk tacked on to the tail of one Nate that suddenly pulled up into an Immelmann and he found himself coming head-on to the nimble little fighter. "I held my fire," reported Newkirk, "and put a short burst in his engine at 300 yards. A cylinder came off and the plane went down smoking." [49]

After diving away in his first unsuccessful attack, Charlie Bond climbed back to 18,000 feet. He observed that there was "no formation, just a swarming mass of a few P-40s and a hell of a lot of Japs twisting, turning, diving, and maneuvering to get into position." [50]

Bond came in on the rear quarter of one Nate, who promptly dove away into the clouds. Leveling off in anticipation of the Japanese fighter coming out again, his patience was soon rewarded as the Ki-27 popped out of the clouds directly in front of him. Opening up with all six guns, Bond watched his tracers smash into the Japanese pilot's cockpit and engine. Overtaking the stricken fighter rapidly, he banked the P-40 sharply to barely miss the flaming cockpit of the Japanese plane. Bond could scarcely suppress a cry of exhilaration that he had gotten "one for Hoffman." [51]

By this time, the Japanese fighters were turning back eastward toward their home base at Lampan. Bond pushed the throttle to the firewall and slowly closed in on a couple of isolated stragglers for one more attack. Suddenly the pair pulled up sharply into an Immelmann and Bond, like Newkirk, found himself on a head-on course with the two Nates. As he squeezed the trigger, he was dismayed to find that only his .30-caliber wing guns fired – his .50s had jammed! As luck would have it, he scored some hits on one of the Japanese fighters and saw it trailing smoke as it disappeared in a dive through the clouds.[52]

With the Japanese retreating and a thick under-cast beginning to form, the A.V.G. pilots decided to call it a day.

One-by-one, the Tomahawks pitched out to land. Charlie Bond made a pass down the runway and did a victory roll to celebrate his first two victories. The results for the day were very satisfying; Sandell was credited with three of the fighters and Bond with two. Noel Bacon, Jack Newkirk, Frank Lawlor, John Dean, Bob Little, Tex Hill, and Bob Prescott each got one. The only P-40 damaged (miraculously!) was Matt Kuykendall's. He took hits in both wings and one slug through the oil tank, which forced him to land early. Another bullet grazed his forehead and tore away his goggles.

At dusk, after refueling and rearming, the P-40s were flown out once again to the outlying dispersal strip. There was one tense moment when George Burgard almost let his P-40 get away from him as he touched down in the dusky twilight. A burst of power straightened the narrow-geared, veering P-40, and he took it around for an uneventful landing. An hour's ride over 36 miles of weaving, dusty road brought the pilots back home – filthy, bone-tired, but happy.

Early the next morning, British intelligence reported Japanese troops west of the Thai border, approaching Moulmein. The British asked the A.V.G. to launch a strafing mission against the Japanese troop concentration. Sandell refused.[53] Strafing aerodromes was one thing, but going against troops in the field with no armor plate protection against concentrated ground fire was something else again.

Second Squadron Vice Squadron Leader Jim Howard, after talking it over with his pilots, volunteered for the mission. Taking Ed Rector and Tom Cole with him, he took off at 1115.

Flying beneath a low ceiling, they flew over Kawkareik, 40 miles east of Moulmein, just inside Thailand. Checking for enemy air activity, Howard and Cole spiraled down to reconnoiter the area which Rector flew high cover.

With Cole on his wing, Howard saw a number of trucks on the road about two miles north of Kawkareik. He made one pass across the trucks, holding his fire, then pulled up for a second pass. Unable to identify the trucks positively as Japanese, he made another dry pass. Deciding that he didn't see anything worth attacking, Howard pulled up for home. As he looked back he saw Cole gliding down towards the trucks as if he were going to strafe.

As Howard reported later:

> *"The next thing I saw was an explosion and a flame shot up 100 feet into the air with black smoke. At first I thought that Cole had hit an ammunition truck, but as I passed over the target I realized Cole had crashed because there was a path cut through the tall grass which bordered the road and bits of plane were visible scattered all over."* [54]

Howard and Rector circled for several minutes looking for ground fire. Then, sick at heart over the loss of their squadron mate, they climbed out for home.

The invasion of Burma began in earnest on January 31. Moulmein fell to the advancing troops, and there were several attempts to cross the Salween River. The Japanese continued to send nightly bombing raids against Rangoon in a persistent attempt to knockout the A.V.G. and R.A.F. at Mingaladon – the only real resistance to their invasion remaining in Burma.

The British rushed one division of Indian troops to Rangoon on February 1 to defend the west bank of the Salween River. But Rangoon's days were numbered.

The A.V.G., both pilots and ground personnel, were wearing down, along with their P-40s. Thirty days of almost constant harassment from the Japanese had taken its toll on men and machines.

By February 1 the A.V.G. at Rangoon counted only 18 of 33 total aircraft as being combat-ready, and that number only the sheer grit, determination, and ingenuity of the crew chiefs. No sooner had a Tomahawk landed but it was crew chiefs and specialists to refuel, rearm, and fill the oxygen bottles. Engine or airframe repairs were begun immediately.

In addition to expected operational wear and tear, the crew chiefs had to contend with dust so thick that carburetors had to be cleared every five flight hours. Oftentimes dirt cakes a half-inch in diameter could be found clogging the oil sump screen.[55]

Aircraft that couldn't be repaired in the field were disassembled and trucked to the CAMCO factory at Loiwing. Additionally, aircraft that had crash-landed in the jungle or rice paddies of southern Burma were cannibalized for any usable spare parts, and several badly damaged P-40s were restored to flying condition by this "jungle parts store."

A radiogram from Chennault to Newkirk and Sandell arrived on February 2 advising that the remainder of the First Pursuit Squadron (7 pilots and 19 ground personnel) would be sent from Kunming to Rangoon as soon as transportation became available. Newkirk and his Panda Bears would be returning to China by C.N.A.C. All but two of their P-40s would remain at Mingaladon, those two would be flown back for overhaul.

As the Japanese moved westward on the ground, so did their air war. Toungoo was attacked for the first time at 0930 on February 3. Robert B. "Buster" Keeton, one of the Second Squadron pilots left at Toungoo, was preparing for a test hop on a repaired P-40, when he heard the low roar of distant aircraft. Looking up, he saw a formation of nine bombers about 11,000 feet high and five miles east. "Jap bombers!" Keeton shouted to line chief Harold Walker. "Let's get the hell out of here!" Sprinting for the hangar, Keeton called to crew chief Dan Keller, "Get the car started!"

The three jumped into the Chevrolet sedan and headed for the gate, hoping to get away from the field before the bombs fell. They had gone only a hundred yards before the ground erupted ahead of them. Another explosion sent the car into a ditch. Walker and Keller jumped out and dove for the ditch. Keeton, however, was stuck in the back seat, his door jammed shut. Keeton recorded in his diary that night:

> *"The bombs were whistling and then hitting, making a tremendous sound. I knew they were dropping awfully close from the impact and the car was being filled with dirt. I could hear the shrapnel falling and whizzing by sounding like machine gun bullets. They must have dropped about 30 bombs on the runway and they were all within a radius of 75 yards of where we were parked."* [56]

After the first wave of bombers passed, Walker and Keller pushed the car out of the ditch and the three drove off the field, and waited for a second wave. After a decent interval of quiet, the trio returned to find that one Burmese soldier had been killed and two others injured. Although the bomb craters had bracketed the armament shack, no buildings had received serious damage.

The base resumed "business as usual" following the raid and Keeton took his delayed test hop at 1100. At 1530 the air raid siren sounded again. Preparing to test his second plane, Keeton jumped in the cockpit and took off, just ahead of the bombs.

Circling the field until he reached 20,000 feet, he was surprised to find that his oxygen bottle hadn't been filled. Determined to stay up and search, however, he continued to orbit for 30 minutes, looking for the elusive Japanese bomber. Beginning to feel the effects of oxygen starvation, he decided to go back and land. Then, as luck would have it, "I just nosed the plane over and started down when I saw a twin-motored Jap plane practically flying formation about 1,000 feet below me. I switched on my guns and dove on him from above and the left side with all six guns going full blast. Passing under him and looking back I thought he was smoking but not certain. I came on down and landed because I knew without oxygen, I couldn't stay at that altitude and do any good." [57]

Although Keeton wasn't sure of his victory at the moment, later confirmation by the British gave him credit for the destruction of the bomber.

The Japanese returned to Toungoo the next morning at 0530. In the first of two passes, three bombs hit the runway. Circling the field slowly, they made another run, also from the east. This time they destroyed the operations building, one hangar, and one Blenheim bomber that was parked close to the runway.[58]

In a repeat of the events at Rangoon following the first December bombings, the cooks fled into the hills, leaving no one to help with evacuation preparations – nor even to cook. That afternoon Keeton and the remaining few at Toungoo began loading ammunition onto waiting trucks for eventual convoy to Kunming. They managed to load four trucks before dark; then, after a supper of canned food, they loaded five more before falling exhausted into bed.[59]

After two days of bombing Toungoo, the Japanese switched again to Rangoon, hitting the besieged city and airfield once on February 5 and four times in the early morning hours of the 6th. Two Hurricane night fighters got off and shot down two of the raiders. Upon landing, however, one of the British fighters dropped its wheels in a bomb crater and was severely damaged.

Later that morning, Bob Neale led a flight of six P-40s in a scramble following an air raid warning. Two Hurricanes joined the flight as they climbed out.

High above in loose formation were 25 Ki-27 Nates from the 77th *Sentai* at Lampan, headed for Rangoon. They didn't seem to see the Allied fighters until the battle was joined.

During the next 30 minutes, Neale and his First Squadron pilots knocked down seven Japanese fighters, without a loss to themselves. Every one of the A.V.G. pilots had scored – Bob Little and Pappy Boyington got two each, and Neale, Bob Prescott, and Billy McGarry each got one. Charlie Bond got a probable, and the two Hurricane pilots knocked down three more.

The seven victories scored by the First Squadron proved to be a milestone. Later that day Air Vice Marshall D.F. Stevenson, Commander, R.A.F. Burma, sent a telegram to Chennault which read in part:

> *"Today the First American Volunteer Group destroyed its one hundredth Japanese aircraft in the defense of Rangoon. I take this opportunity of expressing to Colonel Chennault, their commander, squadron leaders Sandell, Newkirk, and Olson, and to the fighter pilots and maintenance crews the deep admiration of the R.A.F. Burma for this remarkable piece of air fighting..."* [60]

For some unexplained reason, the message wasn't received by Chennault until four days later. Nevertheless, it recognized the A.V.G.'s combat achievements and Chennault had it pinned on the bulletin board at Kunming. Upon reading it, one of the men surmised that, "When a Britisher admits that Americans are good, then we are good, by damn!"[61] The Japanese retaliated with four raids that night, destroying a Blenheim and a Hurricane on the ramp and setting fire to two buildings.

The next morning, Saturday, February 7, Squadron Leader Sandell got down to the line early. He wanted to test-hop his No. 11 – repaired after being struck by the Japanese Nate that crashed nearby on January 28 – before the expected mid-morning Japanese raid. Second Squadron line chief Harry Fox and his mechanics had worked for two days repairing the damage caused by the Japanese pilot that dived onto the field.

Smiling at Fox's admonition to "take it easy," Sandell buckled on his parachute and climbed into the Tomahawk's cockpit. Thirty minutes later he was back over the field. The mechanics and several of the pilots were watching as Sandell approached the field at 2,500 feet. He began a slow roll, continuing on over until he was inverted. There the roll stopped and, to the horror of the spectators on the ground, the P-40s nose dropped and the aircraft dove straight into the ground. The Tomahawk hit with such force that only the tail wheel and right main gear could be salvaged.[62]

In a flurry of radiograms between Mingaladon and Kunming, Newkirk reported Sandell's death, noted that they had 17 airworthy aircraft, and requested five generators and drive gears. Chennault responded with the movement order for the Second Squadron to return to Kunming "in accordance with plan previously issued," but that "all supplied, personnel, and equipment necessary for effective operation of First Squadron will be transferred to that unit." The First Squadron would inherit all the P-40s of the departing Second.

To Bob Neale, he radioed:

> TERRIBLY SORRY TO LEARN SANDY'S DEATH STOP ENTIRE GROUP SENDS DEEPEST SYMPATHY TO FIRST SQUADRON STOP YOU ARE APPOINTED SQUADRON LEADER EFFECTIVE THIS DATE STOP CARRY OUT INSTRUCTIONS ALREADY ISSUED FOR SANDELL STOP RECOMMEND DEPUTY LEADER FOR APPOINTMENT END [63]

Sandell was laid to rest the next morning, the 8th, and shortly thereafter the remainder of the Second Squadron, except for Squadron Leader Jack Newkirk, departed for Kunming. The few remaining at Toungoo were busy as well as Ed Goyette carried out plans to evacuate the A.V.G.'s "kindergarten." Very little remained at the Kyedaw aerodrome, except for three P-40s that were awaiting three parachutes and a pilot from Mingaladon to fly the third one back to Rangoon. When Fritz Wolf arrived with four mechanics and one parachute, Goyette was furious. It would be four more days before the last two P-40s were ferried to Mingaladon.

When Newkirk left for China, via C.N.A.C. transport to Calcutta on the 10th, Bob Neale was officially in charge of the A.V.G. in Burma. He was left with 19 Tomahawks (plus two at Toungoo) and 22 pilots to hold off the Japanese advance.[64]

CHAPTER 12

Evacuation of Rangoon

Almost as soon as Jack Newkirk turned over control of the A.V.G. in Burma to Bob Neale, a lull settled over Rangoon. Indeed, there would be no more raids on Rangoon or Mingaladon for nearly four weeks. Although an uneasy calm had descended over Rangoon, one possible reason was that the Japanese forces were tied up elsewhere. The Japanese air campaign in Burma shifted from Rangoon to the support of the Salween front. The Japanese Army had crossed the un-bridged Salween River just ninety miles east of Rangoon, forcing the defending British and Indian forces to fall back to the Sittang River. They were also advancing steadily down the Malay Peninsula, and on February 15 Singapore, Britain's great citadel at the southern tip of the peninsula fell to the Japanese – the greatest military disaster ever suffered by a European nation in the Orient.

The Adam & Eves settled into a routine of alert duty, reconnaissance patrols and flying escort for R.A.F. bombing missions. There was even time for an occasional golf game on the Mingaladon golf course, or swimming at the posh Kolune Club in Rangoon. Several evenings were spent partying at the Silver Grill or at one of the British homes that had been turned over to the A.V.G.[1]

If Rangoon was calm, Kunming was absolutely peaceful. All the Third Squadron pilots were presented with Chinese wings on the January 31 and two days later, in an impressive ceremony at the aerodrome, Chinese Air Force Chief General Mao Pan-Tsu presented eight of the pilots elaborately embroidered silk bedspreads. Afterward, Squadron Leader Olsen led a six-ship flyby. As the other five peeled off to land, R.T. Smith put on a short demonstration of low-level aerobatics. The Chinese loved it.[2]

The Chinese may have loved it, but, according to Frank Losonsky:

> *"Chennault had a serious morale problem. Folks were unhappy. The ground crew felt the pilots were getting all the glory and awards. No decorations for us, no silk bed coverings, and little press coverage."*[3]

In early February Chennault received word that the long-awaited replacement aircraft – Curtiss P-40Es – would be delivered to the west coast of Africa. On Sunday, February 15, six pilots from the Third Squadron – R.T. Smith, George McMillan, Link Laughlin, P.J. Greene, Chuck Older and Tommy Haywood – were picked to make the trip. R.T. Smith wrote that night, "Don't know how long it will take but it should be ten days at least – we hope. We're all packed and rarin' to go."[4] He underestimated a bit. The trip actually took 35 days.[5]

Leaving Kunming the following morning, the six took a DC-2 to Calcutta, India. From there they boarded a B.O.A.C. flying boat that took them to Cairo, Egypt, via Karachi. India, and Basrah, Iraq. Arriving in Cairo ten days after leaving Kunming, they climbed aboard a U.S. Army C-53 contracted to Pan Am and flew to Accra, on the African Gold Coast, by way of Khartoum, Sudan and Kono, Nigeria. The return trip took twenty days, but was made without undue problems, and on March 23 the A.V.G.'s aircraft strength was bolstered by six new P-40E Warhawks.

A second group – Erik Shilling and Lew Bishop from the Second Squadron and Catfish Raines from the Third – left for Cairo in early March, where they would be met by six Pan-American pilots, ferrying the second six Kittyhawks from Africa. First stop was Calcutta (described by Shilling as "...one of India's largest cities and probably the dirtiest I had the misfortune to visit..."), where they spent two days. When they reached Cairo, they learned that they had a two-week "vacation" while they awaited the P-40Es from Accra. Shilling, designated leader of the group, went to a Cairo bank, cashed $3,000 of their letter of credit and gave each of the group $500 for expenses and spending money. With hotel rooms costing $2.00 a night and cheap food and liquor, they lived "high on the hog," according to Shilling.

After two weeks, MacMillan's group started dribbling through from Accra with stories of mechanical failures, most of which were fouled spark plugs. One of Mac's P-40Es was still in Khartoum awaiting a replacement generator. Similar mechanical problems plagued the incoming replacement aircraft for Shilling's group, delayed a group departure. Departing Cairo in two-ship flights, Shilling, Bishop and Raines accompanied by three of the Pan Am pilots, straggled back to Kunming on April 4.

Mingaladon by this time was a shambles; most of its buildings were either destroyed or badly damaged, and the A.V.G. and R.A.F. dispersed their fighters nightly in three outpost strips laid out in a horseshoe to the north of the city to preserve what little air defense force that was left. Downtown Rangoon was almost a ghost town. Convicts and lepers were released to the streets from their institutions and looting was rampant. With the loss of Singapore, even the most optimistic British residents of Rangoon realized that their city couldn't be far behind. The British businessmen who had hosted A.V.G. pilots in their homes turned over their keys to the Americans and promptly left heading north to Hsi Pau.

The nearly-deserted docks on the Rangoon waterfront were still piled high with hundreds of thousands of tons of Lend-Lease war materiel destined for Chiang Kai-shek's forces in China: nearly a thousand unassembled GM trucks, jeeps, staff cars, and guns – from rifles and submachine guns to antiaircraft guns – with accompanying thousands of rounds of ammunition, hand grenades, 5,000 truck tires, canned foods, hundreds of cases of liquor, medical supplies, aircraft parts, and even several new Brewster Buffalos still in their crates!

The tragedy of the situation was that only a small portion of it was moving. About a thousand tons a day were moving north to the Burma Road, but that barely touched the bulk of the glut piled high on the docks. Few Burmese laborers were left to load the goods onto trucks, and many of those were so afraid of Japanese attacks that they refused to work.

The U.S. Army Mission in Chungking (AMISCA) had only a few men in Rangoon to coordinate the shipment of war materiel to China. These few were given the responsibility of disposing of the great mass of materiel stacked up on the docks. In this atmosphere of chaos and panic, the A.V.G. was invited to come to the docks and take everything that they needed or wanted.

Many A.V.G. opportunists made several visits to the docks, loading up with whatever they could carry away. As Charlie Bond recorded in his diary:

> *"On the docks at the port are all kids of equipment, merchandise and food that are beginning to pile up. Some of the men have begun to go down to the docks, drive their truck up to a pile of whatever they are interested in, speak in a commanding voice, and direct the Burmese laborers to load the stuff onto their truck. They merely drive off with it. Not only are our men moving it, but so are local people. The men say that is where they learned the trick. The men have caches of canned fruit, rifles, shotguns shells, soap, radios and whatnot. John Dean picked up three more jeeps for the men yesterday the same way. Hell's bells! No wonder war costs so much.*
>
> *"Organization in the city seems to have gone haywire. It must stem from administrative officials evacuating. No one seems to be in charge of the docks. Thievery is rife all over the city. Of course, the locals do not condemn us for picking things up at the docks. They say we should have it rather than letting it pile up for the Japs to get later on."* [6]

Paul Frillman, the A.V.G. chaplain, had volunteered to go to Rangoon with the Third Squadron in December. Now he took charge of the salvage operation on behalf of the A.V.G. With whatever ground crew that could be spared, he loaded trucks with aircraft and vehicle spare parts, tires, tools, radio equipment, guns and ammunition. As Frillman wrote of it later:

"We tried to round up a three-month supply of canned goods and liquor for the whole A.V.G. When time ran short, we just snatched whatever crates were handy and portable, with no idea of their contents." [7]

Frillman and his men had to travel in convoys for protection against sniping by Japanese agents or Burmese nationalists. The streets were very nearly deserted by this time, the only living creatures a few dogs, crows and buzzards, plus an occasional figure would dart from an abandoned shop, arms full of looted wares. On one of their last trips from the burning docks, as they passed the deserted Buick automobile dealership, Frillman noticed three new, but undamaged, Buick sedans on the showroom floor. With the Japanese only 30 miles away, they stopped to "liberate" all three, Frillman driving one and two of the ground crew taking the others. One driver managed to lose his within minutes when he slid broadside into a tree while racing through the city.

By February 18 the Japanese had crossed the Bilin River, halfway between the Salween and Sittang Rivers and 30 miles closer to Rangoon. The R.A.F. was making plans to evacuate 240 miles northwest to Magwe on the Irawaddy River. British officials had posted 48-hour warning notices throughout the city suggesting that civilians evacuate Rangoon.[8]

That day, George Burgard and Charlie Bond, who had been staying at the home of Mr. Jensen of the East Asia Co., LTD, were told by Mr. Jensen that he and another British friend, a Mr. Mather, were leaving. After quickly loading their cars with all they could carry, they gave the keys of the house to the two A.V.G. pilots and departed. Quoting George Burgard, "Both were jittery as heck, and I can't blame them. Mr. Jensen left us the house and the servants are staying with us. CRB and I are the lords of the manor now..." [9]

At Mingaladon, several badly damaged Tomahawks deemed repairable were being loaded onto flatbed CAMCO trucks to be ferried to Bill Pawley's aircraft factory/repair depot at Loiwing. Twenty flyable P-40s remained. The R.A.F. could muster 20 Hurricanes, five Blenheim bombers, and nine high-winged Westland Lysanders, a slow, ungainly aircraft normally used for artillery spotting, courier work and the like, but which were now being pressed into service as dive bombers.[10]

The British authorities formally ordered the evacuation of "all non-essential personnel" on Thursday, February 21. With bureaucratic efficiency, the British announced a schedule of departures, establishing a priority list of military members, civil servants and dependents. The wealthier Indian and Chinese merchants loaded their families and goods aboard boats and planes and began moving out. Those not so fortunate began clogging the roadways and trains. The throngs of people crowding the streets and roads following the exodus were described by Charlie Bond:

> *"The citizens of Rangoon are leaving by the thousands. Our hosts at the billets tell us to take anything we want when they leave. They consider it lost and would rather we have it than the Japanese...All the convicts, lepers and insane are being turned out of their institutions and given freedom. Looting is out of all reason...Last night Bob (Smith) started to stop on the road to give some white hitchhiker a ride, but his host told him to ride on – the two were lepers."* [11]

Meanwhile, on the 21st at Mingaladon, the R.A.F. commander sent four of his five Blenheims on a strike against the Japanese troops threatening the British and Indian forces northwest of the city of Bilin, 86 miles northeast of Rangoon. Bob Neale directed Charlie Bond to lead an escort flight of six Tomahawks.

Bond and his flight – Jim Cross, George Burgard, John Farrell, and Robert H. Smith – cruised over the enemy lines at 16,000 feet, gradually dropping to 12,000 to keep the lower flying bombers in sight. Coincidentally, a Japanese force was on its way to Mingaladon at that same time. Sixteen Ki-48 Lilys from the 8th *Sentai* and fifteen 50th *Sentai* Nates had taken off from Nakon Sawon, Thailand, 275 miles southeast of Rangoon. Twenty-three more Nates from the 77th *Sentai* at Lampang, 250 miles northeast, rendezvoused with the bombers en route to Rangoon. The opposing forces met over Bilin.

Smith was one of the first to spot the Japanese formation, to their right and 2,000 feet above. He broke to the left and climbed. At 18,000 feet, with the sun at his back, he dived on the Japanese who, he described, were "just milling around." He got a piece of one of the Nates as he flashed through their formation, then zoom-climbed to altitude for a second attack. He damaged another fighter on the next pass and then joined with another P-40 for his third pass. Falling in behind a lone fighter, he drove in to 300 yards before opening fire. This time he scored; one of the Nates wings folded over the cockpit and the fighter spun down in a twisting dive.[12]

George Burgard tried to warn his flight leader, but Charlie Bond's two-bit, light- plane radio had failed again and he was engrossed in watching the Blenheims drop their bombs on Japanese troop concentrations north of Bilin. He glanced up just in time to see a group of enemy fighters diving in his direction. Bond pushed the stick to the left, snapping the Tomahawk into a half roll and pulled into a violent split-ess as he fire-walled the throttle. The Japanese fighters didn't follow.

Climbing back to altitude, Bond engaged another fighter head-on. The Nate veered off to the right, and Bond thought he had gotten him, but he was "too busy to follow up."

After Burgard called out in vain to Charlie Bond, he, too, dived away from the attacking Nates. "It seemed," he wrote later, "there were more than a million Jap planes all over the sky." Describing the action, he continued:

> *"After a number of attacks I gained a favorable position on top of an I-97 that was maneuvering to make a diving attack on a P-40 below it. As the enemy straightened out to fire, I began firing and in an attempt to make a sharp pull out the I-97 pulled directly into my line of fire... Parts of the motor flew out and the ship fell apart as it spun down."* [13]

The bombers finally turned back toward the east and the Nates followed. The six Tomahawks, now widely scattered, broke off the attack and headed for home. In their first air-to-air combat in over two weeks, they had downed four fighters with no losses to themselves. Burgard got two, while R.T. Smith and Johnny Farrell scored one each.

Bond was still writing out his combat report when Bob Neale rushed in to tell him to get his flight back in the air – they were to fly top cover for Neale's flight of six on a return strike at Bilin.

From his vantage point of 20,000 feet, Bond watched the Blenheims dump their load on the Japanese troops moving along the road west of Bilin. Then Neale and his flight dropped down from their 15,000 feet orbit to strafe the mechanized columns. This time no Japanese aircraft appeared, and the P-40s were able to exhaust their ammunition on over 50 trucks lining the jungle road.

What Charlie Bond didn't know, nor did any of the others at Rangoon at the time, was that the "Japanese" column they strafed on the road to Kyaitko was actually members of the 17th Indian Division approaching Bilin. It was a horrible tragedy of what later became known as "friendly fire." Faulty intelligence had placed a Japanese force moving along the Kyaitko-Bilin road. In reality, the Japanese were slowly moving through the jungle and rubber plantations to the north. Over 160 Commonwealth troops died in the two attacks.

Bond had scarcely climbed out of the cockpit after landing from his second mission when he looked around to see Bob Neale walking rapidly toward him.

"Charlie, take three men and get the rest of those trucks," Neale ordered.

"What about top cover?" Bond asked.

"None," Neale replied.

Picking Bob Prescott, Jim Cross and Bob Smith to go with him, Bond ran for his plane. Prescott's engine failed to start and Cross had to abort shortly after takeoff because of engine trouble. Bond and Smith climbed out toward Bilin.

Flying at tree-top level, Bond and his wingman raked the survivors of the previous attacks by the Blenheims and the strafing by Neale's flight. As he pulled up from his last pass, Bond's aircraft started shaking violently. Throttling back, yet maintaining a gentle climb, he nursed the Tomahawk back to friendly territory. With his canopy rolled back, and oxygen mask and helmet removed, however, he was ready to ditch or bail out.

As he neared the field, he found an rpm setting at which the vibration smoothed out considerably and continued on to a smooth landing back at Mingaladon. Taxing into his parking space, Bond stop-cocked the throttle and switched off the ignition. Then he leaned back to watch the big three-bladed prop wind down to a stop, physically and mentally exhausted after three back-to-back missions. It was only then that he saw the cause of his aircraft's vibration – a neat, one-inch hole through his prop blade, the result of an unsynchronized, .50-caliber round during his last strafing pass.[14]

On their way back to their quarters that night, the men of the Adam & Eves could see the towering column of smoke and flames from the docks on the Irawaddy River in downtown Rangoon. The British had given up trying to salvage the huge stocks of lend-lease supplies lying on the docks and were destroying what they could before they were forced to evacuate. Charlie Bond recalled:

> *"We passed several Indian infantry units on their way to the front and many trucks filled with supplies... I wondered how long we would hold out."*[15]

The answer was... not long. The remaining A.V.G. members were put on a one-hour notice. Japanese troops were moving toward the Sittang River from Kyaitko and were threatening the bridgehead 20 miles northwest at Mokpalin, forcing the British to destroy the bridge and to fall back along the west bank of the Sittang. By the 23rd, the R.A.F. had all but abandoned Rangoon, evacuating their forces: six Blenheims from No. 13 Squadron, four Buffalos from No. 67 and four Hurricanes from No. 17 to Magwe, 240 miles north-northwest from Rangoon. One squadron of Hurricanes remained to cover the retreat.[16]

The same day, Chennault ordered Neale to send a flight of six Tomahawks to Magwe. Working under R.A.F. direction, they were to patrol the Prome Road during the ground echelon's evacuation to Magwe. Those aircraft remaining would be used to carry out attacks on the advancing Japanese, either as bomber escorts or ground strafing.

Upon arriving at Magwe, the six pilots found that conditions at the field were primitive at best. The single, hard-packed dirt runway sent up clouds of blinding, choking dust with each takeoff and landing; there were no revetments or protection for the parked aircraft; the "alert shack," such as it was, was a tent, as were the sleeping and eating facilities for the ground crew. Without a radar capability, the only warning of a Japanese attack was an R.A.F. Blenheim bomber that was maintained on airborne alert. As George Burgard described it in his diary:

> *"This is a hell of a hole... No quarters on the field so we drive to Yenanyuang, a town in the oil fields north of Magwe, where we stayed with an American colony."*

Flight Leader Fritz Wolf, who arrived with the First Squadron contingent, flew to Kunming the next day with the following report:

> *"...The planes at Rangoon and Magwe are in an almost unflyable condition. The tires are hard and baked, and blow out on us continually... We are short on batteries, the battery plates are thin... There is no Prestone available down there at all. And there is no oxygen whatever... We were completely out of auxiliary gear shafts... Fresh foodstuffs of any description are completely lacking in Rangoon. We are living on canned goods entirely. Water is difficult to obtain, most of the water in town has been cut of... The dust from Magwe and Rangoon fouls up the ship's engines considerably. It clogs the carburetion system to such an extent that it is difficult to increase the manifold pressure of the Sharks as the engine quits cold. The entire systems are cleaned up on the ground but in a day's time is back in the same condition. This tendency of the engines to quit makes it impossible to dogfight or strafe, especially at low altitude, as the engine is liable to quit without warning. Out of 8 planes which were started off on the air raid alarm two days ago, only 5 were able to leave the ground. When we tried to contact the AOC at Magwe today to see about getting some gas, he told us he was too busy with evacuation problems to take care of us. He said for us to see if we could find the Duty Officer."*[17]

Back at Rangoon, the Third Squadron's thirteen aircraft were reduced to twelve when Dick Rossi lost power on takeoff and purposely ground-looped his Tomahawk at the end of the runway to keep it from being destroyed in the overrun. No. 13 was a "lemon," as one of the pilots described it, continuously out-of-commission. That left eleven at Mingaladon to carry out the orders to stem the Japanese tide.

Over the next two days, the P-40s and Hurricanes flew one escort mission after another – to Kyaitko, to troop concentrations on the Sittang River, and to the airfield at Moulmein, eighty-five miles due east of Rangoon across the Gulf of Martaban. After the Blenheims dropped their bombs, the fighters descended to make strafing runs.

On the 24th, five Tomahawks returned to Moulmein where they destroyed three bombers and a fighter on the ground. Charlie Bond led four more on a Blenheim escort mission to Kyaitko. Four Hurricanes flew top cover. After the Blenheims dropped their load, Bond and his flight hit a small Japanese supply column.

"We practically obliterated them," Bond recalled. "Some oxen were pulling two artillery caissons, and after our first pass the road was smeared with blood, dead oxen, and riddled caissons. God, what a mess!" [18]

Fog rolled in during the early morning hours of the February 25 – fog so thick that the First Squadron pilots could barely see half way down the sod strip at John Haig, the dispersal field northeast of Rangoon. Somehow the Hurricanes taking off at one end managed to miss the Tomahawks rolling from the other end as they took off into the fog to return to Mingaladon. "I still don't know how we did it without killing someone," remarked Charlie Bond, "but we did." [19]

So began the day that would bring the A.V.G. their greatest number of victories in a single day in their entire six-month war. At mid-morning, 44 Ki-27 Nates – 23 from the 77th *Sentai* at Lampon and 21 from the 50th *Sentai* at Moulmein – plus two of the new Ki-44 Shokis (Allied code name Tojo) out of Hlegue droned toward Rangoon from the east at 18,000 feet.[20] In their first daylight raid on Mingaladon in several weeks, the Japanese were hoping to strike a knock-out blow to the weakened Rangoon defense force. At 1030 the alarm sounded at Mingaladon and six First Squadron pilots raced for their parked Tomahawks. Minutes later they were airborne and climbing out to the east, joined by Hurricanes from No. 67 Squadron.

The P-40s split up into two flights, with Bob Neale, Bob Prescott and Bill McGarry in one and Bob Little, Charlie Bond and Dick Rossi in the other.

As he climbed through 15,000 feet, Little rocked his wings, signaling he had seen the enemy, and rolled away in a dive. Bond and Rossi followed. Neale and company, in their own patch of sky, continued climbing. Fifteen miles east of the field, Neale caught sight of the Japanese formation below and to the left. With a signal to his flight, he peeled off in a screaming dive.

Bob Prescott, following Neale, described his combat:

> *"I picked up one on the edge of the bunch, gave him a long burst, dead on, at close range. Don't think he ever knew what hit him. He started for the ground, I followed to make sure he wasn't playing possum. Watched him crash, climbed back up and couldn't find the flight."* [21]

Diving through the "target rich environment," Neale picked off two fighters and Bill McGarry got one before they streaked for home.

Meanwhile Bob Little's flight was still searching for the enemy. His "tally ho" had proved false, and they never made contact with the Japanese.

Their morning raid broken up, the Japanese returned in force that afternoon. This time 12 Ki-48 Lilys from the 8th *Sentai* at Nakhon Phanom, Thailand, rendezvoused with the same fighters that had made the earlier raid.

Bob Neale picked up the telephone at 1600 to hear, "Large enemy formation approaching from the southwest." [22] This time he ordered every available aircraft into the air, and eleven Adam & Eve pilots ran to their planes. At the opposite end of the field eight Hurricane pilots were strapping in and starting engines.

Climbing out to the west in three flights of three (Prescott and Blackburn were still waiting for their planes to be refueled), Neale had Red Flight, Little was White Leader and Bond led Blue. Since Bond had the best radio of the three, he acted as air commander. About 50 bandits were headed towards Bassien, 85 miles west of Rangoon, reported "Orphan," the R.A.F. ground control.

Bond relayed the call and continued climbing toward Bassein. Then Orphan called, "Bandits southwest of field – twenty miles."

Bond turned his flight back toward the east, but in his excitement forgot to relay the call to Red and White Flights. Neale and Little, by this time several thousand feet above Blue Flight, caught sight of the enemy bombers and Neale called out, "Here they are."

Bond spotted the Japanese formation almost at the same time, twelve bombers "in perfect 'V' formation," with the "usual cluster formation of Japanese fighters directly above the bombers."

Pulling up and into the fighters, Bond took on the nearest one head-on. The Nate caught fire almost immediately, but continued on at Bond's Tomahawk. Yanking back on the stick, Blue Lead watched the burning Japanese fighter flash past scant feet beneath his airplane.

Turning hard, his head swiveling in search of another target, Bond glimpsed another Nate boring in on him in a second head-on attack.

"The Jap fighters were as aggressive as we were," Bond recalled. "We headed straight at each other, each with all our guns blazing. At the last second he pulled up and I did the same thing." Then, to avoid a certain mid-air collision, Bond instantly pushed the stick violently forward, the negative "G"-force driving him up against the top of the canopy despite his cinched-down seat belt.

"We shot past each other," Bond continued, "him above me. I rolled and pulled violently to the left, shoving my head over backwards to pick him up. He, too, was in flames." [23]

By now, Bob Prescott and John Blackburn ("Blackie") were climbing out full throttle for the fight. Leveling off at 16,000 feet, Prescott spotted both the Japanese formation and the attacking Tomahawks. With a gaggle of Japanese fighters above and another below, Prescott chose the latter, rolled and dove. As Prescott related it:

> *"The action was so fast I didn't get to signal to "Blackie. Had only time for one short burst at the plane I picked out and didn't see what happened to him. I dived on away, caught my altitude and was coming back into the thing when I saw a Jap on a '40. I shot way out in front of him, and very luckily hit him. He started burning and went for the ground."* [24]

The air battle swirled above Rangoon for the next forty-five minutes before the Japanese formation turned back southeast and headed out across the Gulf of Martaban. The Japanese fighters had been successful in protecting the formation of Lilys ("The bombers were damn well protected," remarked Bob Neale)[25] – only one bomber was lost, brought down by George Burgard. Protected by the umbrella of fighters, the Ki-48s did severe damage to Mingaladon; two people were killed, 10 Blenheim bombers were hit and one Lysander was destroyed. And when the Tomahawks and Hurricanes landed, they had to weave around numerous bomb holes in the runway.

The Japanese paid a price for their successful raid, however. Twenty-four escorting fighters fell to the combined guns of the A.V.G. and R.A.F. – eighteen for the American volunteers and six for the British. Bond and Little got three each; Prescott, McGarry, Neale and Burgard got two; and Bob "Snuffy" Smith, Dick Rossi, Johnny Blackburn, and Joe Rosbert all scored one each.

The Rangoon defenders, in turn, lost two – the R.A.F. lost a Hurricane and pilot and the A.V.G. lost Ed Liebolt. Dick Rossi had been the last to see Liebolt. Patrolling near Bassein, he saw Liebolt roll his canopy back and shortly after begin a slow descent. His problem was never determined. No radio call was heard and no one saw him again.

After landing at Mingaladon, the Adam &Eve ground crews swarmed over the ten Tomahawks, refueling, rearming and performing minor maintenance. As soon as the first six aircraft were ready, their pilots dispersed them to John Haig.

The First Squadron relaxed over dinner that evening at the ground crews' hostel, nicknamed "The Eighteen Mile Ranch," and quietly celebrated their big day, their enthusiasm considerably dampened by the loss of Ed Liebolt.

It had been a big day; every pilot who was in the battle (except poor Ed Liebolt) had scored, and five had achieved ace status with their victories for the day – Charlie Bond, George Burgard, Bob Little, Bob Prescott and Billy McGarry. During the evening, Bob Neale announced that he had received word from Chennault that they would evacuate on March 1. Only three more days!

That night the Japanese struck back with a vengeance. Twelve unescorted Ki-21 Sallys from the 31st *Sentai* at Phitsanulok dropped 13 tons of bombs on the darkened base at Mingaladon. Few bombs found the runways, but two more Blenheims were but charred ruins the following morning.[26]

On the 26th, the Japanese began their final air assault on Rangoon. Only nine flyable P-40s and six R.A.F. fighters were left to defend the city. Early that morning, Bob Neale gathered six of his pilots for another strike on Moulmein Airport. As they were taxiing into position for takeoff, however, the air raid alarm sounded. Neale gunned his Tomahawk onto the runway and took off, followed by Prescott, McGarry, Rossi, Little, Burgard and Rosbert. After twenty minutes of futile searching, Neale judged it another false alarm and turned his flight toward Moulmein.

They hit the satellite field south of the airport first, destroying two Nates on the ground in a fast hit-and-run attack, then headed for the main field. Attacking in a string, the seven P-40s streaked down on the field, with Neale in the lead and Prescott bringing up the rear as "tail-end Charlie." Down below, Moulmein Airport was bustling with activity. The Japanese had been caught as they were preparing for a morning mission – several planes were warming up and others taxiing into takeoff position.[27]

Suddenly, Neale turned away from the field. "I thought he was going to strafe the town instead," recalled Bob Prescott, "when suddenly straight up from out of nowhere shot a Jap fighter…I caught one taking off. He was almost 500 feet high. I shot full deflection at him – a couple of bursts."

Three Nates had just made it off the field as the shark-nosed Tomahawks arrived. At 400 feet over the field, Neale opened fire on the lead Nate. Pulling up sharply to avoid Neale's fire, the Japanese pilot inadvertently turned directly in front of George Burgard, who, in Burgard's words, "cold-cocked him with one grand burst." [28]

Burgard climbed out to engage the fighters overhead, while Neale and two others came around to make another strafing pass on the field. Burgard caught another Nate with a short burst, but broke off when his guns jammed and turned for home.

Meanwhile, down below, Neale had picked off two Nates as they took off but was bounced in turn by another pair of the little nimble, fixed-gear Japanese fighters. Barely fifty feet off the ground, and unable to dive out, he had little alternative than to hunker down behind his armor-plated seat, push the throttle to the firewall and run for home. The Japanese chased him out into the Gulf of Martaban, but the Tomahawk had a good 50-mph top speed advantage over the Nate and Neale soon out-distanced his pursuers.

Back at Mingaladon, all had landed except for the squadron leader. Charlie Bond said that the last he saw of him was on the deck over the water with a couple of Japs on his tail. J.J. Harrington, Neale's crew chief, recalled Neale's return: "While we were standing around pulling our thumbs, here came this P-40 screaming over the field. Of course, we all breathed a sigh of relief, and here came Bob in for a landing, and he made a beautiful landing even under those circumstances." [29]

As Neale taxied in to his parking spot, Harrington jumped on the wing to help Neale off with his parachute. "The Japanese can't hit the side of a red barn," he told his crew chief. Harrington said nothing as the two climbed into the jeep and drove Neale back to the alert shack. Two hours later, after patching seventeen holes in Neale's Tomahawk No. 7, he drove back over to the alert shack and invited Neale to take a jeep ride with him. Driving down the flight line, Harrington stopped at the newly-repaired P-40 and said, "Those Japs might not be able to hit a red barn, but they sure do know how to hit a P-40." "God Almighty!" was all the comment Neale could manage.[30]

Charlie Bond and Dick Rossi took off at 1030 to make another search for Ed Liebolt's aircraft. Flying out 40 miles west of Mingaladon, they flew a 30-mile radius search pattern. After an hour of fruitless searching, they turned back for Mingaladon.[31]

As they made their approach to the field, they saw people running toward the parked P-40s – obviously they were scrambling for an air raid warning! Circling the airfield until the P-40s got off the ground, Bond and Rossi tacked onto

Bob Little's flight. As they climbed through 18,000 feet, Bond spotted the Japanese formation – 12 Ki-48 Lilys of the 8th *Sentai*, escorted by 20-to-30 Nates from the 50th and 70th *Sentais*.[32] As Charlie Bond recorded in his diary:

> *"I pulled up to Bob Little and waggled my wings, pointing off to the left. Away we went! We had to get a little more altitude to come in from above the fighter escort. I took a quick peek to my rear to see if anyone was with me. Hell, I was by myself!*
>
> *I attacked, and was head-on with one of the fighters. We barely missed each other. I turned to see if he was on fire. He wasn't, and he was easily turning inside of me. Immediately, two others were on my tail. I half-rolled and went into a vertical dive – everything forward. Pulling out and up again, I got into position to make another attack. He saw me coming in and went into a left turn. I led him a lot and squeezed the trigger. His ship straightened up a little, and then started off in a steep spiral with smoker trailing. I pulled up into a shallow climb to follow him.*
>
> *My speed had fallen off, and when I looked back I saw two Jap fighters smack-dab on my tail. Again, quickly into a split-S with full throttle. They followed. All I could do was hunch up behind my armor plate and wait for my acceleration to pull me out of their firing range. None of the bullets hit my fuselage, but I got a few in the left wing tip. When I got out of range, they pulled away."*[33]

The battle raged for over an hour, during which time the First Squadron knocked down 13 Japanese aircraft – 12 fighters and a bomber (once more, the Japanese fighters put up a superb defense for their bombers). Dick Rossi and Joe Rosbert got two Nates each, and Bob Neale, Charlie Bond, Johnny Blackburn and George Burgard were credited with one each. Robert H. Smith brought down the sole bomber that was destroyed. Total for the day, including the raid on Moulmein, was 20.

Again, the Adam & Eves returned without a loss. Jim Cross, however, took a burst of 7.7mm Japanese slugs through the windscreen of his No. 78, holing his oxygen bottle and exiting out the left side of the cockpit. The bullets missed him by mere inches, but he was still showered with flying glass and shrapnel.[34]

After the P-40s were refueled and rearmed, the pilots climbed back in to disperse the ships to John Haig, their satellite field to the north. From there they drove to the Ranch for supper, where they listened to the first wave of Japanese bombers making their nightly foray to Mingaladon.

Charlie Bond recorded the hopeless feeling of most in his diary that night:

> *"We are out of spare tires; if one blows, then that ship is lost. The last oxygen is in the bottles in the ships, and that's it. Other spares are practically gone, and some have already been sent up north. We are just hanging on… The men are really getting edgy."*[35]

Earlier in the day, J.J. Harrington had been to the R.A.F. side of the field to get a battery charged and oxygen bottle filled. To his surprise, as he commented later, "Everything was scattered helter-skelter. I couldn't find anyone." Searching for Bob Neale to report the situation, he found him in the ready shack near the flight line.

Harrington asked Neale, "Bob, did those Limeys say they were going to pull out and leave?"

"No," Neale replied, "I haven't heard anything about it."

"Well, they have," said Harrington. "There ain't nobody down here but us."[36]

That evening, Bob Neale had his regular meeting with the R.A.F. field commander, Group Captain Broughall, who casually mentioned that the A.V.G. might have a bit shorter air raid warning time tomorrow, since the R.A.F.'s radar warning equipment had been disassembled and shipped to Magwe.

Dismayed at the news, since with no warning system the A.V.G. would literally be sitting ducks for the Japanese, Neale woke the sleeping pilots at 0130 to announce that they would begin evacuating Mingaladon at daybreak.[37] Although Chennault's orders had been to evacuate on Sunday, March 1, Neale decided that a 48-hour wait wasn't worth the risk.

Bond, Sawyer and Burgard dressed quickly and drove to the Ranch to tell the ground personnel to gather their personal belongings and prepare to convoy north. Several mechanics went to the John Haig satellite field to prepare the P-40s. Working in the pre-dawn darkness, they had all ready to fly, including Jim Cross's bullet-riddled No. 78, by dawn.[38]

At daybreak on the 27th, Neale formally ordered the evacuation of the A.V.G. Four P-40s, led by Bob Little, took off first, climbing out for Loiwing. Just across the northern Burmese border into China, the base served as the headquarters for Bill Pawley's CAMCO, the major depot for battle- and accident-damaged P-40s. Red Probst and Matt Kuykendall were off next, with Kunming as their destination.

At mid-morning, Neale met with Paul Frillman and said, "You're it. You're taking the first convoy up the road."

"You're out of your mind," Frillman replied. "I'm the chaplain. I'm not taking the convoy anywhere."

Brandishing a radiogram from Chennault, Neale retorted with, "The Old Man says so. Convoys start this morning. Frillman takes the first." [39]

Frillman took an hour or so to load his newly-acquired Buick sedan "to the roof" – including two Siamese cats he had promised to take to one of the R.A.F. officers who had left earlier. He then drove to the end of the field where his convoy was impatiently waiting for him. There were twelve jeeps and eighteen trucks, their stake-and-board enclosed beds piled high with crates and boxes covered with tarpaulins. On the seat beside the driver in most of the vehicles sat a young Anglo-Indian girl, dressed as though ready for a dance. Deciding this was no time for a sermon, Frillman drove to the head of the column and led the way to the north.[40]

After a lunch of cold, canned pork and beans, Charlie Bond, George Burgard, Johnny Blackburn and Bill "Black Mac" McGarry took off for Magwe. En route, they patrolled the Prome Road, flying low over the evacuation convoys. Only Bob Neale and Robert H. "Snuffy" Smith remained at Mingalodon.[41]

Neale, still upset at losing Ed Liebolt, was reluctant to leave as long as there was any chance of his showing up. After all, hadn't several A.V.G. pilots walked in after days in the brush? Maybe Ed would, too. Neale and Smith waited for another night for Liebolt. Neale had even put a blanket into the baggage compartment of his Tomahawk for Liebolt to wrap up in during the flight to Magwe.[42]

As the sun rose in the Burmese sky on the 28th, however, there was still no sign of the missing pilot. Reluctantly, Neale gave up the vigil. Tossing a couple cases of scotch into the empty baggage compartment, he pointed to nose of his Tomahawk down the pock-marked runway of their erstwhile base for the last time; with Snuffy Smith on his wing, he climbed out for Magwe.

CHAPTER 13

Chennault Was "Plenty Steamed"
– *Frank Losonsky*

Smoke and haze covered southern Burma on the morning of February 28 as Bob Neale and Robert H. "Snuffy" Smith took off from Mingaladon headed for Magwe. With neither of the pilots carrying maps of the area, and haze up to 15,000 feet, they overflew the airstrip at Magwe and Neale lost his wingman about 20 miles north of Magwe. After trying vainly to find identifiable landmarks, and with 10 gallons of fuel remaining, he circled a peanut field near the village of Singu, lowered gear and flaps and put the P-40 down in a successful emergency landing.[1]

Meanwhile, after losing Neale in the haze, Smith continued on north. As his fuel gauge neared empty, he attempted a landing in a river bed near the village of Myingyan, some 100 miles past Magwe, bending his prop in the otherwise successful landing. Calling back to Magwe, he reported to Charlie Bond that he had damaged the prop and landing gear, but that he could probably fly it out if he could get replacement parts.

Shortly after, Bond got another telephone call. Walking in to the village of Singu, Neale had found the "head man" and made a telephone to call Magwe for fuel. Answering the telephone, Bond heard Neale's voice, "Charlie! Charlie! That you?"

"Yeah, Bob! That you?" Bond replied.[2]

After listening to Neale's explanation of his situation, Bond promised that he'd have gas on the way the next morning. After he hung up, he smiled, recalling how Neale had commented earlier on Bond's concern that they had no maps for the evacuation flight to Magwe – "Goldarn, Charlie, you can't get lost. Hell, it's right on the river."

Neale spent the night under the wing of the Tomahawk on a bed brought out by the village head man, and by 1500 a truck with 40 gallons of avgas arrived from Magwe. Running the power up while holding the brakes, he took off downhill and was thrown into the air by a ridge on the east side of the field. A few minutes later, he landed at Magwe, commenting, "Probably never live this down." [3]

Meanwhile, as Neale was bedding down beneath the stars, at the A.V.G. headquarters in Kunming the Generalissimo and Madame Chiang Kai-shek were hosting a party in honor of the American volunteers. On their way to attend an inter-Allied conference at Lashio with Air Officer Commanding (AOC) Burma, A.V.M. Donald Stevenson, and Tenth Air Force commander, Maj. General Lewis Brereton, the Chinese leader and his wife had stopped overnight at Kunming. At the banquet, after introductions and opening remarks by Chennault, the Generalissimo gave a lengthy speech in Chinese (translated for the benefit of the Americans), telling the assembled group that "...You have written in the history of this world war a remarkable page, the memory of which will live in our minds forever." Madame Chiang followed, calling the assembled A.V.G. members "my boys" and "angels with or without wings." [4]

During the ceremonies, the Chiangs bestowed the Order of the Cloud Banner on several, including Ken Jernstedt, Fritz Wolf, Robert P. "Duke" Hedman, Ed Overend, Chuck Older, Parker Dupouy, Arvid Olson, Tommy Haywood, and George McMillan. "Sandy" Sandell and Louis Hoffman were given the decoration posthumously.

The festivities were marred somewhat, however, by two "angels" who didn't hear Madame's words. First Squadron pilots Greg Boyington and Percy Bartelt had their own party just outside the banquet hall, supplied by Boyington's personal liquor stash (Chennault had prudently closed the Hostel's bar prior to the dinner). Before the end of the speeches and entertainment, the two errant pilots stumbled noisily into the banquet hall, knocking over chairs as they entered. It was just one more black mark for Boyington.[5]

Following the three-day conference at Lashio, their DC-2 escorted by Olson, Brouk, Hedman and Jernstedt, on their way back to Chungking the Chiangs stopped over once again at Kunming. At yet another banquet that night, they brought word that the $500 promised bonus for each Japanese aircraft shot down would be extended for all aircraft destroyed, including those on the ground from strafing attacks.[6]

On Thursday, March 5, just before the Chinese leaders' departure, Harvey Greenlaw (as Bus Keeton remembered it) "got a hare-brained idea to put on an airshow… told Howard to get six stunt pilots to zoom three times and on the last time to do a slow roll." Led by Frank Lawlor, the six pilots (Lawlor, Bob Layher, Greg Boyington, Red Probst, Gil Bright and "Hank" Geselbracht) gave an impressive display for the Chiangs' entertainment. At one point, Bob Layher's inverted pass came so close to the reviewing party that the Chiangs and Chennault were forced to "hit the dirt." As Lawlor made his roll, his baggage door came open and he had to abort the flight. Passing flight lead to Boyington, the five remaining P-40s joined up on the C.N.A.C. DC-2 transport after it took off to escort it part way to Chungking.[7]

Dropping off from the Chiangs at Changyi, Boyington turned the flight southwest (or so he thought) for Kunming. Unfortunately, he had no area maps, his compass was malfunctioning and his radio was out. Relying on his defective compass, he led the flight south instead of southwest. Finding no recognizable landmarks, their fuel finally exhausted, Boyington spotted what appeared to be a small, flat field at the top of a plateau. Boyington, along with Bright and Probst, put their planes down in the small field (which turned out to be a cemetery!), two attempting to land wheels-down (with disastrous results) and the third skidding along gear-up. Layher put his P-40 down twenty-five miles further north, and Geselbracht crash-landed a few miles further on.[8]

As the pilots climbed out of their wrecked aircraft, they were immediately surrounded by a crowd of Chinese peasants. Fortunately one spoke passable English and informed the Americans that they were near the village of Wenshan, 150 miles southeast of Kunming, and only 40 miles north of the Yunnan-Indochina border. Finding one operable radio in the three wrecked planes, Boyington called Kunming, reporting that no one was injured, but it appeared that only two of the P-40s could be repaired.

Glen O. "Tex" Blaylock, Third Squadron line chief, was tasked with the job of salvaging as much as possible of the wrecked aircraft. On March 6, Bill Reed flew to Wenshan with Blaylock in a BT-9. When they arrived, Blaylock learned that he would have to have the approval by the village "head man" before proceeding further. After meeting with the "funny little guy with a wispy mustache," and after long negotiations accompanied by countless mutual toasts with the strong local rice wine, Blaylock was given the promise of as many coolies as necessary to assist in the salvage operation.

Meeting briefly with Boyington, Probst and Geselbracht, who were still there, they found that two of the aircraft had attempted a wheels-down landing, and the third, Reed's No. 75, had landed gear-up. Blaylock determined that two of the damaged aircraft could be suitably repaired to be flown out. After obtaining Chennault's permission to go ahead, they returned to Kunming, with Blaylock (a licensed pilot before the war) flying their BT-9 part way back.[9]

Blaylock assigned Third Squadron crew chiefs Frank Losonsky and Stan Regis to depart for Wenshan to "salvage the ships before the natives cut the remains up for pots and pans." Checking out a 6x6 flatbed truck, a jeep and assorted spare parts, including props, the two left Kunming on March 8 for Mengtzi, where they loaded the truck on a railroad flatcar and boarded the train to Mengtzi. From there, they unloaded the truck from the flatcar and were joined by a group of Chinese who escorted them the rest of the way.

Arriving at the crash site on March 11, Losonsky commented, "Two aircraft were wrecked beyond repair. The props were smashed and the wings badly damaged. The other two ships were flyable, but needed lots of repair." Over the next three weeks, the two worked their mechanical magic, replacing landing gear, rebuilding battered wings and hammering bent props back into reasonable shape.

After staying a week as guests of the "mayor" of Wenshan, with the aid of an American-made truck and Chinese driver, the stranded pilots made their way back to Kunming some three days later, as Boyington commented, to a "cool reception." [10] However, after returning to Kunming, Greg Boyington volunteered to fly out any of the Tomahawks from his flight that could be salvaged. (To quote Losonsky: "I'm not sure why he volunteered to fly the aircraft out. We could have easily 'hauled' them back to Kunming.")[11]

It took 50 Chinese coolies only two days to hack out a usable, 1,200-foot airstrip from the thick vegetation, extending right through the cemetery, by a rice paddy, down an embankment and off the end of a cliff. Finally, the aircraft's armor plate was removed, along with all but 30 gallons of fuel, just enough to get to the auxiliary air field at Mengtzi.

Boyington arrived via another BT-9 flight on March 15. After the Tomahawk was towed to the very end of the short airstrip, Losonsky advised Boyington to "pour the coals to it and fly the aircraft 'gear down' all the way." (He didn't trust the hydraulics.) Lowering himself into the cockpit, Boyington started the engine and advanced the throttle to full power. After the brakes were released, the Tomahawk slowly gained speed, lifting off just before the end of the strip.[12]

Boyington would return on April 2 to fly out the remaining salvageable aircraft. The two crew chiefs remained to take the guns out of the unsalvageable aircraft, pulled out what they could from the wrecked fuselages and took off the damaged wing. While coolie laborers physically lifted the fuselage two feet off the ground, they pushed a skid beneath it and then pulled it 200 yards to the waiting truck. A fourth aircraft was hauled up from its more distant crash site, and the loading process was repeated. From there, they returned to the train station at Mengtzi, loaded their truck, P-40 fuselages and wings on flatcars for the trip back to Kunming, arriving there on April 15.[13]

And the accidents continued. Scheduled to replace the First Squadron at Magwe with elements of the Third, Chennault had initially sent a flight of four – Parker Dupouy, Clifford Groh, Freddy Hodges and Moose Moss – to the fall-back base. On March 10 (while Glenn Blaylock was en route to Wenshan to recover the two "lost" P-40s), he dispatched five more Tomahawks to Magwe, via a fuel stop at Lashio. Departing Kunming at 1130, the flight included Third Squadron pilots "Duke" Hedman (as flight leader), Bob Brouk, Ed Overend, and Bill Reed, with First Squadron pilot Bill Fish.

Running into hazy weather about an hour and ten minutes out as they approached Lashio, Hedman miscalculated his position and, thinking that he had overflown the air field, turned the flight back 180 degrees. Fifteen minutes later, with Lashio still not in sight, he led the flight in a weaving pattern, searching for recognizable landmarks.[14] At that point, Reed, Overend and Fish broke off from the flight to fly northeast, while Hedman and Brouk continued on a southeast heading of 140 degrees. With 15 gallons of gas remaining, the two sighted a long sandbar on the Nam Ting River, near Kol Long, and landed safely. After landing, Hedman radioed the base at Paoshan that he and Brouk were down OK., but didn't have any word about Reed, Overend or Fish.[15]

Met by a group of Burmese natives, they were taken to a construction camp at Kunlong, some nine miles distant from the landing site (and 60 miles northeast of Lashio).[16]

Reed, Fish and Overend weren't so lucky. According to Reed:

> *"... Fish, Overend, and I picked a valley and decided to attempt forced landings, Fish tried first wheels down, but he went right over on his back. He got out and waved, so after looking awhile more, Overend tried one with his wheels down. He, too, went over on his back, and when I flew by to look I saw that his cockpit and canopy were buried. I landed immediately, wheels down like a conceited fool, and luckily stayed right-side up. I had to dig Overend out, but he was alright. We stayed with three Americans – Hall, a sanitary engineer; Wetzell, an Army captain; and Haas, a Lt. Col. Medical. Had dinner with above-mentioned three and two Chinese nurses, and some Chinese engineers. Almost got a tiger cub, but his owner didn't want to let him go."*[17]

The following morning, Reed and Overend set out for the small island where the two had landed. Picking up Fish from the small village he was staying in, the three crossed the Nam Ting River and located the two P-40s of Reed and Overend (they gave up on Fish's aircraft because it was too far from the road). Working until early afternoon, they were able to strip the guns from the aircraft before heading toward Kunlong.

As the three were making their way to the construction camp, back at Loiwing George Burgard was directed to fly over the area and try to locate the five downed aircraft.

Lord, what a mess," Burgard commented. "10 airplanes in less than one week. The Colonel will die." Flying over the Nam Ting valley, he sighted all five, two on the sandbar and the other three further up the valley. Returning to Loiwing, he reported the status of the five to Kunming and then, not knowing the five had been "rescued," took off in the CAMCO Stinson with food and supplies for the stranded pilots. Finding he couldn't get over the mountains with the little Stinson, he sent Bill McGarry to drop a message. Taking off late in the afternoon, however, McGarry got lost, ran out of gas and he, too, bellied in near Kutzai, about half-way between Loiwing and Lashio.[18]

On Thursday, the 12th, two R.A.F. personnel arrived at the camp in a truck with five 20-gallon drums of gasoline. Accompanied by most of the camp staff, 20 coolies to carry the gasoline drums, and curious natives, the five pilots set out for the hot, three-mile trek through the jungle to the sandbar. After refueling the planes and pacing off the 2,000 feet take-off distance, Hedman and Brouk easily made it off and headed for Lashio, 90 miles away. The next day, a Chinese camp engineer provided a car and driver for the three remaining pilots for their road trip to Lashio.[19] Frank Losonsky commented that Chennault was "plenty steamed" at the loss of the aircraft. In the space of five days, eleven pilots had gotten lost, ran out of gas and been forced down – losing more aircraft than the A.V.G. had lost in combat over a similar period of time.

Meanwhile, while the multiple debacles of lost A.V.G. pilots was going on, Paul Frillman's convoy was making its way toward Maymo, 103 miles southwest of Lashio, the southern terminus of the Burma Road. The first day the convoy shared the road with only a few scattered British and Chinese military convoys also heading north. By the end of the day they had managed to get nearly 300 miles from Rangoon, despite several unscheduled stops to jump for a ditch as Japanese aircraft flew overhead. Frillman surmised that they must have been reconnaissance planes, however, since they were never attacked. After the first day, they weren't bothered.[20]

That first night, Frillman's group camped by the road beneath the stars, each taking a turn at guard duty. British drivers they had met along the road had warned them of incidents of Japanese agents or Burmese nationalists slashing tires and stealing engine parts.

After breaking open a few unmarked crates from the randomly gathered cargo, they were surprised to find box after box of shoelaces, a crate full of kitchen cleanser and enough facial tissue to last a year. Finally they came upon a crate of canned asparagus, which they washed down with bottles of Bristol Cream sherry.[21]

On the road again early the next morning, the convoy made good time until just south of Mandalay, they encountered the first mass of refugees fleeing Burma – mostly poor Indian and Chinese shopkeepers fleeing to China. As they approached Mandalay, the throngs of people trudging northward gave way, as Frillman described:

> *"... a mad jam of cars including new ones of the well-to-do, all in a cloud of heat, dust and flies. On the flats outside the city a swarm of vehicles was packed, among them taxis, buses, delivery wagons, ice trucks, fire engines, and hearses from Rangoon. Tents and cooking fires dotted the parking lots. Many refugees hoped the Japanese would stop at Rangoon and were delaying to see what happened next."*[22]

From Mandalay, Frillman's convoy headed north through wooded valleys toward Maymo, the rear echelon headquarters of the British forces in Burma. After nearly two weeks on the road, they rolled into the city, half-way on their trip to Kunming.

As they made their way through the crowded streets, Chaplain Frillman was surprised to meet two Americans in uniform, Army sergeants with U.S. Signal Corps brass on their collars. On questioning the two, Frillman learned that Lt. Gen. Joseph W. Stilwell, newly-arrived in China, had just arrived to set up his headquarters in Maymo.

Frillman seized on the opportunity to stop at the American enclave. Here they could wash the two-week dust and grime from their bodies, perhaps rest overnight in clean, comfortable beds and eat a real American-style meal. He had met Stilwell years earlier when he was a missionary in Hankow and Stilwell was a colonel serving as military attaché to China. He remembered him as being "serious and somewhat austere."[23]

Stillwell had been sent back to China by Army Chief of Staff General George C. Marshall as "Commanding General of U.S. Army Forces in the China-Burma-India Theater, supervisor of Lend-Lease and U.S. representative on any

Allied war council," and to "equip and train the Chinese forces in China." [24] Lt. Gen. Hugh A. Drum, the U.S. Army senior line officer, and at the time commander of the First Army, had originally been proposed for the job. A "pompous man of large pretensions,"[25] he seemed to be ideally suited for what Chiang Kai-shek had in mind – "someone amenable, of imposing rank and influence at home, but not so knowledgeable as to be critical of Chinese requisitions." [26]

When Drum learned that the position would not include command of American troops, he argued strongly that it wasn't worthy of an officer of his rank and experience. Unhappy with Drum's attitude about the pending assignment, Secretary of War Henry L. Stimson withdrew his name from nomination and invited Maj. Gen. Stilwell to his home for an interview. No less reluctant than Drum to serve under Chiang Kai-shek, Stilwell nevertheless told Stimson he would "go where I'm sent."[27] Discussions between Stimson, Marshall and T.V. Soong followed on February 13, 1942.

Joseph E. Stilwell, newly-promoted to lieutenant general, departed for China by way of Calcutta. A month later, establishing his headquarters in Maymo, Burma, he was given nominal command of the Chinese 5th and 6th Armies and initially provided with a staff of 35 officers and five enlisted men. Later to follow would be 400 technicians and instructors to aid in training the Chinese in the use of American and tactics.

With British guards waving them through, Paul Frillman led his motley caravan through the gates of the military complex. Pulling up in front of a red brick house covered with purple bougainvillea,[28] they were surprised to see a tall, thin American in a crisply pressed khaki uniform with three stars gleaming on each shoulder – they had stopped at the former Baptist mission house that had become Stilwell's residence.

The contrast was appalling – Frillman and his men with several days' growth of beard, dressed in a mixture of rumpled and dirty military and civilian clothing. And Stilwell was appalled. As he stared at the array of scruffy A.V.G. drivers, Anglo-Indian passengers and two mating Siamese cats that jumped from Frillman's dusty Buick, Frillman got out of his car to introduce himself. As Frillman later described:

> *"In the next few minutes, I learned how Stilwell earned his nickname, 'Vinegar Joe'. I was glad he didn't recognize a former young Hankow missionary as chief of what he saw as a gang of bums, but he didn't need the extra ammunition. Before I finished trying to tell him who and how many we were, and what billets we hoped to find, he was launched…"* [29]

Even as Stilwell was blasting the group as being a "disgrace to the U.S. Army," a station wagon roared down the street and skidded to a stop beside Frillman's Buick. When the bleary-eyed driver stepped out of his vehicle, Frillman recognized him instantly as a mechanic who had left the A.V.G. earlier to follow a more lucrative career of black marketing. His passengers were a fat and aging Anglo-Indian woman, her gaunt and toothless mother and a carload of dark-skinned, raggedly-dressed children. Waving a half-empty bottle, the man staggered toward Frillman and Stilwell.

With scarcely another word, Stilwell turned on his heel and went back into his house, slamming the door behind him. "So much for a shower and clean bed," muttered Frillman. That night they camped again beside the mountain road.[30]

Several days later, the caravan reached the Chinese border. Just north of the border lay the Loiwing headquarters of William Pawley's operation. Here, beside the airstrip, was his aircraft assembly plant and repair facilities. Beyond were bungalows that housed the employees and a clubhouse, complete with an American housekeeper.

Frillman's group received a much better welcome here than they had found at Maymo. First order of business was a hot shower, followed by a good, American-style meal. That evening they relaxed in the clubhouse lounge listening to jazz records played on an imported Wurlitzer jukebox.[31]

The following day, they began their long, twisting 717-mile journey to Kunming along the Burma Road. Frillman commented that, "I was very aware that I was back in China… It was a relief after the Burmese who were unreadable and might be hostile." [32] A hundred miles out of Loiwing, the convoy began to climb the ridge of the Salween Gorge. As they looked down upon the Salween River, 6,000 feet below, they marveled at the ingenuity and industriousness of the thousands of Chinese laborers – men, women and children – who built the Burma Road with primitive tools and methods in just over a year's time.

Countless hairpin turns and switchbacks along the route – required to avoid natural obstacles, as well as national monuments, rice fields, temples and sacred cemeteries – made the trip along the narrow road both slow and hazardous. The convoy had to frequently stop to repair overloaded vehicles. Those that couldn't be repaired were stripped, unloaded and were pushed over the side as the men watched with fascination as the vehicles bounced and tumbled into the Salween below.

To negotiate the hairpin turns, the larger trucks had to stop in the middle of the road and back up to the edge of the unfenced shoulder before being able to make the turn around the bend. To complicate the situation, southbound traffic – mostly elements of the Chinese Army – to save precious gas would shut off their engines before coasting downhill. As they careened toward the northbound Frillman convoy, both groups would shout and wave frantically for the other side to get out of the way. On one occasion, the Chinese fired rifles and submachine guns over the heads of the A.V.G. convoy. Frillman's men retaliated with their own weapons at the next encounter.[33]

The second convoy from Rangoon, which left at noon on February 28, finally caught up with Frillman's group before they arrived at Paoshan, half way between the Salween and Mekong River gorges. Headed by Harry Fox, the group included Charlie Sawyer, 1st Squadron flight leader and Doc Richards, the flight surgeon. Their vehicles, added to the first convoy, made up a party of some 50 vehicles. Frillman noted that one vehicle was conspicuously missing, however – the second of the three new Buicks he had commandeered before leaving Rangoon. It had been pitched over the side of the gorge after developing mechanical trouble. The large contingent of vehicles made for a very conspicuous target from the air, and Frillman was understandably nervous about staying in one spot too long. Indeed, a Japanese reconnaissance aircraft had been observed overhead while they were on the road (likely a reconnaissance Mitsubishi Ki-51 from the 71st *Sentai*). Parked at a Chinese military airstrip outside the city, they had a scare when 18 Japanese bombers appeared overhead. Ignoring the military target, however, bombers dropped their loads on the defenseless city.[34]

Hastening to "get out of Dodge," the convoy continued their way north to enter the Mekong gorge. Along the way, they encountered another obstacle – thievery. Several ex-A.V.G. personnel – some who had earlier resigned, some who had been dishonorably discharged – had remained in Rangoon to follow the black-market trade. Several of these shady entrepreneurs left Rangoon on the heels of the A.V.G. convoys and had been shadowing the group the whole way.

The renegades, all of whom were known to the A.V.G. drivers, would gather to visit with the convoy members when the convoy stopped for the night. As the convoy continued on toward Kunming, Frillman noted that, "…I couldn't be sure, but the cargo seemed to be shrinking."[35]

17 days after departing Rangoon, the convoy pulled into Chennault's headquarters compound just outside of Kunming around ten at night. After nearly three weeks on the road, they fell into bed in the comfortable quarters the Chinese had arranged for the A.V.G. there in a former university building .

Reporting to the Old Man after breakfast, Frillman was surprised to learn that Chennault had already heard about the missing cargo and loss of some of the vehicles in the gorges. According to Frillman, although Chennault had greeted him cordially, he had some "sharp questions about the lack of an inventory… and the black-market intrigues."

Later that day, to redeem himself, Frillman offered Chennault the last remaining Buick he had brought from Rangoon – which was accepted without comment.[36]

CHAPTER 14

Attacks on Magwe

Third Squadron leader Arvid "Oley" Olson officially assumed command of the A.V.G. contingent at Magwe on March 12, when he led a flight of three from Kunming.[1] The following day a C.N.A.C. DC-2 brought in twenty more Third Squadron pilots and ground crew, including pilots Dick Rossi and Frank Swartz. On the 13th, the remaining First Squadron personnel boarded the DC-2 back to Kunming. "Kunming looked like heaven after dusty, humid Magwe," Charlie Bond commented. Later that day, he went to Hostel No. 1 to pick up the new A.V.G. insignia designed by artist Hank Porter of the Walt Disney Studios in Hollywood – a winged tiger leaping out of a blue "V" for Victory." The insignia had been requested by the China Defense Supplies office in Washington back in October. "A clever little insignia," said Bond.

When Bill Reed arrived on March 14 from Lashio in Duke Hedman's P-40 No. 92, Oley Olson had a total of twelve pilots – Dupouy, Jernstedt, Groh, Hodges, Reed and himself from the Third Squadron; augmented by Rossi, Prescott and Wolf from the First Squadron, and Moss and Swartz from the Second. While the pilots stayed in an abandoned house, which also served as a communications station, the ground crews traded their earlier tents for another house near the field.

While the Adam & Eves enjoyed relative inactivity during their two weeks at Magwe, with the rapid advance of the Japanese into southern Burma, Olson's Hell's Angels almost immediately began providing more escort for R.A.F. bombing missions and patrolling Japanese troop movements – essentially reconnaissance missions – after which they would land at Toungoo, where the Chinese 200th Division still had their headquarters, to report any usable intelligence.[2]

The Third Squadron began operations at Magwe on March 14 with a reconnaissance mission to Kyaikto, 70 miles northeast of Rangoon. Led by Olson, the flight of four had been tasked to determine the damage done to the railroad bridge across the Sittang River, some 22 miles north of Kyaikto, which had been hit by R.A.F. Blenheims earlier. Finding the bridge still down, Olson led the flight on a strafing run on 20-30 vehicles in the Japanese truck and armored-car compound before heading to Toungoo for refueling and intelligence reporting.[3] Although the flight was unscathed during their strafing attacks, upon landing back at Magwe, Ken Jernstedt hit an unoccupied jeep that was parked on the runway, putting the P-40 on its nose.[4]

The A.V.G.'s first aerial combat since February 26 took place on the 15th, when Parker Dupuoy led a flight of four to Martaban, where they strafed Japanese troops and vehicles near the village of Nyaunglebin. As they headed for Toungoo to make their report, Ken Jernstedt spotted a flight of Nakajima Ki-27 Nates to the east and 1,000 feet above. Attempting to warn his flight leader, he found his radio "dead" (a common occurrence with the cheap, commercial radios supplied to the A.V.G.). Firewalling his throttle, he pulled ahead to signal Dupouy visually.

Cliff Groh, Jernstedt's wingman, had fallen back due to a sluggish engine. Puzzled at his leader's sudden action, he was oblivious to the Japanese Nate that was now moving up behind him. Fortunately, moments before, Moose Moss had spotted four Nates above and behind the one at Groh's six-o'clock. Calling out a warning on his radio, he climbed into the upper four and fired a long burst at the furthest one to the right, which caught fire and dived away. The surviving three immediately turned to attack Moss. Firewalling his engine, Moose soon outdistanced the slower, fixed-wing Nates and later recovered at Toungoo. Meanwhile, finally seeing the Nate behind him, Groh also dived away, successfully pulling away from his pursuer.[5]

On the morning of the 16th, Olson took a flight of five – with Prescott, Hodges, Wolf and Reed – to join with six Hurricanes on a Blenheim escort mission down the Irawaddy River to Prome. Reed later commented, "We watched them drop their bombs, but they didn't hit anything. All five of us following each other 'in trail' strafed a river steamer, a big barge, and then on our way back up the river some smaller craft." [6]

On the 17th, Olson directed Bill Reed and Ken Jernstedt to take a look at the Japanese base at Moulmein and reconnoiter the traffic up to Kyaikto.[7] The mission would also give Third Squadron armorer Chuck Baisden an opportunity to test his new experiment. Since the P-40B Tomahawks had no under-wing bomb racks (unlike the later P-40Es that had both under-wing bomb racks as well as shackles for an external centerline fuel tank), but did have two flare racks to accommodate emergency flares, the 18-year-old armorer had the bright idea of loading incendiary bombs into the flare racks. Obtaining a cluster of British incendiary bombs from the R.A.F. stocks, he modified three to fit into the flare racks and got Jernstedt's approval to give them a try on his ship. At the same time, he installed Canadian concussion hand grenades in Bill Reed's P-40.[8]

Taking off at dusk, the two flew to Toungoo where they refueled and spent the night. Their old training base had been hit hard by the Japanese in two bombing raids the previous day; but, although the village of Toungoo was devastated, the runway at their old training base at Kyedaw was largely undamaged.[9]

Taking off half an hour before dawn on the 18th, Jernstedt and Reed headed south for Moulmein. Flying over the Gulf of Martaban, they observed no shipping, so then turned overland toward their planned target. Less than five minutes later, Reed spotted a satellite field lined with thirty or forty planes. Signaling for an attack, Reed dove with Jernstedt following.[10] As reported by Jernstedt:

> *"... Planes were parked rather close together on either side of this one runway. We each made six passes. On my first pass, I set a fighter on fire and scattered some lead up the line of planes. On my second pass, a transport plane folded up on the ground after receiving the main part of my fire. On my third pass I hit several planes but nothing happened to them except that pieces flew from same. My next pass resulted in one bomber being set afire and other planes being hit. On my fifth pass a fighter crumbled to the ground with parts flying from the engine. During my last pass on this field, I saw at least 5 large fires. We then proceeded towards Moulmein and saw the main hangar with at least 30 more planes on it, mostly bombers. I passed over the hangar on my approach and tried to hit it with an incendiary bomb, but I missed the hangar and hit an Army 97 bomber parked in front of the hangar. This ship was soon ablaze. Reed made one more pass. AA guns shot at us here, the shells bursting as low as 200 feet. Each of us had a lone bullet hole in our planes, but where we picked them up, I don't know. After Reed had made his last pass I saw two planes taking off. I turned and headed west. We left his field with 5 planes ablaze. I claim: 1 transport, 2 bombers, 4 fighters. It would be hard to estimate the amount of damage done to other aircraft, but I feel it was considerable."* [11]

On his six passes, Reed claimed the destruction of one transport, five fighters, and two bombers. On the way back, they strafed a boat, a Japanese staff car and some boxcars on a railroad siding, and Jernstedt dropped Baisden's remaining incendiary bombs on a Japanese compound. Landing at Toungoo once again for refueling and leaving an intelligence report with the Chinese headquarters, they returned to Magwe. Tallying up the day's effort, they counted two transports, nine fighters and four bombers destroyed. Bill Reed recorded in his diary that night, "A pretty good day's work, so Jernie and I took the rest of the day off." [12]

The next day, Parker Dupouy and Cliff Groh flew to Pyu and strafed several trucks and five tanks. The R.A.F. followed early on the 21st with a major strike force to Mingaladon – nine Blenheims from No. 45 Squadron, escorted by ten No. 17 Squadron Hurricanes. Although the Japanese intercepted them some 40 miles north of Rangoon, all the attacking aircraft made it through to the target, accounting for the destruction of sixteen aircraft on the ground at Mingaladon and two fighters shot down by the Hurricanes. One Hurricane was lost in combat on the way back to Magwe.[13]

Even as the R.A.F. aircraft were returning to their base, the Japanese were mounting a huge retaliatory surprise raid designed to completely eliminate the Allied air strength in Burma – over 150 bombers and fighters from nine *sentais* and six different bases. The Japanese had learned on March 9, from high-flying photo-reconnaissance aircraft, that the R.A.F. and A.V.G. had evacuated to Magwe. But it had taken nearly two weeks to gather sufficient strength for a knock-out raid on the base.[14]

The R.A.F. Blenheims of the returned No. 45 Squadron were already refueled and loaded with bombs, while their crews were in their ops tent briefing for a second mission. At 1245, a call came in from a ground station east of Magwe, "Bandits from the southeast at angels fifteen!"[15] The R.A.F.'s sole radar had been pointed south toward Rangoon. A minute later, Dupouy, Jernstedt, Groh, Prescott and Hodges raced to the available Tomahawks, joining six R.A.F. Hurricane pilots of No. 17 Squadron cranking their engines, while Bill Reed and several ground personnel ran for the slit trenches.

As Dupouy and Jenstedt climbed out, reaching 13,000 feet, Jernstedt was the first to spot the incoming Japanese force – 26 twin-engine Mitsubishi Ki-21 Sallys, escorted by 14 Nakajima Ki-43 Hayabusas . By that time, however, the bombers were already over the field and dropping their loads with surprising accuracy, considering the cloud cover over the field. Too late to prevent the raid, the pair continued to climb to 23,000 feet and chased the bombers for 20 minutes before they were able to make an attack. Then, as described in Dupuoy's combat report:

> *"...I made an underneath rear quarter attack on the end bomber without noticing any noticeable effect. After the third similar attack, I saw a lone fighter Model "O" circling in behind me. I headed for him and then saw seven more Model "O" slightly above me starting down. I turned toward the nearest one and fired at him head on. His engine caught fire before he passed me and one .30 calibre from his gun entered my cockpit directly below the wind shield on the left side. I dove straight for the ground and the others followed for a few thousand feet. I saw the Model "O" on fire crash land in the trees about 30 miles northwest of Toungoo."* [16]

But they weren't "Model 'Os'." This was common error reported in quite a number of pilot's combat reports. They were similar in appearance – both were single-seat fighters with radial engines and bubble canopies; and they were dimensionally similar – wingspan for the Mitsubishi A6M5 Model 0, 63' 1;" for the Nakajima Ki-43 Hayabusa (Oscar), 35' 6 ¾." Both were 29' in length. However, the Japanese Navy's Mitsubishi didn't appear against the A.V.G. during their existence, while the I.J.A.A.F.'s Ki-43 was one of the Tiger's main opponents.

During the battle, Jernstedt was hit by a Sally gunner, shattering his canopy. With shards of Plexiglas in his eyes, he was out of the fight.[17] Turning back, "with more luck than judgment," as Jernstedt later put it, "I hit Magwe on the nose on my return trip." Bleeding from glass slivers in his face and left eye, he landed his Tomahawk successfully amid chaos and confusion on the field. Pulled from the cockpit by the ground crew, he was rushed to the hospital to be attended to by Doc Richards. Unable to write out a formal combat report, he did claim two bombers shot down. He was flown to Kunming the following day for treatment.

Meanwhile, the other three A.V.G. defenders – Groh, Prescott and Hodges – were climbing out toward the southeast to intercept the 28 Sallys of the 98th *Sentai*. They met the Japanese southwest of the field and then chased the bombers for 30 minutes toward Toungoo before engaging. Groh and his wingman, Prescott, made two passes, damaging one Sally, before turning for home.[18] Flying at 4,000 feet, Groh spotted a lone, southbound Ki-27 Nate four miles to the east. With a wing-over to the left, Groh was soon behind the Japanese fighter for positive identification. From his combat report:

> *"...I got behind it without being seen and saw the markings on the wings...I dived underneath the Jap and turned towards the left...I gave a deflection shot and fired several bursts before he turned towards me...I kept firing burst after burst until just before I passed him...I turned and saw the Jap crash in the vicinity of the fires (forest) that is SE of the Magwe airport about 15 or 29 miles."* [19]

The Japanese attack came in three waves, over a nearly two-hour period. The first wave, twenty-five 98th *Sentai* Sallys , escorted by 31 Nakajima Ki-27 Nates from the 1st and 11th *Sentai*s struck from the southeast. The second wave – twenty-six 12th *Sentai* Sallys, with their escort of fourteen 64th *Sentai* Hayabusas – came from the northwest. During that attack, a Hurricane pilot, with his engine on fire, attempted a landing on the field. As he skidded to a stop, his plane burst into flames. Crew chief Johnny Fauth and Henry Olson jumped from the jeep they were riding in and ran to the blazing Hurricane. After pulling the injured pilot out of the cockpit, they laid him in the waiting jeep and Olson drove away. Just as Fauth was running to a slit trench for cover, a Hayabusa came across the field on a strafing run, and he was hit in the shoulder by a 7.7mm slug. Frank Swartz, jumping from his trench, ran to Fauth's aid just as a bomb hit not five yards between them. The shrapnel shattered Swartz's jaw and nearly tore Fauth's arm off. Crew chief W.R. Seiple, who was in a nearby trench, suffered internal injuries from the concussion of the blast. All three were rushed to the field hospital between bomb runs, where Doc Richards did all he could for the injured men.[20] There was little help for Johnny Fauth, however; the gravely-injured crew chief died at 0430 the following morning.

Ground-to-air defenses at Magwe were virtually nonexistent – save for the improvised return fire of several people on the ground. Following an alert on the 20th, Chuck Baisden set up a British Bren gun anti-aircraft mount in a shallow indentation in the ground just off the runway, commenting later, "I was prepared if a fighter came in strafing the field." As the first wave of bombers hit the next day, Baisden continued:

> *"I grabbed my Bren and jumped into a Jeep drive by 'Buck' Rogers... The bombers dropped their load, but I saw a Jap fighter coming in to strafe. He was within one-hundred yards of me, but all I could do was hold the Bren up and fire a full clip at him because I had left the anti-aircraft mount on the other Jeep with all my extra ammunition. He fired his machine guns, dropped some anti-personnel bombs and flew off."* [21]

Other A.V.G. ground crews jumped out of their slit trenches and blazed away with whatever weapons they had, from Colt.45s to Thompson sub-machine guns.[22] Joe Sweeney, Keith Christensen and Manny Wakefield all fired on a single Nate as it flashed by. It seemed to falter, and Wakefield later claimed a "probable." [23]

The bombing and strafing lasted until 1430, when the last Mitsubishi Ki-30 Ann single-engine light bomber departed the area, heading back to Moulmein. The Japanese had scored big – runway cratered; all buildings destroyed or badly damaged; hangars obliterated, one with seven R.A.F. mechanics inside; and eleven Blenheims parked on the edge of the runway, loaded with fuel and bombs, were completely destroyed, some of the munitions in their still-smoldering carcasses cooking off hours after the raid. Every aircraft on the ground was damaged to some extent.[24]

Oley Olson and Duke Hedman arrived from Lashio just before dark, landing between bomb craters. Already the A.V.G. crew chiefs were hard at work repairing what they could of the Third Squadron's dwindling fleet. They estimated that, working all night, they might have six flyable P-40s, including the two that Olson and Hedman brought in, ready to go by morning. That evening, Olson was handed a radiogram:

> LOOK OUT FOR FOLLOW UP RAID TOMORROW STOP DID YOU HAVE ANY WARNING OF RAID TODAY QUESTION CHENNAULT [25]

The maintenance crews were cleaning up the debris and filling in holes in the pock-marked runway when, as Chennault predicted, the Japanese returned at 0845 on the 22nd. The force of 23 bombers – a combination of Kawasaki Ki-48 Lilys and Mitsubishi Ki-30 Anns – followed by 61 Nates struck in two waves, fifteen minutes apart. This time, with the R.A.F.'s unreliable RDF unit already dismantled for evacuation, the first warning those on the ground had was the sound of a storm of radial engines overhead. With no time to even attempt a takeoff, the A.V.G. pilots and ground crew raced for any available shelter they could find, or jumped into whatever vehicles could be driven away.[26]

In the lull that followed, Johnny Fauth was buried in the nearby Catholic cemetery, and a C.N.A.C. Douglas DC-2 arrived to transport Frank Swartz and Wilfred Seiple to Calcutta, India, for further treatment for their wounds. Frank Swartz died a month later in the hospital at Poona, India, on April 24, of a "compound fracture of the jaw and acute encephalitis." He was buried in the cemetery of St. Sepulchris in lot BC7. The DC-2 got off the ground just in time; a few minutes later, at 1230, the third wave of Japanese bombers and fighters – 53 Sallys accompanied by 23 Nates and 18 Hayabusas – arrived overhead. As with the first wave, there was no warning, and no A.V.G. aircraft got off the ground. The raid completely finished off the Allied air strength in Burma. Every aircraft that survived the previous raids was hit again. Virtually all of the remaining Blenheims of No. 45 Squadron were destroyed and only eight Hurricanes were salvageable. One remaining P-40 went up in flames and four more were badly damaged.[27] Olson reported:

> RAID ON AERODROME 1230 STOP 80 HEAVY BOMBERS 20 FIGHTERS STOP TOO SHORT WARNING NO SHARKS IN AIR STOP ALL SHARKS HIT STOP POSSIBLY FLY THREE DAMAGED STOP PROCEEDING WITH CONVOY TO LOIWING STOP RADIO WILL OPERATE UNTIL 10PM TONIGHT STOP END OLSON [28]

That message was soon followed by confirmation from Chennault to "begin movement squadron to Loiwing without delay," but to "move radio station last." [29]

Armorers Chuck Baisden and Keith Christensen began removing the machine guns from the unsalvageable P-40s, including R.T. Smith's No. 77, which Dick Rossi had crash-landed following engine failure on takeoff. After loading the rest of the armament equipment onto trucks, they joined the first contingent of the 60-vehicle convoy north on the Burma Road to Loiwing. The first group stopped fifty miles up the road.[30]

Meanwhile, as the evacuation of Magwe was beginning, back at Kunming the first four brand-new P-40Es arrived. George McMillan, R.T. Smith, Link Laughlin and P.J. Greene had made the 20-day ferry trip from Accra, on the Gold Coast of Africa. Chuck Older and Tommy Haywood would follow with two more in the next few days.[31]

Working throughout the night, armorer Joe Poshefko and line chief Frank Van Timmeren managed to get four aircraft flyable for the 300-mile ferry flight to Loiwing. At dawn on the 23rd, Fritz Wolf, Bob Prescott, Freddy Hodges, and Moose Moss took off for their new base. After the ships were safely airborne, Poshefko and Van Timmeren joined the convoy to Loiwing.[32] Olsen put Bill Reed in charge of the convoy, while he and Parker Dupouy waited to "chase up the stragglers," as Olsen put it. He described the first day's journey in his diary that night:

> *"...Had some trouble with breakdowns but most of us made Maymyo by night. Dupouy, Overend, Rossi and I stayed at the Baptist Mission which was being used by members of Gen. Stillwell's staff... Talked to Gen. Gruber about what happened. Drained a bottle of Crème de Menthe I bought in Yenangyaung with the mob before I left."* [33]

CHAPTER 15

"Hell of a Nice Mission..."
– *Robert Neale*

In the deadly tit-for-tat of aerial warfare, it was now Chennault's turn. Recovering from another bout of bronchitis (Chennault was life-long chain-smoker), he began devising a retaliatory raid against the Japanese even before the follow-up Japanese raid on March 22. From reports of earlier reconnaissance missions into northern Thailand, aided by R.A.F. intelligence reports, Chiang Mai, 165 miles due east of Toungoo, and Lampang, 50 miles southeast of Chiang Mai, were ripe targets.[1] (Unbeknownst to Chennault at the time, two of the *sentais* that took part in the raids of the 21st were launched from the two bases – the 12th *Sentai* Sallys from Lampang and the 64th *Sentai* Nates from Chiang Mai).

At 0800 on the 22nd, Chennault called Bob Neale in to give him the order he had prepared:

> *"Strong concentrations of enemy aircraft have been reported in the Chiang MaI/Lambhun/Lampang area of Thailand...One flight of five airplanes each from the First Pursuit Squadron and the Second Pursuit Squadron will attack Japanese air bases at Chiangmai and Lambhun, Thailand, on the morning of March 23rd, 1942. Objective: Enemy aircraft on the ground and/or in the air at Chiang Mai – Lambhun. Squadron Leader Neale of the First Pursuit Squadron, leading "A" Flight is designated Group Leader of this mission and will issue detailed instructions covering same..."*[2]

Leaving Chennault's office, Neale thought, "Hell of a nice mission, as Chiang Mai is 165 miles inside Jap lines." Looking in on Charlie Bond, he told him to get ready, "We're going to make them pay." Soon the ten ships were made ready at Kunming's Wu Chia Ba airport. At 1300, even as the Japanese were hitting Magwe once again, including Neale, with First Squadron pilots Greg Boyington, Billy McGarry, Charlie Bond and Bill Bartling, along with Second Squadron Leader Jack Newkirk and Second Squadron pilots, Ed Rector, Hank Geselbracht, Bob Layher and Bus Keeton, took off for Loiwing. Even though he was the nominal mission leader, he graciously passed flight lead to Newkirk, since he had never flown to Loiwing and thought Newkirk was familiar with the CAMCO airstrip. Even so, there were some problems, as described by Charlie Bond in his diary:

> *"We wandered all over the sky until we hit the Salween River. At this point Jack turned south when he should have turned north. I was convinced Jack was wrong and I checked everything I could see. Bob, Bartling, and I came into Loiwing after trying to get Jack to turn back north."*[3]

Landing at Loiwing at 1515, Bond came in too fast and ran off the over-run at the end of the landing strip, blowing out his tail wheel tire. That was just the first of their problems, though. Expecting a short refueling stop, they found no refueling or maintenance personnel on hand at Bill Pawley's CAMCO-operated airfield. Jack Newkirk and his errant flight straggled in later and Neale decided to stay at Loiwing overnight, postponing the mission by 24 hours.[4]

The A.V.G. pilots made the most of their delay. Eight miles from the airstrip was Pawley's CAMCO operation. There, where perhaps 30 Tomahawks were in various stages of repair or rebuilding, Pawley had built a nine-hole golf course, a club house complete with a wood-burning fireplace, juke box, motion picture projector and a well-stocked bar with refrigerator. A Mrs. Davidson, whom the residents called "Ma," was in charge and ran a smooth operation. Charlie Bond wrote, "We sit in overstuffed chairs and eat great food. There are three huge windows which overlook the beautiful river valley and provide a panoramic view of the distant mountains."[5] Bob Neale called Pawley's headquarters "a wonderful place to stay. Had some cold beer from States." [6] That night they enjoyed another treat – real Simmons mattresses![7]

The next morning, the pilots were up at 0530 for their pre-dawn ride out to the airstrip. With the fog and drizzling rain, Bond thought, "at least we don't have to worry about a Jap strike in this weather." Arriving at the field, Neale put his flight on alert. Later that morning, despite the poor weather, the four surviving P-40s from Magwe came in under the ceiling and landed safely.[8]

Mrs. Davidson drove out to the field around noon, bringing the waiting pilots "a swell lunch," as Bus Keeton put it. Neale kept the flight on alert until 1515, when they took off for the R.A.F. base at Namsang, 220 miles due south. Charlie Bond, flying on Neale's wing, still remembering the navigational problems on the flight to Loiwing, said he, "kept my thumb on my map for my own navigation." [9]

Arriving at Namsang at 1645, Neale had the aircraft refueled immediately and dispersed the ships around the field in two groups, one at the end of each of the two intersecting runways. He then arranged with the R.A.F. to provide five trucks on the left side of each of the two runways, with kerosene lanterns interspersed between the trucks, to provide lighting for the planned pre-dawn take-offs.[10]

After a dinner at the R.A.F. mess, Neale gathered his pilots for a last-minute review of their plans. "We'll take off at 0545," Neale began. "I'll be the first off, with Charlie on my wing, with Bill and Pappy as three and four. Ed, you and Billy will follow as the top cover element. Then Jack, your flight will follow Ed and Billy. We'll join up at 10,000 feet over the field. I'll lead both flights to Chiang Mai, on a heading of one-five-zero. When we reach Chiang Mai, Jack, you'll break off and lead your flight of four on a low-level strafing attack at the airport at Lambhun, just twenty-five miles due south.

Turning to Rector, Neale said, "Ed, you and Billy will hold high as top cover, in case any enemy fighters show up. If none, come on down at your discretion and join the fun." Turning once again to Newkirk, he said, "Jack, if you don't find any aircraft at Lambhun, come on back to Chiang Mai and join the rest of us. After you've expended your ammunition, take your flight back to Loiwing via Heho. I'll take my flight back via Namsang. Any questions?" [11]

There were no questions, but Charlie Bond had had a long discussion with Neale the day before regarding the necessity of taking two of their six for top cover at Chiang Mai. He felt that, with the low-level dawn attack, the element of surprise would negate the need for top cover, and the two aircraft could be better used in the initial strafing attack. Apparently Neale's plan was set, so he said no more about it.[12]

The mood that evening was somber, however. An R.A.F. sergeant warned the men not to drink the local water, not to even brush their teeth, for it was polluted. With a wry smile, Jack Newkirk muttered that, "after tomorrow, I don't think it'll make any difference anyway." Bus Keeton echoed Bob Neale's earlier reservations for the mission in his diary that night, commenting, "…our mission…to my way of thinking is the most dangerous undertaking the A.V.G. has done, going 120 miles into enemy territory where if you have to force land and don't get you the jungle will." In Bob Neale's last diary sentence for the day, he wrote, "Here goes nothing." [13]

Morning on the 24th came early, as a Burmese native woke the pilots at 0405. A few minutes later, an R.A.F. officer burst in with, "All right, you curly-headed fellows, it's time!" By 0525 they were in their cockpits warming up the Tomahawks' engines.

Neale was the first off, at 0545, followed by Bond. As Bond recalled:

> *"I aimed at the truck lights, and felt my way along the grass strip until the wheels felt light. A little bit of back pressure on the stick and I was free of the ground. I couldn't see a thing … A few seconds later I relaxed and looked up for Bob's navigation lights. There his plane was, slightly to the left and above me. I followed and wondered how in the hell those other guys will make it to rendezvous over the field at 10,000 feet."* [14]

In fact, they didn't make it to the rendezvous. Taking off from the intersecting runway, Newkirk's flight of four leveled off inexplicably at 6,000 feet. Whether it was because of the haze, pre-dawn darkness, or the 4,000 differential in altitude – or because their two-bit radios weren't operating – they never made contact with Neale's flight. Then, with a few minutes head start, they headed for Chiang Mai. Unable to contact Newkirk, after circling Namsang for twenty minutes, at 0630 Neale turned his flight on a heading of one-five-zero for their Thailand targets.[15]

At 0710, Newkirk's flight was over Chiang Mai. Although Newkirk's briefed instructions were to fly on to Lamphun, his primary target, he spotted the town's railroad station through the haze and led his flight down for a strafing run. "We followed, but did not fire," Keeton later wrote.[16] Flying low over the railroad tracks between Chiang Mai and Lamphun, they observed a row of 15-20 closely-spaced warehouse buildings and all opened up on the "targets of opportunity," setting them ablaze. Frank Lawlor continues with a description of the attack in his combat report:

> *"We approached Lambhun making a wide circle. We flew over two satellite fields but no airplanes. A third and larger field also had no planes but the area surrounding the field had several barracks. These we strafed and I noticed one was smoking. No enemy fire was noticed during the period up through this attack. Immediately after leaving the area of the barracks Squadron Leader Newkirk headed back up the road to Chiang Mai. He went into a dive to strafe what appeared to be an armored car and I saw his plane crash and burn."*[17]

Bus Keeton, following Lawlor, reported:

> *"...Jack dove and fired on a R.R. station. We followed but did not fire. We then proceeded south and dove on some barracks. All fired but me. I didn't think the target warranted the waste of ammunition which might come in handy later. We then dove on some store houses probably setting them all afire, was about 20 in all, and smoke was coming out of quite a few when I made my run. As I pulled up to the right, I noticed a large flame of fire burst upon a field to the right of me." The fire spread along the field for 100 or 150 yards,"*[18]

Although it's impossible to determine whether Newkirk was a victim of "target fixation," or whether he was hit by ground fire from the armored car, his shallow diving angle caused the flaming wreckage of his disintegrating Tomahawk to spread the length of almost two football fields along the Chiang Mai-Lamphun road when he struck the ground.

Meanwhile, at 0714 Charlie Bond was the first in Neale's flight to recognize the towering mountains that marked the Chiang Mai area. Bond remembered them from an earlier reconnaissance mission that the air field was at the southeast corner of the mountain. Maintaining radio silence, Bond moved the throttle forward and rocked his wings, signaling to Neale that he would take the lead. As Neale dropped back, Bond led the flight in a gentle left turn, and at 6,000 feet they spotted the field through the haze just as dawn was breaking. Continuing their descent, as the flight passed through 5,000 feet, Rector and McGarry leveled off to assume their pre-briefed role as top cover.

On the air field, the Ki-27 Nates of the 64th *Sentai* were parked wingtip-to-wingtip in neat rows along the field, with mechanics running up some of the engines and pilots beginning to climb into cockpits in preparation for another early morning mission. Commenting later, Bond said, "Hell, I hadn't seen this many aircraft in years. Seemed like the whole Japanese Air Force had tried to crowd into this one little field. I couldn't miss."[19]

Diving down to 1,000 feet, Bond led the string of Neale, Boyington and Bartling on their first strafing pass, raking the parked fighters – and three twin-engine types – from south to north. Following the first strafing pass, the flight broke up into individual strafing attacks. Banking sharply to the left, Bond made a 270-degree turn to line up on a second row of Japanese aircraft parked perpendicular to the first line. Boyington, number three behind Neale, reported, "After the initial dive I was unable to follow the flight of the other three allied planes due to poor visibility." As the last man in the string, Bill Bartling made a quick wing-over, reversing his path for a second attack. Describing the action in his combat report, he wrote:

> *"At the end of our first pass, I could see a fire which had all three transports in it, also 3 other fires, 2*

of which had burst into flames from my incendiaries, also one Model "O" I had riddled and put completely out of commission ... I made a quick wing-over, passing 500 feet alt. and came right back over the line. Noticed another line at 90 degrees to it and strafed down that line. At end of those two passes I saw 7 fires, 5 in the first line and 2 in the other line. There were 2 Model "O"s in each of the two fires of the 2nd line. Also one Model "O" riddled but not on fire. I made a quick turn and noticed 4 planes on other side of field, apparently I-97's, set on fire, and riddled another..." [20]

On Bond's second pass, which he made from west to east, he said, "I was so low that it seemed as though my prop tips would clip the heads of Jap pilots who by now were trying to get into their cockpits." As he looked back, he could see three more aircraft in flames. Neale followed Bond in the west-to-east attack on the aircraft parked on the north side of the field, then banked sharply to return to the first line of Nates. Bond's third pass took him to the southwest corner of the field where he riddled another Nate with a several-second burst of his guns.[21]

By this time, the field's anti-aircraft defenses had begun to open up. Bond later reported, "The anti-aircraft defenses had come to life, but they still were frantic and disorganized."[22] Puffs of black smoke began to appear above the field. Ed Rector, flying with Billy McGarry at their 5,000-foot assigned top cover altitude, reported, "A.A. shells were bursting in ever-increasing numbers on mine and McGarry's level, so I began flying a zig-zag course as some were bursting uncomfortably close." [23]

While Bond was starting his fourth pass, Bob Neale noted that "...the anti-aircraft and machine- gun fire from the ground batteries was so heavy I rolled my wings in an effort to indicate completion of the attack." Climbing to 5,000 feet, he circled at five miles southwest of the field, waiting on the flight to rendezvous.[24]

At 0725, when none of the rest of the flight appeared, he struck out on a course for Namsang. En route, Boyington joined him before they arrived at Namsang at 0820.[25]

During the eight-minute attack – from 0716 to 0724 – Bond had noticed only one other P-40. As he came off his fourth pass under a hail of anti-aircraft fire, he saw three P-40s above him – Rector, McGarry and Bartling. Climbing out, he joined the other three at 2,500 feet and assumed the lead, with Bartling on his wing. Rector and McGarry were close behind.[26]

As Bond's flight of four was climbing out, Frank Lawlor and his flight were approaching Chiang Mai from the south. After flying past the flaming wreckage of Jack Newkirk's P-40, Lawlor had assumed lead of the flight and headed north toward Chiang Mai. By this time, Neale's flight had departed and was on their way to Namsang. Over-flying the airfield, Lawlor noted several large fires burning. Flying through a barrage of heavy anti-aircraft fire, he turned his flight north toward their preplanned refueling stop at He-Ho, arriving at the R.A.F. auxiliary base at 0830.[27]

Meanwhile, as Bond's flight approached the Salween River, Billy McGarry's P-40 started falling back and slowly losing altitude. Keeping his head on a swivel as he watched for pursuing Japanese fighters, Bond noticed that his "tail-end Charlie" was dropping behind and rapidly rocking his wings. Bond and Bartling immediately circled around and flew alongside the stricken Tomahawk. Smoke soon began to appear from his engine as McGarry rolled the canopy back. (In a post-war interview, McGarry told historian Robert Andrade that his engine had been hit by ground fire during a strafing run, and when the cockpit became too hot, he rolled the canopy back and prepared to bail out.) Suddenly the aircraft rolled over and McGarry dropped out. His parachute opened immediately and he landed in a small glade, only 300 feet from where his P-40 nosed in.

Apparently OK., McGarry was on his feet and waving as the three remaining P-40s circled overhead. Bond drew a circle on his map to pinpoint where McGarry came down – approximately 60 miles northwest of Chiang Mai – and, making a low pass over the area, dropped the map to him. Bartling also dropped a map and Rector dropped a candy bar. (McGarry later said that he found neither, since the foliage was so thick.) After circling once again, noting that McGarry had disappeared into the foliage of the jungle, they climbed out on a heading to Namsang.[28]

It would be almost three years before any of the A.V.G. learned of McGarry's fate. In early 1945, an O.S.S. team rescued him from a Bangkok jail. After his bailout, he had wandered in the Thai jungle for four weeks. "I knew I was going around in circles," he reported later to historian Bob Andrade, "since I kept coming across the same Buddhists." He was finally picked up by a roving Thai police patrol and taken to Chiang Mai, where he was personally interrogated by Lt. Col. Tateo Kato, the Japanese fighter ace and commander of the 64th *Sentai*. Inexplicably, after the interview, Kato turned him back over to the Thais. McGarry would spend the next three years interned in Bangkok as a "guest" of the Thai constabulary.[29]

Neale and Boyington landed at Namsang at 0820, followed by Bond's flight fifteen minutes later. Bond briefed Neale on the loss of McGarry, and Neale in turn briefed the R.A.F. commander on the field, instructing him to pass the report on to the Chinese in the area for their help in organizing a recovery.

With the birds refueled, the six P-40s took off and headed west for He-Ho, expecting to meet Newkirk's flight. Seeing no P-40s on the field (Lawlor's flight had already refueled and taken off for Loiwing, arriving there at 1100), Ed Rector broke off to fly to Lashio, where he would report McGarry's bail-out to the British Army headquarters there. The remaining five flew on to Loiwing, where they made a low-level victory pass across the field before landing at 1140.

Upon landing, Neale was briefed by Frank Lawlor on the results of the raid on Lamphun and learned of the tragic death of Jack Newkirk. Shortly after, they were in the CAMCO clubhouse enjoying lunch prepared by Mrs. Davidson. Bob Neale sent a message to Chennault, reporting the loss of McGarry, the death of Newkirk and the approximate results of the raid:

> *"...Wing Man McGarry was unable to stay with the formation and was losing altitude. At approximately 0741 and at 1,000 feet above the terrain, he abandoned ship, landing safely. His aircraft crashed in flames...Upon arriving at Loiwing, I was informed that Squadron Leader Newkirk had been killed while executing a diving attack on a ground target south of Chiang Mai. Due to the nature of the existing circumstances of the mission, the exact extent of the damage to the enemy is impossible to determine. However, from my own individual observation, together with a study of the pilots' reports, I am of the opinion that the following damage to the enemy was accomplished:*
>
> *Aircraft burning – 15*
> *Aircraft riddled – 9*
> *Many other aircraft damaged to various and unknown extent.*
> *Many buildings and structures and equipment were damaged or left burning."*[30]

Whatever the final number of Japanese aircraft destroyed, it was a pyrrhic victory, as Charlie Bond noted that night in his diary:

> *"But can the A.V.G. afford to lose such men as Jack Newkirk and Mac McGarry and two P-40 aircraft in these times? We wonder..."*

Daily combat gives little time for reflection on individual losses, as Bond continued:

> *"Soon we were in the bar – the whole gang – celebrating the day's work. It's surprising how quickly one gets over the horrible loss of his buddies in wartime and revels in the successes of the day. Everyone was laughing and enjoying the moment. Yet Jack was gone and we weren't sure of Black Mac. It makes one wonder about the nature of human beings."*[31]

After a hurried stopover in Lashio, Third Squadron Leader Oley Olson arrived at Loiwing with his convoy that evening and received the briefing of the day's activities from Bob Neale. Neale was awaiting further orders from Chennault, but, as Olson recorded, "I had to completely organize this field and accommodations so I will not possibly maintain an alert. So in order to not risk the ships I sent them back."[32] Neale led his remaining flight of eight back to Kunming the next day.

CHAPTER 16

Action at Loiwing

By the latter part of March, the Japanese Army had continued their advance into northern Burma, capturing Kyedaw Airfield, the A.V.G.'s old training base at Toungoo, and pushed the British Commonwealth forces out of Prome, 75 miles west of Toungoo. In the week following the Chiang Mai raid, however, a lull fell over the air war between Chennault's Tigers and the Japanese *sentais*.

Chennault, expecting "the next shoe to fall" on Loiwing, began reinforcing the defenses of the vital CAMCO repair facility. He established a number of outlying radio stations, providing radio and telephone communications with the main station at Loiwing, augmented by Chinese Army observers in the surrounding hills. On March 26, Chennault directed:

> *"This Group from Point 'X' [Kunming] will send two flights of four planes each to Loiwing, China, to reinforce the Third Pursuit Squadron… Squadron Leader Olson, in command of this mission will, upon arrival at Loiwing, China, assume command of the Third Pursuit Squadron and will issue detailed instructions covering operations. All planes of this mission will be attached to the Third Pursuit Squadron Unit until further notice."* [1]

Oley Olson had just arrived at noon from Loiwing for a conference with the "Old Man," but was off again for a return trip at 1430, leading his flight of eight, including R.T. Smith, Link Laughlin and P.J. Greene, arriving there at 1600.

That evening, Chennault called a meeting of his pilots in Hostel No. 1. Explaining that he was under no pressure at this time to have the A.V.G. inducted into the U.S. Army Air Corps, he nevertheless wanted to know how many would be willing to transfer to the U.S.A.A.C., should the A.V.G. be absorbed into the U.S. Army. There was a resounding opposition to the idea, with many voicing the opinion that they would resign immediately or certainly go back to the States after their contract was up. Indeed the majority of the original 107 pilots who were recruited came from the U.S. Navy or Marines (52 Navy and seven Marines, in contrast to 48 from the Army), and most intended to rejoin their parent services. The discussion went back-and-forth for an hour before Chennault concluded with, "I will fight to the end to keep us a volunteer group."[2] Later he fired off a radiogram to Madame Chiang:

> CAREFUL CHECK MAJORITY MEMBERS OF A.V.G. REVEALS THAT NINETY (90) PERCENT WILL TERMINATE CONTRACTS RATHER THAN ACCEPT INDUCTION INTO U.S. ARMY AIR FORCES IN CHINA STOP THIS ATTITUDE FIRM DESPITE MY EARNEST PLEAS MAINTAIN GROUP REGARDLESS STOP MEN NOT UNPATRIOTIC BUT ALL HOPE GO HOME BRIEF VISIT BEFORE FINAL ENTRY U.S. ARMY OR NAVY STOP [3]

Operating out of an alert shack with "Olson & Co., Exterminators" above the door, Olson had eight ships on alert at Loiwing before dawn on the 27th. At noon, the alarm sounded – enemy planes were reported 60 miles south, near Lashio. R.T. Smith, Parker Dupouy, Freddy Hodges, P.J. Greene and Link Laughlin raced to their Tomahawks, took off and climbed to 23,000 feet heading south. Finding an empty sky, Smith called the ground station at Lashio; they reported no further contacts. Returning to the field, R.T. caught sight of two aircraft 5,000 feet below. Diving down, he identified the two as P-40s, flown by Duke Hedman and Bob Brouk who, called from the mess tent while having lunch, had been late in getting off the ground. As R.T. said, "Just a false alarm, I guess." [4]

As the ground war raged in northern Burma and the Japanese steadily advanced against the Chinese Army and remnants of the British Commonwealth troops, Toungoo was at the northern edge of the battle front. Lost and temporarily regained by the Chinese, it was the center of the Japanese assault. Seeking to bolster the morale of his fighting troops holding the line against the Japanese, Chiang Kai-shek exhorted Chennault to have elements of the A.V.G. conduct low-level flights over the Chinese positions on the battle front, so low that the Chinese troops could easily discern the 12-pointed Chinese star emblazoned on the wings of the low-flying Tomahawks. [5]

On the 28th, now with 12 P-40s at his disposal (First Squadron pilots Greg Boyington and Chuck Older had arrived earlier in the day in two Tomahawks to augment Olson's Third Squadron), Olson passed the word to his pilots that they would be going to Burma to patrol over the lines near Toungoo on "special morale missions." With Toungoo over 300 miles from Loiwing, it would mean a refueling stop at HeHo before venturing into Japanese-held territory with their attendant enemy air fields in the vicinity – all for the dubious benefit of "troop morale." R.T. Smith voiced the concern of his fellow pilots that night in his diary:

> *"We're all against it andf it seems senseless, with a good possibility of nobody coming back. This is to be just one day, but it's mighty low grade, Lord, mighty low."* [6]

Early the next morning, Olson led eight P-40s on the first "morale" mission, leaving four on alert with R.T. Smith in charge. They were back an hour later, having encountered deteriorating weather en route. As luck would have it, they couldn't have landed at HeHo anyway – it had been bombed earlier by the Japanese, knocking it out as a refueling stop.[7]

After lunch that afternoon, the alert was sounded that an enemy observation plane had been sighted to the south, heading for Loiwing. Chuck Older and P.J. Greene took off at 1345 and climbed to 27,000 feet. After circling for 10 minutes, Older spotted the Japanese reconnaissance ship to the side and 4,000-5,000 feet below. From Older's combat report:

> *"I waggled my wings for an attack and turned and dove on the enemy and closed to 50 yards. The enemy plane started smoking as I closed in, and burst into flames a few seconds later. One man bailed out of the enemy plane, and the plane started into a dive. The enemy plane disintegrated on the way down."* [8]

Later in the day, a report of another enemy aircraft was relayed to R.T. Smith and Bob Prescott as they patrolled the skies to the south. Visibility was so poor by then – "a haze that seemed to hang like a blanket over Burma," as R.T. described it – that they saw nothing. It was evident, however, that the Japanese were searching for the A.V.G.'s new base of operations, and it would be just a matter of time before Loiwing would be hit.

On Tuesday morning, the 31st, the Japanese hit the R.A.F. base at Lashio. R.T. Smith reported:

> *"This was quite a hectic day. We had an alarm about 9:00 a.m. that bombers were near Lashio and headed this way. We got 8 ships up and patrolled nearby for over an hour, but sighted nothing."* [9]

Shortly after landing, with half the pilots at the mess for lunch, the alert shack radio crackled with the report of a second wave of Japanese bombers over Lashio. Oley Olson and Moose Moss were the first to get off the ground. As they climbed out to 28,000 feet, Moose became separated from his leader, while Olson continued on south. After about 45 minutes, Olson heard over his radio, "many bombers and fighters over Lashio at 25,000 feet." Five minutes later, he saw them – 27 bombers escorted by nine fighters. Discretion being the better part of valor, he climbed into the sun and kept watch on the formation until they turned back south.[10]

Meanwhile, R.T. Smith, Pappy Boyington and three others had rushed to their Tomahawks to follow Olson and Moss. As Boyington advanced the throttle, the Allison engine didn't come up to full power and his P-40 went "sputtering down the runway," as he described later. He barely made it into the air, after using the entire length of the runway, when the engine finally coughed and gave out. With no time to retract the gear for a softer belly landing, Boyington pulled back on the stick to barely miss a 20-foot ditch off the end of the strip and slammed into the ground. Thrown against the instrument panel when his seat belt broke, he suffered severe knee injuries and the gun-sight opened a wide gash along his temple. Crawling out of the smashed cockpit, he watched as R.T.'s flight passed overhead.[11]

Heading south toward Lashio, R.T. and company climbed to 28,000 feet but weren't able to join Olson before the Japanese formation departed. Strangely, the airfield at Lashio was largely untouched by the raid, with the majority of the bombs falling on the city of Lashio itself. After losing his flight leader in the climb-out, Moose Moss recovered on the undamaged field.[12]

The next day, April 1, was, as Oley Olson described it, "April Fools' Day. No fooling around here, however." [13] Indeed, first off, Moose Moss came in from Lashio, but ground-looped on landing. With Boyington's aircraft accident the day before, that was two Tomahawks written off in two days – added to the two lost on the Chiang Mai raid, a loss the A.V.G. could scarcely afford.[14]

Then came a report of an enemy observation ship 60 miles from Loiwing. R.T. Smith and Fred Hodges scrambled to search for it, but only got 20 miles out before giving it up for the poor visibility they had been experiencing for days.[15] (Mysteriously, a combat report signed by Bob Brouk for a April 1, 1942, combat exists in the A.V.G.'s official combat report files. However, Bob Brouk's published diary makes no mention of the report, nor does the official A.V.G. victory credits list.) In the combat report, he stated:

> *"I was leading the second element at 18,000 feet Prescott on my wing. I saw the enemy pass under me flying a course 60 deg. To my right about 4,000 feet below me. I dove and came up behind him. I fired a long burst from 300 yds to about 100 yds., putting his right engine on fire. Prescott continued to chase shooting short burst. The enemy finally spun, caught on fire and burst into flames as he hit the ground. We split the victory."* [16]

A second alarm came in around 1300 that bombers were circling Lashio and were headed toward Loiwing. This time, all eight alert aircraft were scrambled. Although the bombers' initial target was Lashio, they apparently couldn't find the field because of poor visibility and hit another small town to the south of Lashio. Again, the prevailing haze prevented the A.V.G. pilots from finding the invading bombers. Otherwise, Erik Shilling arrived from Kunming in the first of the newly-acquired P-40Es, the number "100" emblazoned on both sides of the fuselage.[17] R.T. Smith summed up the day's activities with, "Turned out to be quite a quiet day too, for a change." Compared to the previous few days, a true statement.

After two days of alert scrambles and aircraft accidents, the Loiwing detachment enjoyed a welcome respite. Two more aircraft were added to Olson's force on April 2 when Frank Adkins and a civilian Pan American Airways pilot flew two more P-40E Kittyhawks in from Kunming.[18]

The highlight of the day, however, was the wedding ceremony between "Fearless Freddy" Hodges and a pretty 20-year-old named Helen Anderson. Helen, the daughter of a British warrant officer, was working in a Rangoon department store when she met Freddy during the Hell's Angels' stay at Mingaladon. The Third Squadron pilot "wined and dined" her at the A.V.G.'s hangout at the Silver Grill before Rangoon was evacuated and asked her to marry him. Her response was "no"; she could "never marry a Yank." Traveling with the A.V.G. convoys first to Magwe and then to Loiwing, after three days at Loiwing she finally accepted Freddy's marriage proposal.[19]

With no minister available, the A.V.G. contingent "elected" CAMCO manager M.D. (otherwise known as "Doc") Walsh as "mayor" of Loiwing, giving him the "authority" to preside over the evening ceremony held in the CAMCO clubhouse. All hands attended, including Pappy Boyington, who, with his bunged-up knees so heavily bandaged by Doc Richards that he couldn't put on his trousers, came dressed in his bathrobe. A.V.G. chaplain Paul Frillman provided a more legal ceremony for the couple in Kunming on May 11.[20]

Later that night, Chennault left suddenly for a hurriedly-called conference in Chungking with the Generalissimo and Madame Chiang Kai-shek, Maj. Gen. Joseph Stillwell and Col. Clayton Bissell, Stilwell's air officer, to discuss the future of the American Volunteer Group. Bissell, a World War I fighter ace, and Chennault had locked horns earlier in their careers over fighter tactics (Bissell had advocated dropping a ball-and-chain device when overflying enemy bombers to foul their engines!) and there was never any love lost between the two. Never fans of the concept of the A.V.G., Stilwell and Bissell insisted on the transference of Chennault's forces to the U.S. Army – and the sooner the better. The alternative would be a halt to any U.S. logistics support – no more spare parts, ammunition, new aircraft or fuel.

Chiang supported Chennault, as could be expected, but finally acquiesced to giving up the A.V.G., provided it would be replaced by a complete U.S.A.A.C. fighter group under the command of Chennault, who would be promoted to brigadier general. Although Bissell wanted the changeover to occur only a month later, on April 30, with Madame Chiang's insistence, it was agreed for a July 4 disbandment.

Chennault reluctantly agreed to the details of the inevitable transition – the American Volunteer Group would be officially disbanded three months hence on the Fourth of July 1942, with the aircraft and remaining personnel being absorbed into the U.S. Army Air Corps' 23rd Fighter Group. Stilwell subsequently sent a message to the War Department with results of the conference recommending that both Bissell and Chennault be promoted to brigadier general, while stipulating that Bissell's date of rank precede Chennault's by one day, insuring that Chennault would be subordinate to Bissell. [21]

Noting Chennault's return to Kunming with the Chiangs on Sunday, April 5, Bus Keeton recorded in his diary, "There is something in the air and it's bound to break soon."

Largely because of the poor flying weather, things were relatively quiet in Loiwing for the next few days, except for an alarm at 0930 on the 3rd. The eight alert birds took off to intercept – nothing. Lew Bishop and five Pan Am pilots brought in another six Kittyhawks en route from Calcutta to Kunming. The next day Bishop, Shilling, Adkins and four of the Pan Am pilots flew the "E's" on to Kunming for further preparation before turning them over to the squadrons.[22]

Easter Sunday, April 5, saw Easter services held at the club house, conducted by an R.A.F. chaplain. Duke Hedman provided piano accompaniment for the hymns, of which R.T. Smith commented, "that alone was worth the price of admission." [23]

Loiwing defenses received a real boost that day, also, when eight R.A.F. No. 17 Squadron Hurricanes and five No. 113 Squadron Blenheims arrived from India. They would use Loiwing as a support base, with Lashio as their operational field in support of the Commonwealth forces in the Irawaddy Valley.

The calm lasted until April 8. Under a bright, sunny sky, devoid of the haze that had lingered for weeks, the day began with a report of an enemy observation ship approaching the field. Chuck Older and Ed Overend scrambled to intercept. Although they sighted it as they climbed out, the speedy reconnaissance bird was already headed back to its base and they couldn't catch it.[24]

Oley Olson, Ken Jernstedt and Bob Little arrived from Kunming at 1240 in three P-40E Kittyhawks. Five minutes later the alarm sounded, reporting "many ships over Lashio and headed this way." Racing to their planes, R.T. Smith and his wingman, Link Laughlin, along with Freddy Hodges, were the first off. Olson ordered Little and Jernstedt to take off immediately to get the new Kittyhawks off the field, despite their not having been refueled. They were followed by Fritz Wolf, with Clifford Groh and John Donovan, and Chuck Older with Ed Overend on his wing. Oley Olson, the last off, followed five minutes later in his new P-40E. Climbing alone to the northeast, Olson radioed Smith and his flight to head toward Lashio to intercept the reported Japanese raid.[25] The four R.A.F. Hurricanes remaining on the field also got airborne, leaving a single Blenheim bomber and two Kittyhawks (one of which was undergoing maintenance) awaiting to be ferried on to Kunming (the Pan Am ferry pilots had gone to lunch and couldn't be found).

The attacking force was the A.V.G.'s old nemesis from Magwe – Col. Kato's 64th *Sentai* from Chiang Mai. Convinced from several reconnaissance ships' reports that Loiwing had become the new Allied base of operations, he split his force that morning, sending 13 of his Nakajima Ki-43 Hayabusas (Oscars) to Loiwing, but only one *chutai* to escort another bombing raid on Maymo.[26]

Smith's flight had reached 22,000 feet and 20 miles out from the field when another call from radioman Harvey Cross came in – enemy ships over the field and strafing. Rocking his wings to signal the flight, Smith wheeled around and dove for the field, pushing the Tomahawk to 330 mph. Passing through 8,000 feet three minutes later, he could see the mass gaggle of Japanese fighters circling and diving on the field in a racetrack pattern as they made their successive strafing passes. With Link Laughlin on his wing, he joined the circling mêlée at 2,000 feet, picking out the nearest Oscar on the left, while Laughlin began firing on the one on the right, 200 yards in front.[27] After firing a two-second burst, Laughlin recalled that, "the E.A. pulled up into a steep chandelle to the right… and I followed it until it rolled over and dove straight to the ground where it burst into flames … only 200 yards south of the field." [28]

Pulling up sharply from his first pass ("I hit the one I fired at," R.T. said, "but don't think it crashed. Saw the one that Laughlin hit go down to the side and crashed."), Smith saw an Oscar above him going into a wing-over. From his later combat report, he stated, "I rolled over and fired at it while on my back, hit it and rolled on out. It pulled up into a stall and dove straight in at the edge of the field." [29]

Turning back into the swirling mass of twisting, turning P-40s and Oscars, he fired at several. Working his way south of the field, he shot at an Oscar that was on the tail of another P-40, noting that it started trailing smoke before it dived away. Spotting another Japanese fighter, he maneuvered into a rear-quarter position and, at close range, gave it a five-second burst. "He immediately went into a vertical dive," Smith reported, "then pulled out a little above level and then went diving straight into the ground... about 2 or 3 hundred yards from the river, near a village."[30]

Screaming down toward the field, following Smith and Laughlin, Freddy Hodges picked out one of the circling Oscars. As he was about to fire, another Oscar suddenly appeared, climbing out and headed straight for Hodges. He related:

> *"I opened all guns on him with a direct head on shot. "I saw no return fire from his guns. As he passed me, I could see black smoke pouring from his engine and then saw him go straight for the ground still smoking where I judged that he hit a little southeast of the field."* [31]

Clifford Groh, flying wing on Fritz Wolf, along with John Donovan, at 22,000 feet heard the radio report that enemy aircraft were strafing the field. Assuming that Wolf had not received the transmission (as mentioned earlier, the Tomahawk radios were notoriously unreliable), Groh rocked his wings to signal his flight leader, then made a diving turn toward the field. Passing through 4,000 feet, he saw several Oscars at the west end of the field. Descending to 2,000 feet, he opened fire on one Oscar that half-rolled away. Doing a one-eighty, he fired on the Oscar again, getting several hits, as another P-40 joined the attack. Moving to the north of the field, he engaged another, but lost him when the Japanese pilot rolled out. Turning south, he rolled in behind another Nate. As Groh described the action:

> *"The Jap didn't know I was behind him and I kept firing burst after burst into him until I was about 10 to 20 yards behind him… He dived towards the ground and pulled out of the dive and came up to about 1,000 feet, then half-rolled again and dove into the ground… about a quarter to half a mile north of the river… south and a little east of the field."* [32]

Following Groh down, Fritz Wolf met several Japanese planes at 6,000 feet as they were climbing off their strafing runs. After firing at two from the rear quarter, with minimal damage noted, he caught the third Oscar as the Japanese pilot half-rolled away. As Wolf reported, "He flipped over and went down to about 400 feet. He pulled up again. He then went down on his back diving for the ground. He crashed west of the field." [33]

Donovan's radio was also out, but when he saw Groh rock his wings and swing back to the field, his immediate thought was "just another false alarm." Following Groh's lead, however, passing 2,000 feet over the field he was surprised to see two Japanese fighters turning into him, their wing guns winking. Describing his predicament that night in a letter to his family, he wrote, "I felt more helpless that a baby and wouldn't have bet ten cents on my chance of getting out." Climbing for a nearby cloud at 3,000 feet to escape the pursuing Oscars, he admitted, "I was thoroughly frightened and I needed a moment to 'wake up'." Coming out of the cloud, he caught an Oscar just pulling up after strafing the ground-bound Blenheim bomber. "I came in from the rear quarter," he continued, "and let loose a burst and immediately he caught fire and dove straight down into the ground." [34]

Meanwhile, Oley Olson had joined the fight. Delayed for five minutes after the alarm sounded due to his Kittyhawk being refueled, with the added weight of a full external center-line tank, he climbed out slowly to the northeast. At 15,000 feet he, too, received the call of Japanese aircraft strafing the field, dropped the belly tank, and dived toward the field. Circling to the west, he picked out an Oscar climbing out and fired a short burst at the Japanese aircraft, too far out to take effect, and shot at another just behind the first. Then, as Olson reported:

> *"I fired a long burst into a third at about 3,000 feet. He turned over shedding a few pieces of airplane and dived straight down into the ground and exploded … in the immediate vicinity of the field slightly to the west or north."* [35]

While Olson was lining up on another Oscar, Fritz Wolf saw two of the Japanese fighters curving in behind Oley (whose aircraft he recognized by the No. 102 painted on the side of the new P-40E). As Wolf described the action:

> *"I saw Oley Olson by the river south of the field with two Japs after him. I shot one off his tail. It started smoking in a glide … I went down again … After firing he again started smoking more intensely … just above the trees and about 100 feet in the air. I then lost him as I didn't want to get too low. It is my belief that he eventually crashed somewhere along the river and foothills south of the field."* [36]

Among the last to get off when the alarm sounded, Ed Overend took off on Chuck Older's wing. As they were climbing through 2,000 feet, however, Older suddenly turned back to the field with engine trouble. Continuing on alone, Overend discovered that he hadn't turned on his oxygen system during his rushed preflight (a requirement in the Tomahawk) and was therefore limited to 12,000 feet. Leveling at that altitude, he orbited for a few minutes before he received the call from the ground station that Japanese aircraft were over the field. As he dove through 4,000 feet, he spotted four Oscars climbing out. Turning into them, one he opened fire, but one tight-turning Japanese fighter rolled in behind him and latched onto his tail. Diving out to escape the Oscar, he fired on another climbing to the southeast directly in front of him. From his later combat report:

> *"I fired from 150 yards until I was almost up on him, at which time he turned violently and I saw the canopy and upper part of fuselage at this point tear away. I passed over him and turning, saw him head down and make a half turn and dive straight down… I saw him hit in the middle of an open paddy field southeast of the field and about ¼ of a mile west of the river."* [37]

Climbing to the east, at 12,000 feet he saw a lone Oscar, obviously retreating from the raid. Pushing the throttle forward, he closed on the Japanese fighter and fired a long burst. The Oscar slowly rolled over and dove straight down through the overcast. Although the stricken aircraft had disappeared through the cloud deck, Overend claimed this one as a "probable," believing that the Oscar crashed six or eight miles in the mountains east of the field.

By 1345 it was all over, with an indeterminate number of Hayabusas straggling back to Chiang Mai. Their successive strafing passes had destroyed one of the two ground-bound P-40Es and the Blenheim and lightly damaged the other Kittyhawk, and two Chinese soldiers were wounded. The airborne A.V.G. defenders came away unscathed. From the after-action reports, they claimed ten Japanese aircraft downed (to a man, identifying the Oscars as Zeros). R.T. Smith and Fritz Wolf claimed two each, while Laughlin, Olson, Hodges, Groh, Overend and Donovan each claimed a single victory.

The exact number remains in contention, however, some researchers claiming that two (or more) pilots claimed the destruction of the same plane.[38] Some contemporary reports give credit for another victory to Bob Little, but the existing file of A.V.G. combat reports does not include one by Little. certainly, in the heat of battle, with a swirling mass of two dozen aircraft fighting it out in a relatively small airspace, it would be easy to miss another friendly aircraft joining the attack. The Japanese admitted to losing four of the 13 aircraft sent on the raid. No matter the actual number of Hayabusas destroyed, however, it was a resounding defeat for the 64th *Sentai*, and the first time that the A.V.G. had met a Japanese force on an equal footing.

As soon as he had finished the debriefing, Olson sent a radiogram to Chennault reporting on the results of the day's air battles. Although overestimating the number of raiders, as well as the aircraft model, it stated:

> ATTACKED BY 20 MODEL O FIGHTERS AT 1315 AERODROME HERE STOP SHOT DOWN TEN CERTAIN AND TWO PROBABLY STOP ONE P-40 BURNED ON GROUND AND ANOTHER SHOT UP STOP ALL OUR PLANES RETURNED SAFELY STOP OUR STRENGTH EIGHT P-40S THREE P-40ES AND FOUR HURRICANES STOP ONE P-40E FOUR HURRICANES AND ONE P-40 DID NOT CONTACT STOP JAPS CIRCLED AND STRAFED FROM 2000 FEET DOG FIGHTS DIRECTLY OVER AERODROME AND EN ROUTE THAILAND STOP [39]

There was much back-slapping at the club house bar that night, as the pilots celebrated the day's victories. R.T. Smith recorded in his diary before turning in, "It was the most thrilling experience I've ever had. The guys on the ground saw it all and are still raving. They say it looked just like a movie only better. Ha!" [40]

The "Old Man," while congratulating his pilots, was not so exuberant. In a radiogram after receiving Olson's report, he asked:

> WHY WERE TWO P-40E'S LEFT ON THE GROUND DURING RAID QUESTION CONGRATULATIONS ON THE FINE WORK BEATING MODEL "O" END CHENNAULT [41]

Above: Upon arrival at Loiwing, Oley Olson commandeered a thatch-roofed, bamboo building as their "Alert Shack." Posing in front of the building, left to right: Herbert Cavanaugh, Tex Hill, John Petach (facing away), Bill Reed, Oley Olson, Moose Moss, Parker Dupouy, Bob Prescott and Cliff Groh. (via Flying Tigers Association)

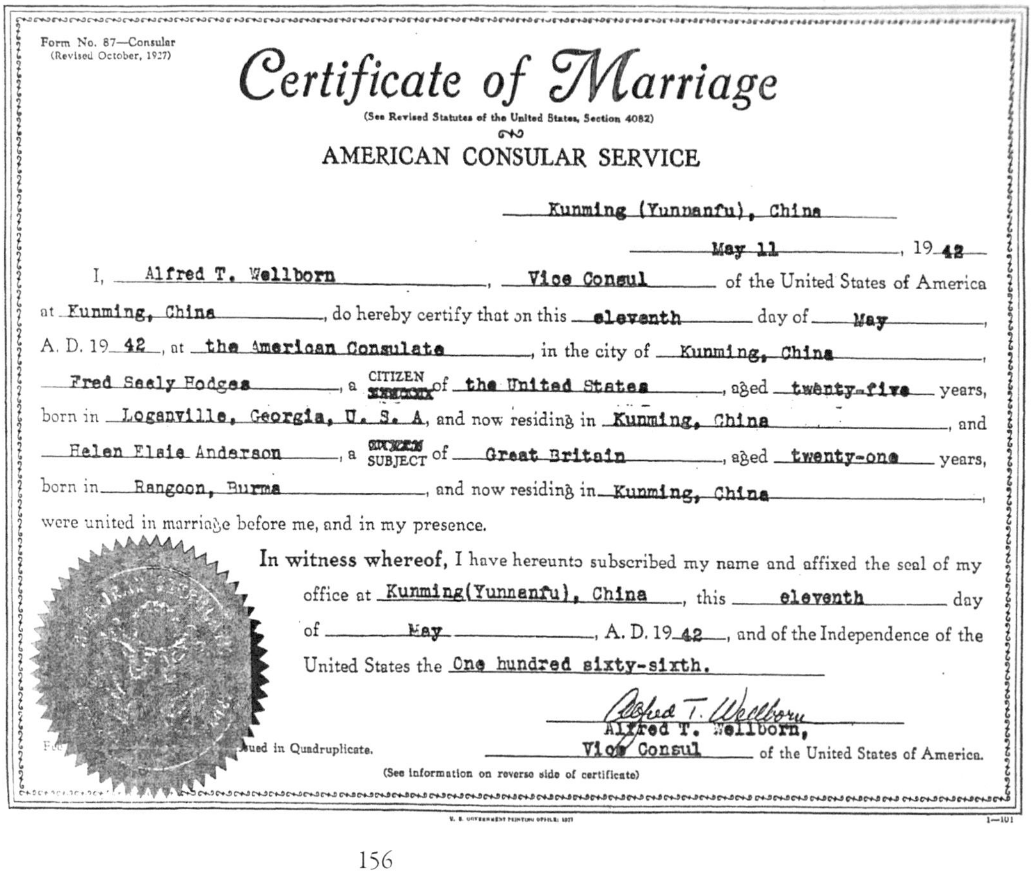

Form No. 87—Consular
(Revised October, 1927)

Certificate of Marriage

(See Revised Statutes of the United States, Section 4082)

AMERICAN CONSULAR SERVICE

Kunming (Yunnanfu), China

May 11, 1942

I, Alfred T. Wellborn, Vice Consul of the United States of America at Kunming, China, do hereby certify that on this eleventh day of May, A. D. 1942, at the American Consulate, in the city of Kunming, China, Fred Seely Hodges, a citizen of the United States, aged twenty-five years, born in Loganville, Georgia, U. S. A, and now residing in Kunming, China, and Helen Elsie Anderson, a subject of Great Britain, aged twenty-one years, born in Rangoon, Burma, and now residing in Kunming, China were united in marriage before me, and in my presence.

In witness whereof, I have hereunto subscribed my name and affixed the seal of my office at Kunming(Yunnanfu), China, this eleventh day of May, A. D. 1942, and of the Independence of the United States the One hundred sixty-sixth.

Alfred T. Wellborn
Alfred T. Wellborn,
Vice Consul of the United States of America.

Fee ... ued in Quadruplicate.

(See information on reverse side of certificate)

Certificate of Marriage between Fred Hodges and Helen Anderson. (Fred Hodges via Bob Andrade)

I, M. D. Walsh, Mayor of the American Community located in Loiwing, province of Yunnan, China, do hereby certify that on this second day of April, A. D. 1942, at the American Community Clubhouse, in Loiwing, China, FRED SEELY HODGES, a citizen of the United States of America, aged twenty-five years, born in Loganville, Georgia, United States of America, and now temporarily residing in Loiwing, China, and HELEN ELSIE ANDERSON, a subject of the British Empire, aged twenty-one years, born in Rangoon, Burma, and now temporarily residing in Loiwing, China, were united in marriage by me, and in the presence of the Rev. A. G. C. Langford, an ordained clergyman of the Church of England, attached to the Royal Air Force, who conferred the blessing.

IN WITNESS WHEREOF I have hereunto subscribed my name at Loiwing, in the province of Yunnan, China, this second day of April, A. D. 1942, and of the independence of the United States the one hundred and sixty-seventh.

Signed in the presence of:

A G C Langford RAF
Rev. A. G. C. Langford, R.A.F.

M D Walsh
M. D. Walsh, Mayor

A E Olson Jr
A. E. Olson, C.O. 3d Pursuit Squadron A.V.G.

SIGNATURES OF CONTRACTING PARTIES:

Fred S. Hodges
Fred S. Hodges

Fred Hodges, his new bride Helen and "Tex" Hill enjoy a relaxing moment at Loiwing. (via Bob Andrade)

CHAPTER 17

Loiwing Under Attack

The Second Pursuit Squadron, now led by David L. "Tex" Hill, held in reserve at Kunming along with the First Squadron, had been idle for nearly six weeks, ever since the Panda Bears had evacuated Rangoon – except for the four pilots who participated in the raid on Chiang Mai, and those few attached to the Third Pursuit at Loiwing. Bus Keeton commented, "Getting sort of restless sitting around."[1]

The natives were "getting restless," to put it in colloquial terms. While many of the pilots busied themselves with checking out in the new Kittyhawks, others, including ground personnel, entertained themselves in other pursuits. Crew chiefs Frank Metasavage and Glen Yarberry were placed under house arrest for a brawl that included Melvin Ceder, George McMillan, Johnny Petach and Skip Adair; and Flight Leader Matt Kuykendall received a $100 fine for "firing his gun in a reckless manner while intoxicated and off duty"[2] ("stewed to the gills," as Charlie Bond put it).[3]

That changed on April 8 when Chennault ordered Hill to take six Panda Bears to "conduct a reconnaissance and strafing mission on enemy air bases in Indochina." After an overnight stop at Mengtzi, 125 miles south of Kunming, they were to hit the Japanese airfields at Gaobang, Ha Giang, Lao Chao and Lao Kay. Additionally, Tom Jones was to make a photo reconnaissance flight over several friendly bases in the Yunnan Province.[4]

Hill took off for Mengtzi at 1730 with Frank Lawlor, Ed Rector, Johnny Petach, Hank Geselbracht and Bus Keeton, arriving at 1815. Trouble started when Ed Rector ran into a ditch on landing, ruining his prop, and Bus Keeton came in with a leak in his Prestone tank. The next morning, with coolant from Rector's disabled plane, Keeton returned to Kunming, while Tex, Lawlor, Petach and Geselbracht took off at daybreak on the planned mission. After hitting two of the Japanese bases with "moderate" success, the four returned to Kunming at 1030.[5] Upon landing, they received orders from Chennault for Hill to take nine aircraft – seven of the new P-40Es and two Tomahawks to Loiwing, where he: "will establish…an operational base in conjunction with the Third Pursuit Squadron… will issue detailed instructions covering operations and will co-ordinate his missions with that of Squadron Leader of the Third Pursuit Squadron." All the P-40Es at both Kunming and Loiwing would be attached to the Second Pursuit.[6] After a hurried lunch, Hill took off with Ed Rector, Frank Lawlor, Bus Keeton, Hank Geselbracht, Freeman Rickets, John Croft, Peter Wright and Moose Moss.

Arriving at Loiwing at 1830, they learned that there had been three alarms that day, with three scrambles of the alert pilots, but nothing was sighted. The seven Kittyhawks and two Tomahawks brought the A.V.G. contingent at Loiwing to 20 aircraft and 30 pilots, leaving approximately 20 First Squadron aircraft and 10 Second Squadron aircraft still at Kunming.[7]

Later that evening, a C.N.A.C. DC-2 arrived at Loiwing with Chennault and several Second Squadron armorers and crew chiefs to round out the force. Previously commanding his dispersed squadrons from his headquarters in Kunming, it would mark the first time that the "Old Man" had come to a forward operating base to personally observe his pilots in action. He gathered them in the club house later to give them the latest intelligence report. As recalled by Keeton, he said: "The Japs are pushing the Chinese and British back along the Burma front and are around Toungoo. We are supposed to fly down over the front and contact the enemy if possible and shoot them down. If no enemy is seen we are supposed to fly low so the Chinese can see us so as to build up their morale."[8] So, the "morale" missions would continue…

The next morning, Friday the 10th, the crew chiefs and armorers had been up well before dawn, readying their aircraft for the early morning missions. By 0600 the pilots were on their way to the field from their quarters just as the sun was coming up. Still three miles out, they saw five Japanese fighters strafing the field in successive passes. "The Japs caught us with our pants down this time," R.T. Smith later commented. There had been no warning at all from the several outlying lookout sites.[9]

The warning net serving Loiwing consisted of 17 operating radio stations equipped with a hand-cranked, 15-watt transmitter and a receiver, with eight more stations being built. Loiwing's radio station was equipped with two Hallicrafter receivers and a 250-watt transmitter. The problem was that most of the stations were located to the east of Loiwing, with only a few in the south. There were none to the west.

It was a return match for the 64th *Sentai*. Col. Tateo Kato personally led the flight of five Ki-43s from Chiang Mai (three more had turned back en route for engine problems). After several runs over the seven parked Kittyhawks, 13 Tomahawks and three Hurricanes, on Kato's signal the Japanese flight climbed out and headed southeast for home.[10] The "all clear" was given at 0700.

Despite the lack of any air defenses that morning, amazingly little permanent damage was done to the grounded fighters. Only nine were even hit, and just one of those so badly damaged that it couldn't be quickly repaired. The only injury reported was a Chinese guard who was wounded in the attack. But, "nothing we could do about it," R.T. lamented. Eight of the surviving Tomahawks were immediately put on alert (Chennault had given the task of defending the field to the Tomahawks, while the P-40Es were to be used primarily for raids against the Japanese positions). In less than half an hour, Tex Hill was airborne with Peter Wright and Freeman Ricketts for Toungoo on a pre-planned reconnaissance/morale mission.[11]

At 1100, the monitoring station at Lashio broadcast an alarm of incoming Japanese aircraft, sending seven P-40s to altitude. With a broken overcast at 15,000 feet by this time, the pilots found nothing, although ground observers reported one or two Japanese "bombers" over the field (maybe above the overcast). After the alert flight landed and was refueled, the "all clear" was sounded once again at 1230.[12]

Chennault launched another recce mission at 1400, sending Ed Rector, Johnny Petach and Bill Fish to survey the Japanese-Chinese lines around Toungoo.[13] Thirty minutes later, the nervous warning net operators at Lashio picked up a flight of aircraft headed north and sounded the alarm at 1445. Seven P-40s took to the air again with Chuck Older leading Duke Hedman, R.T. Smith, Bob Brouk and Bill Reed. Frank Lawlor with Bus Keeton rounded out the defenders. Climbing to 25,000 feet, they patrolled for almost 30 minutes before they sighted nine Oscars 10,000 feet below.[14] Col. Kato was back for round two for the day.

Older and Hedman had become separated from the flight as they passed through a cloud deck on the climb out, leaving R.T. Smith as flight lead. Diving down with Brouk on his wing and Reed close behind, Smith led the attack on four of the Japanese fighters. Passing below them at 14,000 feet, he pulled up and opened fire on the first one as Brouk went after another Oscar.[15] Reed, who had noticed oil leaking from his filler cap on the climb out, got in a shot at an Oscar as it passed 90 degrees in front, then rolled in behind him for another long burst from the rear. With the oil leak getting worse, he engaged two Oscars in a head-on attack, noting his tracers making contact in the Oscar's cockpit area. The stricken fighter "reared like a bucking horse," before it fell off into a spin. "I must have hit the pilot, causing him to jerk the stick back," Reed said later. At this point his oil cap came completely off, drenching his canopy, and Reed was forced to drop away, out of the fight, his visibility "like being in a Link trainer," as he later described it.[16]

Smith observed smoke from the engine of his targeted Oscar, then rolled away to engage two more. After nearly ten minutes of gut-wrenching, high-"G" turns, with each of the three trying desperately to gain an advantage, the more maneuverable Oscars finally gained a firing position, forcing R.T. to dive away. Looking south, he spotted three more in the distance as they were disengaging from the battle and were headed for home. As he reported later:

> *"I chased them and after about five minutes caught up with the last one. Closed to almost 100 yards and opened up. He went into a dive and I followed, shooting on the way down. He crashed in a rice paddy on the side of a mountain about 30 or 40 miles down the valley."*[17]

While his flight leader was busy with his first target, Bob Brouk made three passes on his selected Oscar, observing hits on each pass. As he later described the action:

> *"On the fourth pass, a head-on attack, I hit him and he turned to my left, went over on his back, and flames shot from under his left gas tank. He started down and I followed him for several seconds … while he was in flames."* [18]

After Chuck Older and his wingman, Duke Hedman, came out of the cloud deck, they didn't see another aircraft, friend or foe. After circling for a few minutes, Older turned south in the hope that they could catch some retreating Japanese. Finding nothing but empty sky after flying south for some 20 minutes, Older gave up and turned back toward Loiwing. Almost immediately, an Oscar appeared almost head-on. "I pulled up sharply to identify it," Older wrote in his combat report, "and started back down on his tail. I got in one burst and saw the plane jerk perceptibly." Then began a classic, extended aerial dogfight, from altitude down to tree-top level, with the Oscar maintaining an approximate course of 160 degrees for Chiang Mai, while being harassed by the two P-40s. As Older continued:

> *"After about 20 minutes, during which time Hedman and I both fired many bursts, the enemy plane began giving off a stream of black smoke … and finally saw the enemy go down steeply toward a hillside, apparently out of control, with black smoke streaming."* [19]

The last off the ground when the alarm sounded, Frank Lawlor took off with Bus Keeton on his wing, followed by three Hurricanes. Climbing to 25,000 feet, they circled for an hour before Keeton saw three aircraft coming head-on. Initially thinking they were the Hurricanes, he was a bit surprised when he saw flashes from the wing guns of the oncoming aircraft. Dropping his belly tank, he dove toward the left, then reversed in a steep climb, looking for the Japanese planes. As he flew over the field, an Oscar suddenly appeared coming head-on and smoking, followed closely by a P-40. "Five minutes later," Keeton wrote, "I spotted another Jap. Maneuvered into position and let go one burst. He just rolled over and went straight in. I followed him down until he crashed." [20]

Climbing back up, in-and-out of the clouds, Keeton spotted Bill Reed's Tomahawk No. 59, its canopy completely covered in oil and oil streaking along the sides of the fuselage. Reed had rolled back his canopy to get some semblance of visibility. Pulling up beside the oil-drenched Tomahawk, he signaled Reed to get on his wing. Keeton escorted him back to the field where he made a straight-in approach to the runway. When Reed's P-40 was safely rolling down the strip, Keeton climbed back out and was directed over the radio to proceed to Mengtzi where he was to remain until further notice. As he approached the auxiliary field, however, he received the call to "return to base." [21]

Following the "all clear" at 1640, all seven P-40s and one Hurricane were back on the ground. The A.V.G. had come through the one-hour battle unscathed, although the R.A.F.'s No. 17 Squadron had lost two Hurricanes. One pilot bailed out and John F. "Tex" Barrick, an American volunteer with the R.A.F., bellied in after being brought down by an Oscar, but not before he shot down one Oscar (his fifth aerial victory with the R.A.F.) and damaged another.[22] In the pilots' subsequent combat reports for the day, the A.V.G. claimed a total of four Zeros, while the R.A.F. claimed one destroyed, for the loss of two Hurricanes. Allies – 5, 64th *Sentai* – 2.

After the sunrise surprise visit by the Japanese, Chennault re-initiated the dawn patrol on April 11. Taking off in pre-dawn darkness, the patrol orbited the area for an hour without seeing any other aircraft, then returned to the field, "disappointed," as Bill Reed commented, "when not more than one Rising Sun came over the horizon." [23]

All was quiet until 0955 when the alarm siren sounded. The Lashio radio reported that enemy planes had been sighted (or heard) about 80 miles southeast of Loiwing. R.T. Smith joined a half-dozen other off-duty pilots who raced out to the top of a hill near the club house "to see what would happen."[24] Oley Olson led seven alert Tomahawks to intercept the incoming Japanese, but again, no contact was made. Upon landing, Olsen reported that they "saw one ship but could not catch it. Too high." [25]

Keeping up the pressure, Chennault sent out three more reconnaissance missions later in the morning. One flight – Bus Keeton, Hank Geselbracht and John Croft – was to reconnoiter the lines around Toungoo, but had to abort the mission just past Lashio for bad weather. [26]

Later that afternoon, Tex Hill gave Croft another chance for action when he selected him to join Peter Wright and himself for another trip to Toungoo. Landing at Lashio for refueling at 1600, they took off again for Toungoo. The weather hadn't changed, however, forcing them to return to Lashio, where they remained overnight.[27]

Meanwhile, after dinner that night Chennault called the men into the main room of CAMCO's American Club to give them an update on the induction situation, confirming that the Group would be absorbed into the U.S.A.A.F.'s 23rd Fighter Group. However, he reassured the men that they would be able to finish their contracts and then return to the States, where they could rejoin the U.S. armed forces as they chose, which, as Oley Olson recorded, "The sentiment of most everybody is that after a year they deserve a rest in the States."

Taking off at dawn the following day, Hill and his flight of three set out on a course of 175 degrees for Toungoo. Passing over Pyinmana, 50 miles north of Toungoo, John Croft called out seven bombers passing beneath the flight. Suspecting they were headed for Kyedaw Aerodrome, Tex continued on the southerly heading until they were over Toungoo, where he saw one bomber just landing. Then, as he wrote in his combat report:

> *"I gave instructions for my two wing men to stay up while I went down and strafed and proceeded to go down. On the first pass I burned up one and saw another off to the left so decided to make one more pass, which I did. This plane also burst into flames. I saw about six planes in the same area but I was hit by A.A. on my second pass so broke off the attack. The A/A was very heavy in that area. There was also heavy A/A south of the Swa River which is about 3 min. north of Yodasho. My flight was broke up so returned to Loiwing alone."* [28]

Following their leader after his first pass, Wright and Croft dove through the increasingly heavy anti-aircraft fire to make their strafing run. Wright flamed "a heavy twin-engine bomber on the left side of the runway," as he reported, before climbing out for home. Both made it through without getting hit but became separated as they dodged the ground fire.[29]

Just past Pyinmana, Peter Wright spotted another Japanese single-engine bomber (probably a Mitsubishi Ki-30, "Ann") headed south. Doing a quick wing-over, he chased the bomber, finally catching it 10 miles north of Toungoo, where he, "fired two bursts dead astern from 100 yards. E.A. burst into flames and rolled over," as he wrote in his combat report. Returning to Pyinmana, he refueled for the flight back to Loiwing.[30]

Lunch at the club was interrupted by another alarm. With no previous warning, no ship got off the ground, but several on the flight line reported seeing a high-flying reconnaissance aircraft.

Later that afternoon, Bob Prescott was playing cribbage with the "Old Man" in the small, bamboo alert shack as several other of the pilots lounged around the inside "kibitzing" the game. Suddenly, from just outside the shack someone yelled, "Zero! Zero!" Everyone inside, including Chennault, bolted for the doorway, while others dived through the windows before running to a slit trench. Then, as the radial-engine Zero zoomed overhead in a low approach to the field, it was recognized as the two-place BC-1 trainer used by the pilots to shuttle supplies and people to-and-from Kunming. Upon landing, Erik Shilling explained that his radio was inoperative, preventing him from giving an advance warning of his arrival. R.T. Smith commented later, "Chennault and everyone else chewed his ass out royally, you can be sure." [31]

The last bit of excitement for the day followed the evening's movie, a second showing of "Varsity Show 1937." While cleaning his Thompson submachine gun after the movie, Parker Dupouy accidentally discharged a few rounds, putting a couple of holes in the movie screen. Fortunately, no one was hit and Bill Reed signed off his diary for that night with, "Note: Be careful when handling or cleaning guns." [32]

Japanese reconnaissance flights over Loiwing were becoming more frequent. "Surer than anything they send over one every day," as Bill Reed observed.[33] As if to illustrate Reed's point, at 1145 on the 13th, with no previous warning, a twin-engine reconnaissance ship was spotted over the field. R.T. Smith and Cliff Groh raced to their Tomahawks and were airborne in minutes. By the time they reached altitude, however, the Japanese recce bird had turned south and was long gone. "We couldn't catch it," Smith reported, "or even see it." [34]

The alarm siren wailed at 1320.[35] Lashio reported airplanes overhead headed north. Chennault launched all 15 available aircraft – according to Bill Reed, "the largest number of ships we have ever put in the air." [36] This time the A.V.G. would be ready. But again, no Japanese aircraft were anywhere to be seen.

At 2100 the radio station at Mongma reported hearing "motor noises," and the alarm siren was activated again.[37] Unable to launch at night, several piled into a jeep and drove across the river where they waited for an hour until the "all clear" was sounded.[38]

Another false alarm…

CHAPTER 18

"Our Morale Won't Be Good Fast"
– *R.T. Smith*

The situation for the Allies in Burma was deteriorating rapidly. The Japanese had landed 40,000 troops in Rangoon with mountains of supplies, the Chinese had been pushed back to Pyinmana, 350 miles southeast of Magwe on the Chinese border, and the British were retreating along the Magwe front. Chennault, who had just accepted return to the U.S. Army Air Forces as a colonel (and soon to be brigadier general), was under increasing pressure to throw the A.V.G. into the breach. He continued to send out daily patrols from Loiwing on reconnaissance and "morale" missions for the beleaguered Chinese troops.

Diverted to Lashio for weather on one such mission on April 14, Bus Keeton met with General Lin, head of the Chinese Army in Burma, who implored him to request that Chennault send a reconnaissance flight to Rangoon to confirm his intelligence reports. Before returning to Loiwing, Bus ran into a U.S.A.A.F. Col. Echart, who informed him that they were trying to obtain bombs for the A.V.G. "These missions are silly enough as it is," Keeton commented later, "without going down to bomb with a pursuit plane." [1]

In an attempt to regain the offensive, Chennault planned a second strike on Chiang Mai, this time in concert with the R.A.F. Four P-40Es and four Tomahawks would escort four Blenheim bombers on a raid on the Japanese fighter base. Calling his pilots together on April 14, he briefed them on the planned mission and asked for volunteers. The room was silent until Bus Keeton spoke up, with verbal backing from Hank Geselbracht and Frank Lawlor, pointing out that the Blenheims were too slow, destroying the element of surprise. It would be suicide! The loss of two pilots, as well as aircraft, on the first Chiang Mai mission also weighed heavily on their minds.[2]

As many of the men stared at the floor, with some looking around the room, Tex Hill slowly stood up and, in his low, Texas drawl, said, "We'll take that mission." [3] He was immediately joined by fellow Panda Bears Ed Rector, Frank Schiel, Duke Hedman and R.J. "Catfish" Raines. That night, R.T. Smith recorded in his diary, "It [Chiang Mai] bristles with anti-aircraft and is about 160 miles inside enemy territory… we figure two or three may get back." [4]

That evening, Tex and his flight took off for Lashio, their staging base for the raid. The next morning they received word that the raid had been called off – the Blenheims had become lost, causing an unacceptable delay in the rendezvous.[5]

The refusal of the Loiwing pilots to take the mission was just one indicator of the growing morale problem within the A.V.G. R.T. Smith's diary entry for April 17 provides a good insight into the situation:

> *"Word from Kunming that things are in a bad state there. Several more men quitting to go home. Adair, in charge there, got in touch with the Col. down here & told him he'd better get back up there as soon as possible. To top things off, the Chinese army, & of course the British, are being pushed back in Burma. Now they intend to hold the line & want us to give them air support. Our twelve ships against the whole damn Jap Air Force. It seems mighty futile to all of us & we're wondering what's taking the U.S. so damn long to get something over here. At this rate, our morale won't be good fast! Phooey!"* [6]

A five-day period of relative quiet was broken on the Saturday, the 18th, when the alarm sounded at 1215. Six P-40s scrambled to intercept a lone twin-engine Japanese reconnaissance aircraft. Leveling off at 18,000 feet, Bob Brouk, a two-ship element leader with Bob Prescott, spotted the aircraft as it passed 4,000 feet beneath them on a northeasterly heading. Diving in a wing-over, Brouk slid in behind the unsuspecting Japanese aircrew and, as he reported:

> *"...fired a long burst from 300 yards to about 100 yards, putting his right engine on fire. Prescott continued the chase shooting short bursts. The enemy finally spun, caught on fire and burst into flames as he hit the ground."* [7]

At precisely the same time that the air raid warning sounded at Loiwing, the bomb-bay doors of the lead B-25 piloted by Lt. Col. Jimmy Doolittle opened to drop the first four bombs over the industrial section of Tokyo. Originally planned to recover in eastern China at pre-arranged landing fields, they were forced to launch four hours from the U.S.S. *Hornet* when the fleet was discovered by outlying Japanese picket boats, and thus were woefully short of expected fuel. Although it was a morale victory for the American public, who hadn't seen an American victory since the A.V.G.'s December victories over Rangoon, it did very little material damage to the huge city of Tokyo. All but one of the 16 B-25's aircrews – whose bomber landed in Vladivostok, Russia – were either forced to ditch or bail out near the eastern coast of China. Two crews were captured by the Japanese (one man drowned and three were executed), but the remaining 69 were aided by Chinese villagers and eventually made it to Chungking. [8]

Later that evening, "all hell broke loose," as R.T. Smith put it. Chennault had called another pilots' meeting in the CAMCO club house right after dinner. Stung by so many of his pilots refusing to take the planned Chiang Mai mission, his attitude had stiffened. He began by announcing his selection for promotion to brigadier general in the U.S. Army and that he was to be under the command of Lt. Gen. Stilwell. He emphasized that the A.V.G. pilots would be expected to take any mission assigned – or resign, in which case they would receive "dishonorable discharges." [9]

That brought on a loud chorus of indignant voices. "How can we, under a civilian contract, be given a 'dishonorable discharge' for resigning from the company?" one pilot questioned. Some spoke of the continuing "morale missions," forcing them to fly low over Chinese ground troops that often fired on them. Others cited the deteriorating condition of their planes, with few repair parts and few replacements and the change of their mission, from intercepting the Japanese fighters to being required to escort the slow-flying Blenheims and flying reconnaissance missions deep into enemy territory, all the while being expected to "fight the whole Jap Air Force in senseless, suicidal missions." [10]

Listening to the litany of complaints, Chennault's usually scowling face became even darker. Visibly angry, he told the men, "If you don't want to obey orders as given, you can show the white feather and get out!" With that, R.T. Smith rose to his feet and told Chennault that "there is a hell of a big difference between common sense and cowardice," and that he couldn't believe, with all they'd been through, that Chennault was now calling the men cowards. As heads nodded in assent, Chennault quickly said that was not what he meant. He said no more, however, and turned on his heel and left the room.

Following Chennault's ultimatum, the pilots had their own meeting, voting 24-to-six to resign in protest. A proposed letter was drawn up and Oley Olson read it out: "We the undersigned pilots of the American Volunteer Group, hereby desire to terminate our contracts with the Central Aircraft Mfg. Co. and our service with the A.V.G." [11]

When the sheet was passed around to be signed, Tex Hill took one look at it, then slowly stood up and said, "I'm not going to sign this damn thing." The room became instantly quiet, as all eyes were fixed on the usually quiet, but well-respected Texan. Hill reminded the group that, although they had signed on as civilians under a contract, now the United States was at war with Japan; as Americans, they had a responsibility to carry the fight to the Japanese just as if they were in U.S. uniforms. He defended the aborted mission to Chiang Mai as an effort to strike the Japanese at one of their major fighter bases, dismissing the charge that it was another "suicidal" mission. Continuing, he said, "Hell, I understand how you guys feel, and I don't like some of these missions any better than you do, but I'm not about to resign in protest – I'll fly when and where I'm told, regardless." [12]

Hill sat back down, but he had made his point. Then, as R.T. Smith remembered it:

> *"It was almost like a union meeting, lots of arguments back and forth, leading up to a strike vote. Four or five of the others of the thirty-two pilots present agreed with Tex, but all the rest, myself included, voted to call the Old Man's bluff."* [13]

The resignation letter was formally typed on official A.V.G. stationery and handed to Chennault the next morning.[14] It was signed by F. Ricketts, R.B. Keeton, Frank Lawlor, H.N. Geselbracht, Jr., R.W. Prescott, R.R. Brouk, C.F. Overend, C.H. Older, T.C. Haywood, K.A. Jernstedt, W.H. Fish, Jr., P.J. Greene, R.T. Smith, J.T. Donovan, R.C. Moss, J.S. Croft, P. Wright, F. Wolf, W.N. Reed, J.R. Rossi, P.S. Dupouy, Fred S. Hodges and A.E. Olson, Jr.

Chennault's usual frown didn't betray his inner anger as Oley Olson handed him the letter at 0800 the following morning. But the list came as a real blow to the "Old Man." Of the 23 on the list (Link Laughlin had changed his mind and scratched through his name before the letter was submitted) it included Third Pursuit Squadron Leader Olson and many of his most aggressive pilots. If all left as threatened, he would lose nearly half of his combat-ready pilot force. With no further comment, he said he would meet with the group again that evening at 2200.[15]

At the appointed time, with all assembled, he coldly stated that he could not accept their resignations. If they persisted, he continued, they would be considered deserters and their names would be reported to Washington. Chennault tossed the letter aside and left the room.

The "Loiwing Revolt," as it came to be known, had been quashed. Interestingly enough, however, the pilots were never again ordered to escort the Blenheims on low-level missions deep into enemy territory. Chennault had won the round. Resignation was one thing, but "deserter" carried a far uglier connotation. After all, as R.T. Smith later stated, "Few if any of us wanted to leave, but we did hope that some good might come of our protest." And Bus Keeton summed up the feelings of most when he recorded in his diary that night: "There wasn't much we could do but stay. Here we fought tremendous odds and done a good job. Desertion would ruin everything we had done." [16]

Tex Hill and all but four of his Panda Bears had refused to sign the petition and wanted nothing to do with the revolt. Even as the Sunday evening meeting with Chennault was taking place, Tex, Ed Rector, Duke Hedman and Catfish Raines were on their way to Lashio to refuel for an early morning "morale" mission to Pyinmana in northern Burma. Taking off from Lashio at 0645 on April 20, they arrived over the target area an hour later. Circling at 8,000 feet they encountered anti-aircraft fire and, per Chennault's orders, turned north and climbed to 9,000 feet to continue their patrol. Twenty minutes later, Rector spotted an aircraft 2,000 feet above approaching from the south. Then, as Rector later reported:

> *"I informed Hill by radio then turned toward the plane and started climbing. The plane turned west on sighting us and poured on all guns to escape. I chased him for six minutes at 210 mph indicated then finally started closing on him as he turned south. I gave him three bursts in the turn, one more long one as he straightened out eastward and started in a dive toward the ground. Hedman had meanwhile made a pass at him, and gave him a final burst, setting him on fire. As he rolled over on his back and went in. The two of us then proceeded to Loiwing."* [17]

That afternoon, Chennault sent Bob Brouk, Parker Dupouy, Cliff Groh and Link Laughlin to Namsang, the R.A.F. airfield 150 miles due south of Lashio near the Thai border, to patrol the area around Pyinmana. "All we did," Brouk said, "was dodge anti-aircraft fire which gave off black puffs of smoke about two feet in diameter…" before recovering at Namsang. After landing at 1630, they saw a reconnaissance ship flying about 10,000 feet over the field. Brouk and Laughlin took off in an attempt to intercept the intruder, but Brouk had to return because of an overheat indication on his oil temperature gauge.[18]

The 21st began with an early morning flight of four – Hank Geselbracht, Tom Jones, Freeman Ricketts and John Croft – lifting off at 0630 from Loiwing for a reconnaissance mission along 225 miles of the Burma Road from Lashio, to Mandalay, then south to Yamethin. By 0920 they had just reached the southern end of their flight plan at 11,500 feet, approximately three miles south of Yamethin, when they observed 18 Nakajima Ki-27 Nates three thousand feet above. Four of the little fixed-gear fighters dove on the P-40s. In the ensuing tight-turning combat, Tom Jones managed to get a few hits on one of the Nates, as he described in his combat report: "I closed in… and fired a long burst and saw his plane propeller slow up and plane pull up in front of my guns and fall off…" As more of the Japanese fighters joined in the battle, the four Panda Bears dived away and recovered for refueling at Lashio.

Working into the night at Namsang, crew chief Henry Olson and Bob Rogers prepared three of the Tomahawks for the planned dawn mission on the 21st – Brouk's aircraft, however, had a damaged canopy that couldn't be repaired quickly. After a pre-dawn breakfast, Dupouy, Groh and Laughlin took off at 0630 for their reconnaissance mission as Brouk prepared to return to Loiwing with his bent bird.[19]

"The takeoff was normal," Brouk wrote later, "but the oil temperature started creeping up [again] so I circled the field a few times and when the temperature hit the peg and the pressure dropped, I decided to come in." After landing, he was taxiing toward the hangar when, as he described it, "I saw white flashes and heard noises coming from the right side of my engine…The flashes blinded me, I felt sudden pangs in my legs… I looked up and saw a plane over my head. I knew I had been strafed." [20]

With his Tomahawk still rolling slowly forward, Brouk unbuckled his seat belt and dived over the side of the cockpit onto the ground, narrowly missing the aircraft's tail as it rolled past. Fumbling with his 'chute harness, he threw it off before leaping headfirst into a dugout beside the runway. It was not until then that he noticed "a small hole clear through my thumb below the first joint…and about four gunshots in each leg." [21]

Namsang had been hit at 0730 by ten Ki-43 Oscars of Col. Kato's 64th *Sentai,* now based at the former training ground at Kyedaw Aerodrome outside Toungoo. The high-flying Japanese reconnaissance ships so recently observed over Namsang had obviously reported the A.V.G. activity on the field. Coming in from the east at 10,000 feet, they dropped down to 5,000 feet. As they approached the field, three peeled off to strafe Brouk's landing Tomahawk. The Japanese fighters made four passes each, in string, until the plane began to burn, then climbed out to the west. Surprisingly, no other P-40s were damaged in the attack.[22]

As Col. Kato's Oscars were strafing the airfield at Namsang, back at Loiwing the crew chiefs were busily preparing the P-40s for a late morning mission. With one of the "spares" available, Bill Reed thought it a good time to check out in one of the new Kittyhawks. Taking off in the early morning haze, he flew around the local area for half an hour before he received a radio call requesting him to reconnoiter the area around Lashio.

Turning south, but with no area map (why would he need one, for just a flight around the field?), he was confident that he could find their former base; it was only 100 miles distant. An hour later, however, he admitted to himself that he was hopelessly lost. He had overflown Lashio, hidden beneath a covering of fog and haze, and was now 40 miles south of the town.

With fuel running low, he spotted an open area with a flat pasture, circled once, then put the Kittyhawk down to an uneventful landing. Climbing out of the cockpit, he walked for 10 miles before coming to an outlying R.A.F. early-warning site. There he remained a guest of the R.A.F. station commander for the next two days until he was driven back to his aircraft, miraculously untouched by the Shan tribesmen in the area. As luck would have it, he was able to flag down an A.V.G. fuel truck driven by R.W. "Buck" Rogers, crew chief for the A.V.G.'s Beechcraft "hack," coming down the road on the way back from Namsang to Loiwing. After getting enough fuel to get back to Loiwing, Reed still delayed one more night for the weather to lift before flying back to Loiwing.[23]

To add to the activities of the 21st, Greg "Pappy" Boyington finally resigned, with Chennault giving him a dishonorable discharge before he boarded a C.N.A.C. DC-2 to India. Charlie Bond, who had been a flight leader in the First Squadron, took Boyington's place as Vice Squadron Leader.[24] With the departure of Boyington, the 21st pilot to resign, Chennault's pilot force had dwindled to 54 – half the 107 that had been recruited in mid-1941; out of those, ten were serving in non-combat administrative staff positions.[25]

It was small consolation to Chennault when, on the 22nd, his promotion to brigadier general in the U.S. Army Air Corps was made official (his long-time antagonist, Clayton Bissell, had been promoted one day prior, instigated by

Stilwell, insuring that he would be Chennault's superior). Besides the reduced number of 44 pilots he could count on to fly combat, he had but 20 Tomahawks and eight P-40Es in commission, discounting the 22 being repaired or rebuilt at the CAMCO factory at Loiwing. To add to the problem, later in the day he received word that one more pilot and plane were missing. Cliff Groh, ferrying a P-40E flight to Kunming, had apparently become lost or had mechanical problems and never arrived at Kunming.[26]

Meanwhile, the Japanese army advancing against the collapsing British and Chinese forces were now between Kyidaunggan and Pyokwa, about 15 miles north of Pyinmana. With the deteriorating situation, Chennault ordered the evacuation of Namsang . The trek back to Loiwing began at 1800 on the 22nd.

On behalf of Chennault, Bissell sent a radiogram to the War Department which included the following:

> FOLLOWING FROM CHENNAULT: URGE IMMEDIATE APPEAL BY PRESIDENT TO ALL MEMBERS OF A.V.G. TO REMAIN ON DUTY WITH PROMISE EARLY REINFORCEMENT. GROUP LITERALLY WORN OUT – NERVES AND MORALE SHOT . PILOTS QUITTING. [27]

By April 23, the Japanese were in Loilem, a few short miles from Namsang. The A.V.G. contingent had departed the former R.A.F. base just in time. Early the next day, the Blenheims took off from Lashio to strike Japanese truck convoys reported to be on the roads leading away from Loilem. One was lost to anti-aircraft fire, but the remaining four returned safely to base.[28] Chennault followed up later that morning, sending six Second Squadron P-40Es, each with six 35-pound fragmentation bombs under their wings, with five Third Squadron Tomahawks for top cover, to hit the Japanese convoys. Taking off from Loiwing at 1000, the eleven headed south for Loilem.[29]

Led by Tex Hill, with Ed Rector, Peter Wright, Johnny Petach, Freeman Ricketts and Catfish Raines, a mile-and-a-half north of Loilem the Panda Bears sighted a group of some 30 trucks bottlenecked on the road running northeast of the city. Peeling off in trail formation, the Kittyhawks dropped their loads, the bombs striking 50 feet apart down the center of the road and along the sides, devastating the truck convoy. After the initial run, they circled and swung back around to make a strafing pass along the line of burning trucks. "The trucks were probably carrying gas as they burned readily," Hill related in his combat report. "I counted eight burning." [30]

As he pulled up for another run, Hill saw a Japanese reconnaissance plane (later identified as a Ki-51 Sonia) off to his left. A quick wing-over brought him around in pursuit of the recce bird, which was also being attacked by Wright and Petach. Ed Rector also joined in the attack on the hapless Sonia, as he described in his combat report:

> *"As we started firing, the Jap nosed down, turned 180 degrees and flew up a slight valley. I fired a head-on burst as he headed towards me, then turned sharply to keep from hitting him. He made another turn in the valley and crashed into a hill as one P-40 made a final run on him."*

After dispatching the Sonia, the four returned to the blazing truck convoy and made four more strafing runs each before climbing out, their ammunition expended. Hill ,and his flight made their way back to Loiwing, reconnoitering along the Hsipaw road as they went. Credit for the Sonia was split four ways between Hill, Wright, Petach and Rector.[31]

Coming down from their top cover, Oley Olson led his five Hell's Angels Tomahawks to strafe a line of trucks at Konghaiping. Chuck Older became separated from his leader on the climb out but soon rendezvoused with the other three. Leading them to Kehsi Mansam, he spotted a column of 40-to-50 trucks. Three strafing runs each left the rear one-third in flames. It was a very successful fighter sweep.[32]

With the Japanese steadily advancing toward Lashio, the city's residents, both permanent and temporary, were ordered to begin evacuating the city immediately. The R.A.F. sent their four remaining Blenheims and skeleton ground crew to Loiwing; the small contingent of A.V.G. personnel, including Harvey Greenlaw's wife, Olga, joined the fleeing Chinese on their trek north up the Burma Road.[33]

美志願軍第一大隊

HEADQUARTERS

FIRST AMERICAN VOLUNTEER GROUP

IN REPLY
REFER TO

Loiwing, China,
April 19, 1942.

We, the undersigned, pilots of the American Volunteer Group, hereby desire to terminate our contracts with the Central Aircraft Mfg. Co. and our services with the AVG.

F. Ricketts
R. B. Keeton
Frank Lawlor
H. M. Geselbracht Jr.
R. W. Prescott
R. R. Brouk
C. F. Overend
C. H. Older
T. C. Haywood
E. A. Jernstedt
W. H. Fish Jr.
P. J. Greene
R. T. Smith
J. E. Donovan
R. C. Moss
J. S. Croft
P. Wright
F. Wolf
~~C. M. Laughlin~~
L. N. Reed
J. R. Rossi
P. S. Dupouy
Fred S. Hodges
A. E. Olson Jr.

Original letter of resignation during the "Lowing Revolt" on 19 April, 1942. Note that C.M. Laughlin, who orignally agreed to resign, crossed his name off the list. (via Bob Andrade)

CHAPTER 19

Retreat from Loiwing

Highly pleased with the results of the Kittyhawk's first bombing mission on the 24th, Chennault followed up on the 25th with another 11-plane fighter sweep – the same mix of six P-40Es and five Tomahawks.[1] Taking off at 0950, they proceeded south toward Namsang. One hour later, over Loilem, they saw 25 trucks leaving the town toward the north. Diving on the convoy in string formation, the six Panda Bears dropped their loads of 35-pound fragmentation bombs before circling back for a strafing run.

Coming off their third strafing run, in a repeat of their combat on the 24th, Hill, Petach, and Wright engaged another Sonia.[2] As described later by Johnny Petach:

> *"I saw an Army 97 Recc. Bomber (Jap) to the right of me. I called up that fact on the radio and turned immediately to intercept it. P. Wright and D. Hill also saw it and turned with me so that we were almost in line abreast chasing the E.A. The other planes were catching up quickly but then the E.A. turned towards me. The other planes were firing and I saw the tracers from the planes ahead of me, falling behind the E.A. so I led him plenty. My tracers appeared to go well ahead of the E.A. The E.A. made a slight turn, started to smoke, and crashed into a mountain side and burned."*[3]

As before, after the Panda Bears made their attacks, it was the Third Squadron's turn. Catfish Raines was the last Hell's Angel to make his fourth strafing run. Finding himself all alone as he climbed out, he was surprised to see a Japanese fighter coming head-on, about 500 feet higher. Racking the P-40 around to get behind the enemy aircraft, he spotted a second fighter about a mile behind the first. Then, from Raines' combat report:

> *"I straightened out, flew north till I was nearly even with the second fighter. I turned under him about a 1000 feet and caught him about 10 miles south of Loilem. Pulled up under him, gave him a short burst. He started smoking, rolled over and went straight in, bursting into flames on impact."*[4]

Back at Loiwing, the base was a bee-hive of activity. At 1045 the air raid alarm sounded, with an "all clear" at 1137. Eight minutes later came another that lasted for an hour before "all clear." During this time, five P-40Es came in from Kunming, bringing the A.V.G. force at Loiwing to 17 – the largest concentration of P-40s they had enjoyed since their first combat on 20 December 1941.[5]

The Panda Bears and Olson's Hell's Angels began landing just past noon, with ground crews rushing out to refuel and rearm as the pilots climbed out of their cockpits. With barely time for them to grab a bite of lunch, Chennault ordered a second fighter sweep, this time adding the newly-arrived Kittyhawks. "I was off duty today, supposedly," R.T. Smith recorded in his diary, "but Reed came after me after lunch & said they wanted us to the field. We went out and they were planning another fighter sweep over the lines."[6]

Tex Hill's flight of six – in three elements of two – took off at 1430, followed by the five new P-40Es and Oley Olson's six. As before, Hill's flight would be the primary dive bombers, while the P-40Es would come in at mid-level for strafing runs, leaving Olson's Hell's Angels to fly top cover.

The armada of P-40s had no sooner arrived over Kohnhaiping than two Ki-51 Sonias were spotted. Hill's flight broke up into a ragged string formation and opened fire. As Link Laughlin, third element leader in Hill's flight, described the action:

> *"Two dove down on EA from about 5000 feet, hit him with a burst from about 250 feet. He turned under me and I noticed two or three other P-40s making runs on him. I caught him in a shallow turn close to the ground. Started firing at about 300 feet and closed to about 50 feet...A trail of smoke was coming out of the starboard side of the engine when I turned off...I turned off to the right, the EA turned left. I got behind him and two P-40s. The first P-40 hit him with a short burst and the EA skidded from a 50 feet altitude into the ground, cartwheeled and burst into flames."*[7]

R.T. Smith, one of the Kittyhawks in the middle formation, also took a piece of the action, probably hitting the same Sonia, as he stated from his combat report:

> *"On escort patrol approximately eighty miles south of Lashio we encountered two reconnaissance planes. About six of us attacked and on the first pass I made on one, it caught fire and crashed. I then made two passes on the second one and hit it, but someone else finally shot it down in flames."*[8]

The "someone else" in Smith's report was probably Catfish Raines, as he described in his subsequent report:

> *"The Jap observers sighted just south Kohnhaiping after about six planes had made runs on the one caught fire and dove in. The other started south. I dove down from 3,000 feet, started firing from 200 yards, up to about 50 yards. The Jap burst into flame. I pulled over him about 30 feet. The Jap looked as though he was trying to jump. The plane crashed just about 500 feet from where he caught fire, being about 50 feet up when he caught flame. 5 minutes later we caught some trucks in a gorge...Fired on five armored cars, started fire in one."*[9]

With the two Ki-51s taken care of, the combined flights proceeded with their original mission – to knock out as many of the mass of Japanese truck convoys as possible. From individual combat reports, they were very successful. "Strafed about 50 trucks at Kensi Mansam... I saw about ten in flames" (T.C. Haywood). "Strafed trucks and motorcycles on road about eighty miles south of Lashio" (F.W. Adkins). "Strafed trucks north of Konghai-Ping and patrolled road looking for more columns. 10 trucks burned" (E.F. Overend).

After two hours and twenty minutes, with most of their armament and fuel expended, the flights returned to Loiwing, most with only 30 gallons remaining. All but Link Laughlin's P-40E were unscathed. His Kittyhawk had been hit eight times by the Sonia's rear gunner. Upon landing, his crew chief found large holes in his right elevator, a broken landing light, two holes in the fuselage, four holes in the right wing and one hole through the prop. With so many making multiple passes on the two Sonias, with attendant claims of damage, equal shares of the aerial victories were given to Link Laughlin, Catfish Raines, R.T. Smith, Peter Wright, Johnny Petach and Freeman Ricketts.

That night, Chennault sent all available trucks from Loiwing back to Lashio, under the supervision of Third Squadron crew chief Henry L. Olson, to bring out whatever foodstuffs and other supplies that remained. What they couldn't load on their trucks, the men destroyed as much as possible. Still, large numbers of trucks, construction materials, and even fuel stocks, had to be left for the Japanese.[10]

Despite the A.V.G.'s daily missions against the Japanese positions, strafing and bombing the advancing columns of troops, tanks and vehicles, the mass of Japanese men and material overwhelmed the limited resistance that the A.V.G. could muster. By April 26 the Tigers were preparing to evacuate Loiwing. Chennault ordered Oley Olson's Third Squadron to prepare to move to Mengshi, 80 miles north on the Burma Road, while Tex Hill's Second Squadron would remain at Loiwing temporarily for a holding action, then move to Paoshan.

Early that morning, the 26th, three air raid alarms were sounded, with the alert flight rising each time to meet a non-existent Japanese force. Later that morning Chennault sent Bill Reed, John Donovan, Parker Dupouy and Moose Moss to reconnoiter enemy movement on the Lashio-Hsipaw road and other areas further south. Flying over Lashio, they were somewhat surprised to see no activity, contrary to earlier reports that the Japanese were in the city. Continuing to Hsipaw, they observed "many trucks and cars moving toward Lashio." [11]

Minutes later, they looked up to see a mass of twin-engine Japanese bombers, six-to-eight thousand feet higher – two flights of nine and one of ten – heading southeast in a perfect V-of-Vs formation . (They later learned that the bombers had just struck the town of Hsipaw.) Ignoring the bombers, they continued over Hsipaw, which was now in flames. Turning west, they were weaving back and forth over the roads, searching for enemy traffic, when John Donavan spotted three Japanese fighters above (again, Ki-43 Oscars from the 64th *Sentai*, as it turned out) diving on the flight. As Donovan later reported:

> *"I attracted the attention of Reed by diving by him wobbling my wings. I pulled up in a cloud and came out where I thought I would be above the enemy fighters. As I came out I saw several other Model "O"s and fired at one. I was firing at another when I saw still other enemy fighters above and to my right. I saw a Model "O" climbing apparently one that had dived on us…I dove toward him and fired a long burst. After diving away, I returned to Loiwing."* [12]

Also on the 26th, Chennault ordered George Burgard, Jim Cross and Dick Rossi to leave for Karachi, India, to accompany 13 Chinese Air Force pilots in ferrying 16 Republic P-43 Lancers back to Kunming.[13] Two earlier attempts to deliver Lancers from India had ended in disaster. The Chinese pilots were able to deliver only two P-43s on the first mission and only one arrived safely on the second. The three boarded a C.N.A.C. DC-2 at 1230 for Calcutta. There would be two more such ferry missions for other A.V.G. pilots, giving them a welcomed respite from combat.

Monday was relatively quiet at Loiwing – "Another day – same old story," as R.T. Smith observed. The alert flight scrambled at 1020 on an air raid alarm, but, again, no enemy aircraft were encountered.[14] Administrative duties had to continue, however. Over the past two days, Chennault had given dishonorable discharges to three ground personnel – Second Squadron crew chief Harold Walker on the 26th (for attacking a French doctor for refusing to perform an abortion on his wife), operations clerk John D. Grotty and a second Panda Bear crew chief, J. B. Hauser, on the 27th (both for overstaying their leave in Calcutta).

The next day, four more were released for "having deserted the A.V.G. in time of war" – Third Squadron line chief Glenn O. Blaylock, First Squadron line chief Richard E. Graham and armorer Jack G. Jones and Group Engineering Section line chief Mark H. McDowell.[15]

As far back as 1937, Chennault had been aware of the practice of the Japanese military to celebrate the Emperor's birthday with a major military victory. Emperor Hirohito's 41st birthday was only two days away, on April 29. With the rapid advance of the Japanese Army into northern Burma, the capture of Lashio, accompanied by a knockout blow of the A.V.G. at Loiwing, would be a perfect gift for the Emperor.[16]

Accordingly, on the morning of the 28th Chennault sent 14 P-40s – all he could gather – on a fighter sweep to the south of Lashio in anticipation of a major Japanese effort.[17] Taking off at 0930, the A.V.G. contingent consisted of nine from Oley Olson's Third Squadron, flying a mixed bag of Tomahawks and P-40Es, and five Panda Bears in Kittyhawks led by Tex Hill. The Hell's Angels leveled off at 10,000-12,000 feet, while Hill's flight flew top cover at 15,000 feet.[18]

As they took up their combat air patrol, circling 70 miles to the south/southwest of Lashio, at 1015 Hill sighted 27 twin-engine Ki-21 Sally bombers about 15 miles out to his right, slightly higher, heading northeast. When he couldn't raise Olson on the radio, he relayed the information to R.T. Smith, another in the Hell's Angels flight. Then, turning toward the bombers, Hill saw a string of 12-15 Oscar fighters above and in trail of the bombers. With Hill leading one element of two and Lew Bishop leading the other ("My fifth man had wandered off somewhere," Hill said later), they waded into the nearest eight Oscars. As Hill described the action:

> *"I fired four short bursts at my first opponent before I got him. In the 4th burst he started off in a shallow dive smoking and never did change his attitude until he crashed into the jungle. From then on the fight was a rat race with "O"s and P-40s on each other's tails. I got my second "O" on a head-ion approach with him having a slight altitude advantage."* [19]

Lew Bishop continued the narrative:

> *"I attacked from a very steep angle upward, a fighter that had managed to get on Hill's tail. A long burst started his plane smoking and pilot immediately bailed out. I was then thrown into a spin by loss of speed and recoil of guns. Recovered about 3,000 feet below flight. I started back up and noticed Model "O" heading out of fight. Gave chase and closed range to about 150 yds. Long burst cut off left wing and set plane on fire..."* [20]

Fred Adkins, following closely behind Bishop, entered the wide "Lufberry circle" with P-40s chasing Oscars which were trying to gain lead on each P-40 ahead. "I saw a Model Zero on a P-40s tail and gave him a burst," Adkins reported, "but it was short and I gave him another and he rolled over and dove straight down crashing into the ground southeast of Hsipaw." It was Adkin's first combat victory.[21]

As Bishop returned to the battle, he noticed an Oscar going down in flames, which he later confirmed for Tom Jones, then latched onto another Oscar, firing short bursts as the Japanese fighter spun toward the ground. Climbing back up, he found himself in an empty sky; the combat was over as suddenly as it had begun. Separated from their leader, Bishop, Adkins and Jones joined up and headed back northeast toward Loiwing.[22]

When R.T. Smith received Tex Hill's radio transmission, he and his wingman, P.J. Greene, turned toward the bomber formation. Before reaching the Sallys, however, Smith sighted several Oscars circling about four or five miles to his left. Calling to Olson (who continued to have radio problems) that he and Greene were going to break off to engage the Japanese fighters which were "just circling around, not going anywhere," as Smith put it. Continuing, he reported:

> *"The two of us went into the Lufberry on the tail of one Jap. We both shot at it, and then Greene had to break off as another Jap came up. I followed the first one, and when he tried to dive out I caught him. He pulled up sharply in a steep chandelle, and I followed. Got a good rear quarter shot at him and he flew right thru a burst. He caught fire immediately, and crashed out of a long spiral."* [23]

When P.J. Greene turned away to engage the second Oscar, he found himself on a head-on pass with the Japanese fighter. As the Oscar turned away from the deadly game of "chicken" at the last moment, Greene was able to do a quick chandelle, coming out on the Oscar's tail. After several bursts from the Tomahawk's guns, the Oscar caught fire and went down.[24]

Even before Hill's warning transmission, Chuck Older heard Japanese voices on their radio frequency. Shortly after, he spotted the enemy formation and rocked his wings to alert his wingmen. Fire-walling the Allison V-1710, he turned and climbed to intercept the bombers. As he approached the Sallys, he saw two Oscars coming up from below. A quick wing-over and a tight, high-"G" turn brought him behind the two. As he later reported:

> *"The leader waggled his wings as I came down and they separated. I took one of them and saw another P-40 start for the other. The enemy I chose to follow pulled up into a slight turn...I closed in and gave him a long burst. As I passed behind him I saw black smoke and flames coming out under his wing. He started down in a dive and I lost sight of him..."* [25]

At this point, Ed Overend, Older's wingman, entered the battle a half-mile behind Older. (He was the one that

Older described as "another P-40" in his combat report.) As Overend described the action:

> *"I caught up with four Model "O" fighters. One on the left and one on the right pulled up and away from the other two as I approached. I gave the one on the right a long burst at point-blank range. He caught fire on the right side of his engine and wing and pulled straight up and then down…"*[26]

Oley Olson, the self-described leader of the pack ("I was leading a fighter sweep of seven Tomahawks and seven Kittyhawks out of Loiwing to the south," as he stated in his combat report), with his intermittent radio was never able to exert command after takeoff. He did report hearing the same Japanese voices as did Chuck Older and apparently heard R.T. Smith acknowledging Tex Hill's warning. He then called for his flight to form on him and turned to intercept the Japanese formation. When no one followed, and he found that he had become a "one-man air force," it became evident that he had a radio problem. He briefly joined the turning fight that Older and Overend were in and made three ineffective passes before diving out when he was jumped by another Oscar. Pulling out of his 320-mph dive, he momentarily blacked out. When he regained his vision, he found himself alone again in an empty sky. After circling for another five minutes, he turned northeast for Loiwing.[27]

Ken Jernstedt, leading the second element of Older's flight, observed his leader's signal to attack but saw no enemy. As Older pulled far ahead, however, he, too, spotted the bomber formation and turned to intercept. Following Older and Overend, he saw the four Oscars climbing into Older's six o'clock and watched as Older and Overend took on the two that split up. One of the two remaining attackers was climbing so sharply that he stalled out, giving Jernstedt a chance to open fire. With two short bursts, the Oscar burst into flames and headed down.[28]

Flying south at 14,000 feet, Parker Dupouy and Tommy Haywood saw the northbound enemy formation 4,000 feet above. As they made a 180 and climbed to intercept, two of the escorting Oscars turned to engage the P-40s. In the twisting and turning battle that followed, the two Hell's Angels made three passes each at one of the Oscars, taking deflection shots when possible. Finally, as Haywood reported, "the fighter started breaking up and crashed into the jungle out of control." Pulling away from the doomed Japanese fighter, and seeing no other Oscars, the two turned to chase the Sallys, finally catching them as they began their run on Loiwing.[29]

Meanwhile, on the ground at Loiwing, the air raid alarm had been sounded at 1030, just as a flight of Pan Am DC-3s and a U.S. Army C-47, flown by Ferry Command pilots Col. Caleb Haynes and Col. Robert L. Scott, were approaching the field with a load of supplies from Dinjan, India. As the C-47 landed, Chennault ran out to warn them of the imminent attack. Restarting their engines and turning into the wind, they lifted off for China just ahead of the Japanese bombers.[30]

The Sallys appeared over Loiwing at 1105, harassed by Dupouy and Haywood, who made successive beam attacks until their ammunition ran out. Damage to the bombers was minimal, as was the damage to the airfield (perhaps thanks to having the bombers' attention diverted by the two attacking P-40s). Although the main runway was pockmarked with a few bomb craters, and the tail of one of the Pan Am DC-3s (which landed right after the C-47) was blown off, no other aircraft was hit nor any personnel injured. All returning P-40s were ordered to recover at Mengshi.[31]

All except for Oley Olson and R.T. Smith. As Olson approached Lashio, he was joined by Lew Bishop, Tom Jones and Fred Adkins. Just south of Loiwing, the three Panda Bears peeled off and departed for Mengshi after being advised by Loiwing radio that the runway had been damaged and for them to proceed to Mengshi. Unable to receive the radio transmission, Olson droned on to Loiwing and made an uneventful landing on the undamaged short runway.[32]

Following their combat south of Hsipaw, Smith and Greene headed for Lashio which Loiwing radio reported was under attack. Finding no enemy over Lashio, they turned for home just as Smith spotted the Japanese bombers on a return course of 210 degrees after hitting Loiwing. "Saw them in the distance," Smith reported, "but they were too far ahead, and our gas supply was getting low." [33]

When they arrived in the vicinity of Loiwing, they too, were directed to Mengshi. Unfortunately, Smith had lost his map in the high-G turning combat and had to rely on dead- reckoning as he turned for Mengshi. With a third P-40 joining on them, Smith asked Mengshi ground if they could see three P-40s overhead. When Mengshi replied with an "affirmative," Smith began straining to find the airfield. After 10 minutes of circling and futile searching, it became evident that the three aircraft overhead confirmed by Mengshi ground were, in fact, three other P-40s recovering there.[34]

With all his fuel gauges reading "zero," Smith elected to put his Tomahawk down in a hilly pasture below before his engine quit. As Greene, with his gauges showing a few gallons remaining, turned toward Mengshi, Smith lowered his landing gear and made a gliding approach to the pasture. Sailing over some unexpected humps and gullies, he slammed down and skidded to a stop on his belly, his gear washed out and prop bent. As his "landing" site was only 200 yards from the road to Mengshi, he was soon able to catch a ride to the base.

Meanwhile, P.J. Greene, with a bit more fuel, climbed out to altitude and proceeded on to search for Mengshi. Passing over the next mountain range, he found the right valley and soon spotted the airfield at Mengshi. The last A.V.G. pilot into the field, he landed safely on the muddy strip, but his engine died as he was taxiing in. A near thing![35]

After getting the combat reports from his pilots, at 1300 Chennault sent the following radiogram to P. T. Mow:

> FOR GENERAL MOW JAP BOMBED OUR FIELD 1105 TODAY STOP NO DAMAGE PLANES OR PERSONNEL STOP WE SHOT DOWN 11 FIGHTERS SOUTH OF LASHIO WITH NO LOSSES STOP NO WARNING BNG NET IS BREAKING UP STOP SITUATION CAMCO BAD THINK YOU SHOULD COME HERE AT ONCE CHENNAULT[36]

Later that day in Kunming the A.V.G. lost another pilot and plane. First Squadron Wingman Johnny Blackburn, a former Chinese Air Force instructor at Yunnan-yi recruited by Chennault in January, was on a test flight over Lake Kunming, checking a Kittyhawk's guns. Inexplicably, he dove straight into the lake. Bob Neale took off at 1800 to try to locate any sign of wreckage but found nothing and returned just before dark. Both Neale and Charlie Bond surmised that it was either a matter of "target fixation," or the P-40E's reputation for "mushing" in a high-speed pull-out.[37]

With the Japanese Army just a few miles south, Oley's "Hell's Angels" evacuated Loiwing on the 30th, after setting fire to 22 Tomahawks in various stages of repair and assembly in CAMCO's hangars. Chennault sent most of the Tomahawks to Kunming, but ordered five to Paoshan, a Chinese city a short distance from the east bank of the Salween Gorge. He ordered Mengshi evacuated on May 1, and the Second Squadron's cars, trucks and Tomahawks headed for Kunming.[38]

CHAPTER 20

The Salween Gorge Missions

Sunday, May 3rd, dawned with a thick haze and low clouds at Paoshan. An escort mission for a Blenheim bombing raid was called off because of the foul weather.[1] At 1100, however, in an effort to determine the status of the recently abandoned CAMCO facilities at Loiwing, Chennault ordered Bob Neale to reconnoiter the Loiwing area. Taking off at 1420 with a flight of six, Neale observed a motorized unit in the vicinity of Wanting, but no activity or burning buildings in the CAMCO complex.[2] After turning for home, five miles north of Loiwing his wingman called out an unidentified plane flying south below a low cloud deck. With a quick wing-over, as Neale reported:

> *"I came out of a cloud directly behind a Jap single engine observation plane. After the first burst, the enemy plane started a gentle turn to the left. My second burst was at very close range after which he started a diving turn to the left, finally ending in an inverted dive in which attitude crashed on the edge of a small river approximately 10 to 15 miles NE of Lungling."*[3]

The hapless Mitsubishi Ki-51 Sonia was Bob Neale's 15th aerial victory, making him the A.V.G.'s first, and only, triple ace.

On May 4, the Japanese Air Force attacked Paoshan with fifty Mitsubishi Ki-27 Sally bombers from the 12th and 98th *Sentais,* from Mingaladon and Toungoo respectfully. By this time, Chennault's carefully developed warning net had faded away with the fleeing Chinese. The Third Squadron element was caught by complete surprise, with only Charlie Bond and Bob Little managing to get airborne during the raid.[4] Climbing to 19,000 feet, 30 minutes later they caught the second wave of 25 bombers halfway to the China/Burma border, 1,000 feet below in a "V" formation. Bond reported:

> *"I concentrated all attacks on the flank ship. His left engine blew up on about the fourth pass, caught fire and went down through the overcast. I expended the rest of my ammunition in a similar maneuver on the other flank bomber with no results but I know some of the bullets must have hit this ship..."*[5]

Claiming one bomber destroyed and one "probable" before running out of ammunition, Bond turned back to Paoshan. After buzzing the field, he made a slow "victory" roll, circled the city at about 1,000 feet and made his approach to land. As he began to lower his gear, he suddenly felt bullets striking his armor plate. Looking back he discovered three Japanese fighters on his tail as his P-40 was being riddled by the Japanese gunfire. With fire spreading in the cockpit, he slid the canopy back, rolled the aircraft and tumbled out, his chute opening scant feet above the ground. Landing heavily in the local cemetery, he suffered burns to his hands, torso and head; his scalp was lacerated by two machine gun bullets. He would live to fight another day.[6]

Others on the field weren't so lucky, however. First Squadron pilot Ben Foshee, riding through the town in a convoy from Loiwing, was severely wounded by a bomb blast, along with a CAMCO employee and a few Chinese hostel workers. A Chinese doctor attempted to amputate Foshee's left leg, but Foshee wouldn't allow it, demanding that Doc Richards attend to him. The A.V.G. flight surgeon had an eight-mile journey to reach Foshee, however, and was delayed by the mass of refugees clogging the streets in the bombed city. Foshee died from loss of blood before Doc Richards arrived.[7]

The three Oscars that shot up Bond's P-40 were part of 13 Oscars of the 64th *Sentai* that had been escorting the 98th *Sentai's* Sallys bombing the city. Seeing no fighter opposition, they left the bombers to make strafing runs on the airfield. The strafing attacks destroyed two Tomahawks and damaged five others, four of which were quickly repaired and flown out to Kunming that afternoon.[8]

With the city of Paoshan in a shambles, Chennault ordered the field to be evacuated. He sent a U.S.A.A.F. C-47 to Paoshan the next morning to be loaded with the remaining pilots, most of the ground personnel and Foshee's body. Also loaded were the machine guns, ammunition, instruments and any other salvageable parts from the two wrecked P-40s. Tomahawk No. 33 had been repaired after an all-night effort and was flown out by Bob Little to escort the heavily-laden transport back to Kunming. Only Ralph Sasser and his radio crew (BC-5) were left on the field to relay radio traffic.[9]

To provide additional escort for the returning C-47, Chennault dispatched a flight of seven P-40Es from Kunming. Led by Tex Hill, with Ed Rector, Frank Lawlor, Frank Schiel, Freeman Ricketts, Matt Kuykendall and Raymond Hastey, the flight landed at Yunnanyi, halfway to Paoshan, for refueling. However, the message sent the night before to Yunnan-yi requesting refueling preparation was somehow not received, and their refueling was delayed until nearly noon.[10]

The seven P-40s were still on the ground when they received a radio message at 1115 from BC-5 that bombers were headed toward Paoshan. Fifteen minutes later, six took off, minus Tex Hill. After two unsuccessful attempts to get his Allison V-1710-34 started, he had passed the flight lead to Ed Rector and sent them on their way. He finally got his engine started and was airborne at 1155, just as Ed Rector was coming back in. He had aborted the flight due to a fuel cap coming loose shortly after take-off, turning flight lead over to Frank Lawlor.[11]

As Hill was climbing out to 21,000 feet, Lawlor and company had arrived over Paoshan, setting up a patrol over the area at 23,000 feet. At 1240, their radios crackled with the report that 16 enemy planes were circling the city. With vectors from Sasser, the flight sighted 16 Nakajima Ki-27 Nates 5,000 feet below. Peeling off, Lawlor led the flight in string formation to a rear quarter attack. Reporting his first attack, Lawlor wrote:

> *"...saw the plane I shot down burst into flame and pieces of it that were shot off filled the air with debris...I dove down on another I-97 and shot it down. Parts of the plane passed thru my propeller and damaged my undercarriage. I claim two fighters (I-97 type) certain."* [12]

Frank Schiel continued the narrative:

> *"After the first pass, the flight broke up into individual fights, that is, one P-40 against each individual group of Japs. The Japs broke up into groups of 3 to 5. At about 1250 36 Jap Type I-97 bombers approached from the SW at an altitude of 20,000 feet. I continued my engagement with the I-97's. On my first rear attack I got in a good burst. When I looked back the plane was burning in a spin...I made a total of four or five more attacks. On two of them I got good bursts in on a front quarter approach, from close range. I could see the tracers enter the planes. Both rolled over and disappeared."* [13]

Gil Bright, right behind Lawlor, saw Lawlor's first victim fell in flames before he was in range. He then:

> *"...fired at one nearby which half rolled quickly, shedding its right wing in the process. I then chandelled up and after circling for position fired a burst at another that was coming head on. This plane rocked violently and dove steeply under me."*

Bright's three minutes of combat was cut short when his fuel tank cover plate and cap blew open, filling his cockpit with fumes and a fine spray of avgas. With fuel gushing out and choking from the fumes, he turned for Yunnan-yi, arriving there with but ten gallons remaining in his reserve tank.[14]

Raymond Hastey, third in the string behind Gil Bright, raked a Nate "from nose to tail," observing parts flying from the stricken fighter as it went into a steep twisting dive, and out of control.[15]

Meanwhile, Freeman Ricketts was diving on his own target. As he wrote in is combat report:

> *"I got on the tail of an I-97 and raked him, finally pulled up to avoid collision. He fell seemingly out of control but I didn't see him crash. I came back into the fight and attacked a Model Zero who burst into flames shortly..."* [16]

Matt Kuykendall, bringing up the rear, began firing on a Nate from long range before closing to a few yards from the little Japanese fighter. He reported that:

> *"The enemy plane started down out of control with pieces flying from wing and fuselage. I made four more runs through the group of enemy fighters. Three I scored hits on from short range went down seemingly out of control but I did not see them disintegrate or catch fire."* [17]

After a 20-minute delay in getting airborne, Tex Hill leveled off at 21,000 feet and flew a timed course for Paoshan, expecting to rendezvous with his flight there. Arriving five miles south of Paoshan at 1255, he spotted seven Japanese fighters circling around below, which he reported as:

> *"...Seven Zeros milling about 7,000 feet below me, so immediately picked out one and attacked it. He didn't see me until I was in range and started to turn toward me immediately. I was already firing and in about 3 seconds he snapped into a spin and spun all the way down to the ground..."* [18]

Hill made another pass at a Nate in a group of four but couldn't turn with the agile Japanese fighter and saw his tracers falling behind the Nate. With three of the Japanese's flight mates arcing down behind him, he dove out and climbed back to 18,000 feet. With a suddenly empty sky, he circled Paoshan for another 10 minutes before BC-5 relayed that there was no more activity in the area. With that, he turned for Kunming, landing at 1420.[19]

Four of the other P-40s recovered at Yunnan-yi, undamaged save for Lawlor's prop and landing gear hit by flying debris during the air battle. After two hours and forty-five minutes of flight time since leaving their refueling base, most had less than 25 gallons remaining in their tanks. The exception was Frank Schiel, who had a damaged propeller and pitot tube after flying through the debris of an exploding Nate.

With no airspeed indicator, as well as defective radio transmitter, Schiel took up a course for Yunnan-yi at 1328. With fuel running low, he switched to his reserve tank. Shortly after, the reserve tank indicator showed red, indicating only five gallons remaining. Still some distance out of Yunnan-yi, he elected to make an emergency landing while he could. He spotted a deep, narrow valley ahead and made a gear-up landing, which sheared off the oil and Prestone coolers and tore off the propeller. After sliding along the ground for 45 yards, the P-40 came to a halt. Climbing out of his damaged Kittyhawk, Schiel discovered that the Chinese refueling crew at Yunnan-yi had failed to fill the reserve tank.[20] With the additional incidents of blown fuel caps on the aircraft of both Ed Rector and Freeman Ricketts, it certainly appeared that the refueling crew at Yunnan-yi might have some explaining to do!

Despite the problems incurred, the A.V.G. had scored eight victories for the day – two for Lawlor and one each for Hill, Schiel, Bright, Kuykendall and Hastey – "first blood" for the latter two.[21]

By May 5, the same day as the attack on Paoshan, the Japanese Army's 56th Division had advanced to the west bank of the Salween River, 75 miles along the Burma Road inside China. Chinese resistance had melted away, with much of the disorganized Chinese Army fleeing into the mountains and others pushing civilian refugees out of the way along the Burma Road as they raced for the Hweitong-chao suspension bridge across the Salween River gorge.[22]

The Japanese columns, pressing closely behind, advanced rapidly along the road, passing thousands of Chinese refugees trying to reach the Chinese border. The remnants of the Chinese army made it across the bridge just ahead of the Japanese. The Chinese then blew up the bridge, effectively stopping the advance of the Japanese, but also leaving untold numbers of hapless Chinese refugees and a few Chinese Army stragglers on the west bank – as well as A.V.G. First Squadron crew chief Gale McAllister, making his way east on the Burma Road. Had the bridge not been blown, the Japanese would have had a clear, unobstructed route to Kunming. Once Kunming fell, denying the Chinese their last major supply point by air, China would be out of the war. Only two things were stopping the Japanese: a demolished bridge – and the A.V.G.[23]

But now Chennault was faced with a dilemma – the Japanese troops were effectively stalled on the Salween's west bank, but so were the Chinese refugees that didn't make it across the bridge. An attack on the Japanese would undoubtedly kill hundreds of Chinese civilians. On May 6 he sent a radiogram to Chungking:

> GENERAL C.J. CHOW FOR H.E. MADAME CHIANG KAI-SHEK STOP A.V.G. SHOT DOWN 8 JAP FIGHTERS YESTERDAY STOP JAPS BOMBED CITY AND AERODROME STOP ONE A.V.G. PILOT MISSING STOP ALL COMMUNICATIONS WITH PAOSHAN CUT STOP LAST REPORTS STATE JAPS ON WEST BANK OF SALWEEN RIVER 1500 HRS 4 MAY AND BRIDGE DESTROYED STOP JAPS MEETING NO OPPOSITION ANYWHERE AS SOLDIERS CIVILIANS STRICKEN AND FLEEING EAST ALONG ROAD AND BRIDGES ARE DESTROYED AND DETERMINED OPPOSITION IS DEVELOPED STOP DUE TO FACT MANY CHINESE TRUCKS STILL WEST OF SALWEEN PRESUMABLY IN HANDS OF ENEMY REQUEST AUTHORITY H.E. THE GENERALISSIMO TO ATTACK TARGETS IN BETWEEN SALWEEN AND LUNGLING CITY END CHENNAULT [24]

The response from Madame Chiang Kai-shek was quick and positive:

> GENERAL CHENNAULT YOUR TELEGRAM MAY SIXTH RECEIVED STOP GENERALISSIMO INSTRUCTS YOU SEND ALL AVAILABLE A.V.G. ATTACK TRUCKS BOATS ETC BETWEEN SALWEEN AND LUNGLING CITY MADAME [25]

For several months, Chennault's armorers had been trying unsuccessfully to fit workable bomb racks on the P-40B Tomahawks. With the arrival of the newer model P-40Es in late March, however, the problem was solved. The "E" model came with under-wing mountings for six 35-pound fragmentation bombs. Roy Hoffman, the A.V.G.'s chief armorer, and Third Squadron armorer Chuck Baisden devised a rack beneath the aircraft's belly that could carry the plentiful 570-pound Russian demolition bombs supplied to the Chinese. The Second Squadron, made up largely of ex-Navy pilots, provided the dive-bombing experience needed.[26]

On the morning of May 7, Chennault directed Tex Hill to lead a flight of four P-40Es, accompanied by former U.S.S. *Ranger* pilots Ed Rector, Frank Lawlor and Tom Jones, who, recovering from his latest bout with malaria contracted during a tiger hunt in December, pressed Chennault to let him go. Bob Neale would provide a top cover of four First Squadron Tomahawks.[27]

Later that morning, after a refueling stop at Yunnan-yi, the two flights arrived over the target area at 1010 to find Japanese engineers busily working to span the muddy Salween with pontoon bridges. At the juncture of the Salween River and the Burma Road, the road drops sharply over a mile down into the gorge. Japanese armored columns, trucks and troops snaked back east for 20 miles along the narrow approaches to the gorge. As Chennault described later, the Japanese were trapped in the open "like flies on flypaper – a sheer precipice on one side of the road and a rock wall on the other." [28]

Signaling his flight for the attack, Hill peeled off in a 60-degree dive toward the western end of the Japanese column as. Aiming for the sheer rock walls above the convoy, he released his 570-pound demolition bomb. As the bomb found its mark, tons of rock came cascading down on the hapless Japanese troops below, effectively sealing off any retreat back along the Burma Road. Rector, Lawlor and Jones followed in a string, each dropping their own payloads and causing further chaos and panic below.

Hill and company followed up with successive attacks along the 20-mile length of the Japanese column, expending their fragmentation bombs and ammunition from their six .50-caliber machine guns.

Their ammunition gone, Hill and his flight climbed out of the target area to allow the top cover flight to add to the carnage. The four Tomahawks, led by Bob Neale, swept the column with their own .30-caliber guns until their ammunition was expended; then they joined Hill and his flight as they headed west for Yunnan-yi.[29]

The scene below was sheer pandemonium – vehicles were burning, the bodies of dead Japanese soldiers were strewn about like cordwood, and smoke rose thousands of feet in the air.

The A.V.G. had just begun. The next day, Chennault sent another flight of Kittyhawks, each loaded with six 35-pound fragmentation bombs, escorted by seven Tomahawks, back to the Salween Gorge. Tom Jones led the attack flight this time, while R.T. Smith led the top cover of Tomahawks. Chinese General C.J. Mao provided two aging, Russian-built, twin-engine SB-3 bombers and even a few Hawk III biplanes from the training school at Kunming.[30] Lew Bishop, in Tom Jones' flight, gave this report on his return to Kunming:

> *"At 1530, 4 P-40Es equipped with 6 35-lb fragmentation bombs each and escorted by 7 P-40s of the 3rd Pursuit Squadron took off from Kunming en route Paoshan and then to road on south bank of Salween River to bomb and strafe trucks bringing bridge building materials for crossing the river. Flight arrived over objectives at 1700 without interference from enemy aircraft. Circled low over objective and located trucks on open and straight stretch. Lined up close together. Made first pass from W to E and, at an angle of about 50 degrees' speed 300 MPH. Dived to within 100 feet of trucks releasing bombs in string. Continued on down gorge gaining altitude and turned back…noticed results of attack as 4 bombs burst amidst line of trucks other two slightly beyond but close enough to do destruction by fragmentation.*
>
> *Made three strafing passes from same direction and after first attack noticed small fire upon looking back. On third and fourth attacks drew fire from anti-aircraft guns stationed near road. By the time I completed my last strafing attack, I noticed one extremely large fire and two smaller fires amid trucks. Believed to have destroyed about 50 trucks during the course of attacks made by four planes. Estimate 200 casualties among personnel on ground. There was not any interference by enemy aircraft. All planes returned safely to base after raid."* [31]

On May 9, Chennault sent a report to Madame Chiang Kai-shek:

> FOR H.E. MADAME CHIANG KAI-SHEK STOP A.V.G. ATTACKED JAP ARTILLERY TRUCK COLUMN ABOUT 5-10 KILOMETERS WEST SALWEEN BRIDGE SITE WITH BOMBS AND .30-CALIBER GUNS AT 1640 YESTERDAY STOP MANY TRUCKS WERE BURNED AND A GREAT MANY MORE RIDDLED BY BULLETS STOP ALL OUR PLANES RETURNED BUT ONE DAMAGED BY GROUND FIRE STOP NO JAP PLANES SEEN STOP LOW CLOUDS PREVENT AIR ACTION THIS MORNING STOP PLEASE ADVISE ANY CHANGE GROUND SITUATION END CHENNAULT [32]

Chennault continued the attacks with small flights of Tomahawks and P-40Es over the next three days, harassing the small remnant of retreating Japanese as far as Wanting. The drive on the Salween had been stopped. The Japanese were never able to cross the Salween Gorge, and even though they maintained artillery and infantry positions on the Salween's west bank, they were never able to again attempt a river crossing. With the emboldened Chinese on the east bank, the opposing forces faced each other in a stalemate until the Japanese were driven back south into Burma in the summer of 1944. Chennault and the A.V.G. had broken the back of the Japanese invasion of Western China.

Meanwhile, 3,800 miles to the southeast, another historic battle was taking place. The Japanese had decided to invade and occupy Port Moresby, in southeast New Guinea, in an attempt to strengthen their defensive position in the

South Pacific. Alerted by signals intelligence, the U.S. sent two U.S. Navy carrier task forces and a joint Australian-U.S. cruiser force to oppose the Japanese offensive. Beginning on 7 May, the opposing carrier forces engaged in air strikes over two consecutive days. Although the U.S.S. *Lexington* was sunk and the U.S.S. *Yorktown* heavily damaged, Japan lost the light carrier *Shoho* and the fleet carrier *Shokaku* was heavily damaged, enough for the Japanese to recall the Port Moresby invasion fleet. The Battle of the Coral Sea, as it came to be called, was the pure carrier duel, and, coupled with the Salween Gorge victory, a major boost to morale back in the States.

As the A.V.G. was pounding the Japanese at the Salween Gorge, Gale McAllister had finally reached safety. He had had quite an adventure. As he told it:

> *"I was on the Burma Road retreating towards Paoshan when my truck broke down...some Chinese picked me up. We drove all night long. Traffic was a mess. Towards dawn we were getting close to the Salween River bridge...when a Chinese General and his bodyguard stopped us...to turn around and fight the oncoming Japanese. I told him I had to get back to my squadron. We didn't go far before we hit another traffic jam. I started walking towards the bridge... Suddenly the Japanese artillery started shooting over my head...I gave up on the bridge and headed straight for the river bank. I saw some Chinese soldiers trying to swim the river. They never made it. I decided to walk upriver, but hadn't walked two steps before the Chinese blew the bridge...I walked upriver most of the day. At dusk I came to a ferry boat...Eventually we crossed the river. I stopped at a "commercial" establishment. The proprietor let me sleep on the dirt floor with lots of other retreating Chinese. The next morning I started out on a mountain trail towards Paoshan with a group of Chinese soldiers. Soon my shoes fell apart. A Chinaman gave me a new pair. That night we arrived at a little village...Next morning the soldiers and I continued up the trail, but soon we parted ways. I walked all night. By morning (8 May) I was near Paoshan...I came on a Chinese officer who fed me...flagged down a truck and put me on it. Soon I arrived at one of our airfields. Twenty minutes later I was flying back to Kunming on a special C-47 headquarters had sent. My retreat was over and I was home."* [33]

Alone Above Paoshan. *Oil on canvas by Roy Grinell. Painting shows Charlie Bond taking off from the field at Paoshan during the surprise Japanese attack on the city on May 4, 1942. As one of only two A.V.G. pilots (with Bob Little) to become airborne during the attack, it was a day to be remembered for Charlie! (Courtesy Roy Grinnell Studio)*

Left: Charlie Bond, head bandaged following his bail-out after being hit by Japanese fighters, briefs fellow "Tigers" on his return to base. Left to right: Bob Neale, Bus Loane, Bond, unknown and John Dean.

The Decisive Moment. *Oil on canvas by Roy Grinell. Painting shows Tex Hill climbing out of the Salween Gorge on May 7, 1942. (Courtesy Roy Grinnell Studio)*

CHAPTER 21

"The Train Wasn't Worth Bishop"
– Claire L. Chennault

With Paoshan in ruins, the Japanese began to threaten the A.V.G. base at Kunming. Just before noon on May 9, a report came in that a Japanese intruder had been sighted overhead. At 1140 Bob Neale and four other First Squadron pilots were airborne, climbing out to 18,000 feet. Fifteen minutes later, they spotted a Mitsubishi Ki-46 (code-named "Dinah") twin-engine reconnaissance ship over the south end of Lake Kunming, about 7,000-8,000 feet above and two-to-three miles ahead. Pushing his throttle to the stop, Neale slowly closed on the unsuspecting Japanese. Hoping to get the Dinah pilot to turn, he began firing at long range. When this had no effect, he later reported:

> *"My wingman, Bartling, had a fast plane so I motioned him ahead. His first burst caused the rear cockpit to blow up, and the Jap started down making gentle spirals. At this time two other P-40s joined the fight. I circled the fight area and watched the 4 P-40s making individual runs for about 5 minutes at approximately 16,000 feet A burst from one of the P-40s caused the enemy to burst into flames after which the plane went into a spinning dive and crashed about 60 miles southwest of the southern tip of the South Lake…"* [1]

Three days of bombing and strafing the Japanese columns along the west bank of the Salween was having the desired effect, as the Japanese troops were observed retreating on the Burma Road toward Mengshih, 40 miles southwest of the Gorge. On Sunday, May 10, Chennault maintained the pressure, sending Ed Rector to lead a flight of Second Squadron P-40Es, including Freeman Ricketts, Gil Bright and Peter Wright – again escorted by Oley Olson with six Third Squadron Tomahawks – to attack the retreating Japanese.[2]

Taking off at 0705, two hours later they were over Mengshih. Dropping down to look for targets, Ed Rector spotted a reconnaissance plane on the ground, but continued on five miles south to attack a convoy of trucks on the road. Turning back toward the aerodrome at Mengshih, he strafed the recce bird, setting it on fire. Landing at Yunnan-yi for refueling, they were delighted to see P.J. Greene and Stan Regis, who had also been caught on the wrong side of the Salween when the bridge was blown.[3] Frank Schiel and Harry Bolster, in the escort flight, had belly tanks installed on their P-40Es, so were "fat" on fuel when the rest of the flight turned toward Yunnan-yi. They continued on to Loiwing to report on the condition of their old field, finding all but the CAMCO factory buildings in ashes. Turning back northeast, they strafed a column of eastbound Japanese trucks, setting several on fire before returning to Yunnan-yi at 1030.[4]

It was one of Harry Bolster's first combat missions with the A.V.G. A member of the American instructor pilot cadre at the Chinese Air Force training school at Yunnan-yi, when the school was shut down a short while earlier, he, along with Al Wright, Bus Loane, Les Hall and Red Shamblin, elected to stay in China and fly with the A.V.G. They were a welcome addition to Chennault's dwindling pilot force.

At 1545 that afternoon, Vice Squadron Leader Jim Howard took off with Frank Schiel and perennial volunteer Tom Jones to "bomb and strafe enemy trucks and convoys on the Salween front." Given the report that a Japanese column was moving south at Mengshih, they flew all the way to Wanting, 50 miles further south on the Burma Road before they saw a few scattered trucks. Howard and Schiel hit the trucks with individual attacks, expending their 35-pound fragmentation bombs, while Jones flew 15 miles further south where he hit "pay dirt" – 50 trucks – most of which he set ablaze.

Meeting no opposition, they turned for Kunming, landing at 1845.[5]

Meanwhile, Chennault approached Tex Hill with another mission. He ordered Hill and Duke Hedman to accompany a fourth group of ten Chinese pilots to Karachi, India, to bring back a dozen P-43s. After a few days delay awaiting a C.N.A.C. DC-3, they boarded the transport with their ten Chinese "wingmen," en route to Karachi, leaving Vice Squadron Leader Rector in temporary command of the Panda Bears.[6]

On Monday, the 11th, it was reported that "hostile ground forces have retreated along the Burma Road as far as Lungling." Accordingly, Chennault sent Oley Olson with a flight of four Tomahawks to rendezvous with an aging Chinese twin-engine SB-3 bomber over Paoshan for a bombing raid along the Road, after which he was to "lead his flight on a reconnaissance flight as far north on the Salween River as the bridge on the Paoshan -Tengchong Road to observe enemy air activity." [7] It was a "milk run," with Olson reporting no activity on his subsequent reconnaissance, as he wrote in his diary that night: "Made a tree level reconnaissance in the valley for a column of Jap troops reported moving along the West Bank. Saw nothing." [8]

Also on the 11th, one of the first Army Air Corps pilots reported in to Kunming. Capt. Albert J. "Ajax" Baumler had flown for the Spanish Loyalists during the Spanish Civil War. Flying Russian-built I-15s and I-16s, he was credited with destroying 4½ Nationalist aircraft. Rejoining the Army Air Corps upon his return to the States, he resigned his reserve commission in 1941 to join the A.V.G., but was refused a passport due to his previous combat in Spain. Rejoining the Army (again!), he was on the way to Burma on December 7, 1941, when his Pan American Clipper (carrying tires and spare aircraft parts for the A.V.G.) was turned back at Wake Island as the Japanese attacked. Months later, he was assigned to the cadre of the soon-to-be designated 23rd Fighter Group and sent to China. His first job for Chennault was "flying" a desk.

Tom Jones, who had earned the reputation as an "eager beaver" with his fellow pilots due to his enthusiasm for combat, had for weeks been scheming for a strike on Gia Lam, the Japanese-held airfield just outside Hanoi. Indeed, after returning from a reconnaissance flight over the area, during which he observed row-upon-row of Japanese aircraft at Gia Lam, he had discussed just such a plan with Pappy Paxton. When a report came in that a "heavy concentration of enemy aircraft have been reported at Hanoi, French Indochina," he had his chance.

Cornering the "Old Man," he obtained permission to gather a flight of six for the mission. After recruiting five more volunteers – Frank Schiel, Jim Howard and Lew Bishop from the Second Squadron, with Link Laughlin and John Donovan from the Third – he reported back to Chennault.[9] Although Chennault had some misgivings about Donovan – he was the least experienced of the six – he finally approved. Gathering the six into the ready room, Chennault briefed the mission in detail: "Takeoff at 1430, refuel at Mengtzi, depart Mengtzi at 1600 with ETA Hanoi at 1730, attack from the north at 300 feet with 200 yards separation. If the ground fire is minimal, make a second pass." [10]

At 1430 on May 12 the six P-40Es were airborne. After the 40-minute flight to Mengtzi, they refueled and were off again at the pre-briefed time of 1600. Thirty minutes later, however, Jim Howard's engine began to run rough and he was forced to abort. The five remaining proceeded on at 14,000 feet before they were forced down to 3,000 feet by a descending overcast about 50 miles from Hanoi. Nearing their target area, they passed abeam an outlying Japanese airfield where they observed five fighters taking off. Fire-walling their throttles, they soon outdistanced their pursuers and within three minutes were over Gia Lam Aerodrome.[11] Then, as Lew Bishop later described the attack:

> *"I was the last plane in the attack formation when Flight Leader Jones started his dive from the west. The anti-aircraft batteries were already firing... Wingman Laughlin was already in his dive when I started down. As I neared the field I noticed about eight aircraft (fighters) lined up... Four which were lined up together were already on fire... Laughlin was concentrating his fire on a DC-3 type transport... I concentrated my fire on a long line of fighters and then dropped all bombs from about 300 feet and strafed whole line of planes... looked back and saw transport on fire. Four of my bombs burst amidst line of fighters, destroying three by fire... Estimate fifteen to twenty severely damaged. All thirty bombs dropped within target area and were very effective... As the anti-aircraft fire was very intense we all started back..."* [12]

The anti-aircraft fire was indeed "very intense." What Lew Bishop failed to see as he climbed away from Gia Lam was John Donovan's P-40 being nailed by the enemy ground fire. The young volunteer went straight in. The remaining four, after dodging heavy thunderstorms on the return leg, landed at Mengtzi with minimum fuel at 1843.[13]

That night, Chennault sent a radiogram to Chungking:

> GENERAL C.J. CHOW URGENT FOR MADAME CHIANG KAI SHEK STOP ... 5 OF OUR PLANES ATTACKED GIA LAM AERODROME HANOI WITH BOMBS AND GUNS 1730 STOP 15 JAP PLANES BURNED MANY OTHERS DAMAGED MANY PERSONNEL KILLED STOP WE LOST ONE PILOT KILLED BY ANTIAIRCRAFT FIRE STOP CHENNAULT [14]

Going through John Donovan's personal effects later, they found the following letter to his parents: [15]

> *Dear Folks:*
> *You must not feel badly about my death. The small part that I have played in the war, though it has cost me my life, I am glad to give that. Life has meant much to me, but not so much that I am distressed at leaving and neither must you be. I had only a few things planned for the future, one of the most important was a nice home for Momma. It will please me much if she will live in a comfortable home with many flowers and trees. I am happy, and so must she be.*
> *Love to all,*
> *John Junior*

As R.T. Smith bitterly commented in his diary following the raid, "They set fire to about 15 ships on the field, but Donovan was shot down. It wasn't worth it." [16]

After two days of relative inactivity, on the 15th Chennault sent out two missions – one east to northern Burma in the Myitkyna/Bhamo area and the other southeast to reconnoiter the French Indochina border near Lao Kay. First Squadron Wingman Robert H. "Snuffy" Smith, returning from the northern Burma run, reported "no enemy troops or transport of any kind at Myitknya." [17]

On their return, they received the news that the surviving five members of the Hanoi mission had been promoted one grade – Howard to Squadron Leader, Jones, Schiel and Bishop to Vice Squadron Leader, and Laughlin to Flight Leader. (Strangely, John Donovan was not included.)[18]

The promotions were largely symbolic (but did increase their pay $50-$75 per month), since neither of the current squadron Leaders or vice squadron leaders were going to be replaced. Not all were happy about the promotions, however, as R.T. Smith commented: "We are all rather sore about that now. The Col. says it was an order from the Generalissimo, but there is some doubt about that in some peoples' minds." [19] An understandable feeling, perhaps, since no one received any similar recognition following the very effective Salween Gorge raids.

The next day, the A.V.G. lost another pilot. The A.V.G. had constructed a gunnery and bombing range some distance from the airfield. Tom Jones (as usual!) volunteered to be the first one to make a practice dive-bombing run on it. "We loaded him with six Chinese practice bombs made from ceramic material and filled with white powder, probably rice flour," recalled Third Squadron armorer Chuck Baisden. "Tom made several runs. Then, suddenly, he flew the aircraft right into the target. I think he had target fixation, but we never knew because he was killed instantly. It was very tragic and unhappy day for all of us." He left a young wife and new-born baby he'd never seen.[20]

Keeping up the pressure on the French Indochina front, on May 17 Chennault sent six P-40s to hit the railroad south of Lao Kay in an effort to disrupt transportation of vital enemy supplies. Second Squadron Vice-Squadron Leader Lew Bishop would lead the flight of four P-40Es, including Peter Wright, Jim Howard and U.S.A.A.F. Col. Robert L. Scott, recently arrived in Kunming sent to form the cadre of U.S.A.A.F. personnel taking over from the A.V.G. six weeks hence.[21] To gain combat experience, Scott had talked Bus Keeton, originally chosen for the mission, out of his slot.[22] To provide top cover for the Kittyhawk "bombers," Chennault assigned Third Squadron pilots R.T. Smith and Bill Reed.[23]

Arriving at the target area, about 50 miles south of Lao Kay, Bishop found no train and turned the flight north toward Lao Kay. Shortly after, R.T. Smith spotted a train and so informed Bishop. "He said for us to go back and strafe it," Smith related later, "and his gang would bomb the station at Lao Kay. Reed & I went back and made four passes each and blew up the engine." [24]

Continuing on to Lao Kay, Bishop led his four Kittyhawks in a diving, string formation on the freight yard. As he released his six fragmentation bombs, his P-40 was apparently hit by concentrated anti-aircraft fire. With his Kittyhawk belching fire and smoke, he slid his canopy back and went over the side. Peter Wright, flying behind Bishop, described the action:

> *"I was number two to dive and directly behind Bishop who was leading. As I was releasing my bombs at about 500 feet I saw Bishop pull up with about four feet of flame coming out of his tail. He pulled up to the left and rolled on his back and bailed out. His chute opened immediately and his plane crashed into a mountain setting it on fire. Bishop landed in that part of the town of Lao Kay, Indochina that is on the west side of the river. He landed in the town proper and was alive as he disappeared into the trees. In the meantime, I had noticed that his bombs, mine and the plane behind me, flown by Howard, had all gotten direct hits on the three lines of freight cars in the freight yard."* [25]

In a repeat of Billy McGarry's experience months earlier, Lew Bishop managed to evade capture for three days before he was caught by local French residents and turned over to the Japanese for interrogation and imprisonment. Shuttled between several P.O.W. camps over the next three years, he endured torture equivalent to that suffered by American P.O.W.s during the Vietnam War, 25 years later. On May 10, 1945, he managed to escape from a moving train while being transported to yet another P.O.W. camp in northern China. Aided by Chinese guerrillas, he reached the Allied lines 40 days later [26]

Following his return to freedom, in his official report of his mission and subsequent imprisonment by the Japanese Bishop stated that he thought that a premature explosion of his second fragmentation bomb had likely been the cause of his P-40s damage. Third Squadron armorer Joe Poshefko, who knew of the often-defective munitions supplied to the A.V.G., agreed with this premise. Additionally, Peter Wright mentioned no incident of anti-aircraft fire in his subsequent combat report. Nevertheless, the Vichy French army unit at Lao Kay was credited and decorated for downing Bishop's plane.

Despite the destruction of the train and rail yard, the A.V.G. had lost another pilot and plane – the third in five days (and, coincidentally, three of the five Hanoi raiders). Col. Scott, discussing the mission later, said, "General Chennault said the train wasn't worth Bishop…we should have left Lao Kay alone." [27]

Olga Greenlaw's *War Diary* reflects little activity from May 18 through the 22nd, other than weather reports – mostly rain – from Kunming and outlying auxiliary fields. R.T. Smith's diary for May 18 supports the calm:

> *"Same old grind again today. Nothing has happened, as usual. The Jap air force doesn't seem to be very active these days, at least not in our territory." Whether through circumstance or design – the A.V.G. had lost five pilots on ground attack missions in the last two months – Chennault launched no more bombing and strafing missions for four days. According to Bus Keeton, "The Colonel was quite put out about losing 3 pilots in the last week so he called off all missions for the present."* [28]

During the lull, however, the A.V.G. had a visit by Chennault's immediate boss, Brig. Gen. Clayton Bissell, who was also acting as the chairman of the induction committee. On the evening of the 21st, Chennault called a meeting with all personnel in Hostel Number One to hear what Bissell had to say about the impending transfer of the A.V.G.'s assets and personnel to the U.S.A.A.F.'s 23rd Fighter Group.[29]

Opening the meeting, Chennault explained that all who declined induction into the U.S. Army would have their CAMCO contracts terminated as of the 1 July date of the A.V.G.'s disbandment, they would be paid one month's salary and for unused leave, up to one month's salary, and be given $500 travel expenses back to the States.[30]

Then Bissell took the podium, outlining rank structure to be expected for those accepting induction – technical sergeant for most of the ground personnel, and reserve commissions "at an appropriate rank" for the pilots. After some question-and-answer discussion, during which it became increasingly evident that he wasn't persuading many to accept induction, an obviously frustrated Bissell ended by stating that "for any of you who don't want to join the Army, I can guarantee to have your draft boards waiting for you when you step off the gangplank onto U.S. soil!" [31]

That did it! Keeton stated later that "his job was to convince the A.V.G. that the best thing for us to do would be to remain here. Instead of keeping us here, he changed a few minds to the contrary." [32] That opinion was shared by Charlie Bond, who later wrote, "I think the talk hurt more than it helped the cause. It would have been much better to have it explained by the Old Man," [33] and echoed by R.T. Smith, commenting that Bissell "had all the charm of a cobra." [34]

Receiving reports of hostile heavy artillery fire from Japanese gun emplacements on the west bank of the Salween Gorge, Chennault set aside his growing misgivings of using his men for bombing/strafing missions and laid on another such mission on the 22nd. He ordered First Squadron Vice Squadron Leader Bob Little to take four P-40Es armed with fragmentation bombs back to the Gorge, with Bob Neale and four Tomahawks to act as top cover.[35]

Neale, Charlie Sawyer and Einar Michelson (with one other unidentified Second Squadron pilot) departed Kunming at 1245 to refuel at Yunnan-yi, since their P-40Bs didn't have the range of the E-models. They would form up with the Kittyhawks when they were overhead the auxiliary field. Little, Snuffy Smith, Bill Bartling and Carl Brown were off one hour later. Meeting their escort above Yunna-yi, they proceeded west, where they ran into rain showers and decreasing visibility from Paoshan to the target area.[36] As Neale held high, 5,000 feet over the area, Little took his flight down over the demolished bridge that once spanned the Salween. Visibility was now down to one-quarter mile, and they circled for five minutes before they saw a row of tents. Then, as reported by Smith, Little's wing man:

> *"Little peeled off and following him I saw he was diving on the target. At about 1,000 feet, before Little had leveled off, I heard an explosion. Glancing at his plane, I saw a burst of flame off his left wing. He immediately went into a tight spin…half his left wing was missing. He made no attempt to jump, or pull out, and was on fire when he hit the ground…I observed no anti-aircraft fire, nor any tracers from the ground…"* [37]

The death of the well-liked, ten-victory ace, the A.V.G.'s second loss to ground fire in five days, was a personal loss to close friend Bob Neale, who wrote in his diary that night, "…It was all I could do to fly home as I felt his death more than I can express." [38] Although Snuffy Smith's report indicated no observed anti-aircraft fire, suggesting that one of his-wing-mounted fragmentation bombs made have detonated prematurely upon its release, Neale later said he thought he saw flashes of ground fire, which may have struck the bomb.

It had been announced on the 19th that the First and Second Squadrons would soon be moving to Chungking, with the Third remaining at Kunming temporarily as a "rear guard." The First Squadron would move with the Tomahawks, while the Second Squadron would take the P-40Es. Bus Keeton was less than thrilled, remarking that "we will be doing all the bombing and dirty work, or I should say, the risky business." [39] On the 23rd, the orders came for those leaving for Chungking to get packed and be ready for departure on the 24th.

Later that day, George Burgard, Jim Cross and Dick Rossi arrived in Kunming with ten P-43s destined for the Chinese Air Force. They had started from Karachi, India, on 11 May with 16 aircraft and 13 Chinese ferry pilots. Often battling sand storms and heavy rain, six of the aircraft were lost en route due to accidents and mechanical problems.[40]

Early on the morning of the 24th, the men selected to go to Chungking assembled on the field with their bags packed. As they waited for the low ceiling to lift for take-off, around noon a wire was received from Chungking. The Japanese were threatening the river crossing in the Salween Gorge, bringing up reinforcements and shelling the Chinese forces across the Gorge with heavy artillery and the Generalissimo requested immediate A.V.G. support.[41] Chennault ordered the movement to Chungking postponed, except for Charlie Bond and Jim Cross, who departed on a DC-3 when the skies cleared. Bond would be the A.V.G.'s advance man, preparing for the arrival of the rest, while Cross would be the liaison between the A.V.G. and the Chinese Air Force.[42]

"Looks like we will have to carry on with a lot of foolish bombing raids," Bus Keeton commented when he heard the news, adding, "As far as I'm concerned the bombing raids do nothing as far as good is concerned and you are sticking your neck out too far." [43] Bob Neale agreed – "A great way to make a living." [44]

Chennault wasted no time; at 1325 Frank Schiel, Jim Howard and Ed Rector, escorted by four Third Squadron P-40s led by Oley Olson, took off to bomb Japanese positions on the west bank of the Salween. The flight struck a small village and the roadway leading into town, but saw no hostile fire. They returned from the uneventful mission at 1600.[45]

Chennault continued the Second Salween Campaign over the next five days. Chinese forces had counterattacked across the Salween and had some 300 Japanese troops surrounded in a small village just west of the river. Beginning on Monday, the 25th, Chennault sent eight-to-ten P-40s a day to dislodge the enemy positions from the Salween Gorge down the Burma Road as far as Wanting, 60 miles south. The Panda Bears, under the temporary command of Ed Rector while Tex Hill was away in India, took the brunt of the attack missions, with Adam & Eves flying top cover. The missions provided a good opportunity to "break in" the new recruits from the instructor staff at Yunnan-yi; Van Shapard, Lester Hall, Arnold Shamblin and Frank Adkins all received their baptism of fire during the final missions to the Burma Road.[46]

The A.V.G. flew their final mission to the Salween on Friday, May 29. Although surrounded by Chinese troops, the Japanese were still holed up in the little northern Burmese village, putting up a stubborn resistance.[47] Led by Ed Rector, the flight of eight P-40Es included Frank Schiel, Frank Lawlor, Peter Wright, Bus Keeton, Van Shapard, Lester Hall and Harry Bolster. Shortly after the 1230 take off from Kunming, Hall aborted the flight for engine trouble while the others continued to Yunnan-yi for refueling. Then, according to Frank Schiel:

> *"The first four made a bombing and strafing pass with very good results. The next three made a bombing pass most of the bombs falling south of the city. No activity of any kind was seen on the ground. Weather was good and the return trip was uneventful. Arrived Kunming at 1600."* [48]

Thus ended the Second Salween Campaign.

On Saturday, May 30, Chennault switched once more to northern Indochina, sending four P-40Es, led by Gil Bright, to attack the railway between Hanoi and Lao Kay. Bright took "new guy" Lester Hall and two originally-recruited pilots who had little previous combat experience – Curtis E. Smith, who hadn't been in combat since December, and Ed Conant, the original volunteer who had so much trouble checking out in the P-40. After refueling at Mengtzi, they made a fruitless search for reported rail traffic before placing their bomb loads on a section of the rail line, destroying several hundred yards of track. Taking a wide detour to the west to avoid Lao Kay, they returned from their "milk run" to Kunming.[49]

That evening, Raymond Hastey was enjoying a few drinks with Bus Keeton, Van Shapard and Rose Carney in the bar in Hostel Number One. It would be one of their last get-togethers, since Hastey was just waiting for transportation out of Kunming after being dishonorably discharged on May 24 for black-marketeering. As Keeton recalled later, Hastey was "talking about their big business deals and how much they made. It's unimaginable how much money is being made by looters and so on of this group. Hastey is really making a killing trading and selling stuff for members of this group." [50]

The following day at 0829, came a report of nine Japanese bombers striking Luichow, and another at 0920 that 20 were bombing Yunnan-yi and heading for Kunming.

The alert was sounded at Kunming and the Panda Bear pilots sitting alert were airborne in minutes. Describing the action, Bus Keeton wrote,

> *"We took off and headed for Yunnan-yi to intercept them. About half way there heard that Model Zeros were strafing the field. We poured on the coal but nothing was happening when we arrived. Looks as though the Japs are baiting us, which we want them to do."* [51]

A second alert and scramble that afternoon also proved fruitless. The day – and month – ended with a gala Chinese dinner thrown for the A.V.G. by the W.A.S.C.[52]

CHAPTER 22

The Move to Kweilin

The first of the cadre of the newly-formed 23rd Fighter Group (discounting Col. Robert L. Scott, who attached himself to the A.V.G. in April) arrived in Kunming on June 1. Second Lieutenants Romney W. Masters, Vaden J. Carney and James L. Dumas were attached to the First Squadron; Second Lieutenants Martin S. Chick and Dean A. Carter were attached to the Second Squadron; while the Third Squadron received Second Lieutenants Leonard M. Butsch and Lee N. Minor. Although Col. Scott had arrived earlier, Chennault placed 34-year-old Curtis Smith administratively in charge of the new arrivals as Detachment Commander, similar to Smith's administrative duties on board ship en route to Burma in 1941. Also, considering his little combat experience, Chennault felt him to be more valuable as an administrator than as a combat pilot.[1]

With the pending move to Chungking, A.V.G. activity was reduced to alert duty and checking out the new arrivals in the P-40. On Wednesday the 3rd, however, the excitement of the day was the arrival of two U.S.A.A.C. B-25s, of the anticipated six, from Dinjan, India, 400 miles "as the crow flies" from Kunming. One had battle damage, with its radio operator dead, and both landed with hardly enough fuel to taxi off the runway.[2]

The six Mitchells, scheduled to give Chennault his long-desired bomber force, had left Dinjan with full bomb loads and extra cases of ammunition for the A.V.G. Instead of flying directly to Kunming, however, some unnamed official at the last minute directed them to bomb the Japanese base at Lashio on the way, requiring a dog-leg adding 300 miles to their planned route. Over Lashio, they dropped their bomb load and turned to the northeast for Kunming. Only two arrived at Kunming. Whether the others became disoriented because of the poor weather, or simply ran out of gas, is a matter of conjecture. Four of the six were lost because of an ill-planned, last-minute tactical decision. As George Burgard commented, "The man that ordered that mission ought to be court-martialed." [3]

Six A.V.G. pilots left Kunming on a DC-3 for Chungking on June 4 to take the battery of written and physical exams for regular commissions in the U.S. Army. Appearing before Brig. Gen. Clayton Bissell, the board chairman, they were none too optimistic about their chances. Charlie Bond and friend George Burgard were too old (Bond had turned 27 in April and Burgard was only a month younger); Charlie Sawyer failed the hearing test; and Robert "Catfish" Raines, a former Navy pilot, had never graduated from the required Army flight school.[4] Frank Schiel and Ernest "Bus" Loane were accepted without waivers. Recalled Charlie Bond, "After I handed my papers in, I asked General Bissell about my chances of getting a regular commission. I told him the date I graduated from flying school, and he said one word – 'Impossible'." Sometime after returning to Kunming, however, they learned that all had been recommended for acceptance as second lieutenants in the Regular Army, with temporary ranks as captain or major.[5]

Low clouds and drizzling rain had constrained flying operations at Kunming for the last three days, providing an opportunity on June 6 for a long-planned awards ceremony. With a small band to provide martial music and a Chinese Army honor guard assembled in the main hangar, Chinese Colonel Wang Shu-ming presented the 25 pilots and three ground personnel present for duty with appropriate decorations – the Chinese Cloud Banner (4th, 5th, 6th or 7th Class) – for "bravery and outstanding combat performance." Similar decorations were given posthumously to seven pilots who died in combat or flying accidents. (Strangely ignored were M.I.A.s Billy McGarry and Lew Bishop.)[6]

The departure of the First and Second Squadrons for Chungking was delayed once again – the weather for June 7 in Chungking was reported as "ceiling and visibility zero." It was good enough at Kunming, however, to disperse several

P-40s from Wu Chia Ba to the outlying Generalissimo Field, the auxiliary dispersal airstrip. After lounging around a bit during the morning and playing a hand or two of cribbage, Charlie Bond drove out to the field to lead a flight of six to Generalissimo, as directed by Bob Neale. To his surprise, when he arrived, he found that Bill Bartling had already taken off with a flight of four – with three new Army pilots – to the dispersal area. Jumping into the station wagon, he drove immediately to Generalissimo Field, where he found a chaotic sight – three wrecked P-40s!

Arriving overhead at the auxiliary airstrip, Bill Bartling had peeled off for a landing, with his three new charges obediently following in trail. Cutting his power to idle, he set up a fast glide onto final. Touching down long, as he passed midfield he was still going much too fast – he had misread the windsock and was landing downwind. In an attempt to avoid running off the end of the runway, he jammed the rudder pedal in an intentional ground loop. His P-40 swung around, shearing the gear and came to an abrupt stop. Following closely behind, two of the Army pilots plowed into Bartling's bird, but the third pilot managed to skid to a stop just short of the wreckage. Fortunately, injuries were confined to a few scrapes and bruises – and to Bartling's pride. All four piled into Bond's station wagon for the ride back to Kunming where they faced a very angry Bob Neale. In Neale's own words, he "gave the boys a wee lecture." [7]

The weather at Chungking finally cleared on Tuesday the 9th, but by that time Chennault had amended his plans; the Second Squadron would set up shop in Chungking, while after an overnight rest stop, the First Squadron would continue on to Kweilin, 400 miles to the southeast to provide a presence in East China. Reports had come in that heavy concentrations of enemy aircraft were based at Hankow, Hupeh, Hangchow, Chekiang and Canton; indeed, the Japanese had been using the Kwangsi Province capital for bombing practice for some time, much as they used Kunming during the dark days of 1941 before the A.V.G. arrived. Seventeen First Squadron mechanics boarded a C.N.A.C. DC-3 for Kweilin that afternoon, and Chennault would follow on another C.N.A.C. flight the next day.[8]

Ed Rector led the departure to Chungking with the Second Squadron's P-40s. The First Squadron followed an hour later with Bob Neale and George Burgard leading nine and Charlie Bond leading six. The 15 P-40s roared low over the center of Chungking when they arrived, then circled back to do it again, giving the impression (to any enemy spies who might be watching) that a much larger force had arrived. As they climbed out of their cockpits upon landing at the Peishiyi airport, they were met by A.V.G. nurses Emma Jane "Red" Foster and Jo Stewart with the news of the U.S. Navy's victory at Midway, sinking four Japanese aircraft carriers.[9]

In early June, Admiral Yamamoto had sent out the bulk of the Japanese fleet, including four heavy and three light aircraft carriers, with orders to engage and destroy the American fleet and invade Midway. U.S. intelligence broke the Japanese naval code, however, allowing three heavy aircraft carriers of the Pacific Fleet to be stationed 350 miles northeast of Midway, awaiting the advance of Yamamoto's armada. Beginning on June 3, 1942, the battle sea-sawed back and forth, with the Japanese losing four heavy carriers and one heavy cruiser, with the U.S. Navy losing a cruiser and the aircraft carrier *Yorktown*, which had been badly damaged a month earlier in the Battle of the Coral Sea. The battle so decimated the Japanese naval forces, they were never again able to effectively challenge the U.S. Navy. It was a definite turning point in the war in the Pacific.

The planned departure of the Adam & Eves was delayed for two days, as reports came in from Kweilin that the Japanese had bombed the airfield three times a day for the past two days; the runway had suffered direct hits. On June 11, however, Bob Neale's squadron finally took off at 1000 for Kweilin – minus George Burgard, who was to have led a flight of four. Unable to get his engine started, he was delayed for an hour while his spark plugs and leads were being replaced. During that time, Charlie Bond (leading another flight of four) also landed after aborting with a sick engine. By noon, however, all repairs had been made and the two were on their way to Kweilin.[10]

As they approached the field, Charlie Bond was impressed with the remarkable change in the landscape, as he described in his diary that night:

> *"The Kweilin field is surrounded by high mountains that look like inverted ice cream cones. The valley is lush, and the field is the biggest and best that I have seen in China. It is over a mile long and has revetments that can take B-17s. The runway is a hard surface, gravel mixed with clay."* [11]

After landing, the P-40s were dispersed about the field and the pilots were driven to their new hostel, some five miles from the field. There they found equally improved amenities – American furniture, showers, and a pool table in the club room. "This is the best layout yet," observed Bond.[12]

Bond and Burgard weren't the only late arrivals. Matt Kuykendall had engine trouble en route but managed to put his crippled bird down at an auxiliary field where it nosed over on landing and caught fire. Uninjured, he quickly climbed out of the aircraft before it burned up and caught a ride to Kweilin with a group of Chinese.[13]

That evening, Neale called his pilots together to discuss combat formations and tactics for the anticipated third Japanese raid in four days. It had been a full month since the Adam & Eves had engaged in aerial combat, and he felt they needed a tactics review. Before turning in, Neale recorded in his diary, "Tomorrow should be a big day, as I have 11 planes waiting for them should they come over. Here's hoping." [14]

After a short night, the 11 pilots were awakened at 0300 and driven out to the airfield where they began sitting alert. A few were playing cribbage when at 0500 came a report of a lone enemy reconnaissance ship overhead, signaling an imminent raid. Running to their planes, Neale had his eight Tomahawks and three "E"-models airborne by 0530.[15] Circling 15 miles to the northwest, Neale took Bill Bartling, Dick Rossi and 2nd Lt. Romney Masters to 17,000 feet, while Charlie Bond, Joe Rosbert, Al Wright and 2nd Lt. James Dumas leveled at 15,000. George Burgard took top cover at 21,000 feet with John Dean and 2nd Lt. Vaden Carney. At that moment, coming from Canton, 250 miles to the southeast, were five 90th *Sentai* Kawasaki 48 Lilys, escorted by eight 54th *Sentai* fixed-gear Mitsubishi Ki-27 Nates and five twin-engine Kawasaki Ki-45 Nicks, a new Japanese fighter they had not seen before.[16]

At 0555 the voice of radioman Ralph Sasser crackled over their headsets with, "Come on in, boys – the weather's fine!" – alerting the flight that bombers were over the field. Joe Rosbert was the first to spot the five Lilys, which were already retreating to the southeast after dropping their loads. Sliding in close to Bond, he fired his guns to get his leader's attention. Bond, in P-40E No. 129, peeled off to lead the attack, followed closely by Rosbert . After two rear-quarter attacks, firing short bursts at the Lily, five of his six .50s quit firing. As he slid out to recharge his guns, his coolant overheat light glowed yellow and smoke began curling out from beneath the instrument panel. Bond's first thought was that his coolant system had been hit by return fire from the bomber's rear gunner.[17]

Diving down and away, he looked back to find two Nates on his tail. Steepening his dive until his airspeed indicator read 315 mph, he leveled out only 1,000 feet above the ground, dismayed to see that one of the Japanese fighters was still with him, just out of range of his guns. As Bond related later:

> *"I was beginning to feel that my number was up this time. My oil pressure dropped to zero... hunched behind the armor plate, listening for the ping of the Jap's guns while trying to make a decision about bailing out or belly landing. Luckily, the Jap must have thought I was a goner, for he turned his fighter away and climbed back to the battle area. A second later, my propeller stopped..."* [18]

Too low to bail out by that time, Bond elected a wheels-up landing in one of the numerous rice paddies he was gliding over. His Kittyhawk hit the ground hard, bounced out of one muddy rice paddy, sailed over a small dike and came to an abrupt halt in another in another rice paddy, throwing him against the shoulder straps and into the gun sight. Momentarily dazed and bleeding from two deep gashes in his forehead, he slid the canopy back, gathered his gear and climbed down onto the wing of his smoking P-40. Approaching a group of Chinese working in the field, he managed to get to a telephone where he was connected to Catholic missionary, Reverend Herbert Elliott. With help from the Chinese, he made his way to the missionary's village, where his head was cleaned and bandaged, was fed and put on a train back to Kweilin.[19]

Meanwhile, after making the two ineffectual passes with Bond, Joe Rosbert, in P-40E No. 131, wheeled around where he attacked three stragglers. As he reported:

> *"I picked one out and made three or four diving attacks from the rear... I left him after his left engine*

was burning and he was going down. I made three or four passes on the other bomber which made violent sharp turns and dives. Finally I got right behind him and with one burst at short range the rear of the plane blew up and went down in flames..." [20]

Former Yunnan-yi instructor Al Wright, in his first aerial combat, followed closely behind Bond and Rosbert as they made their first passes on the bomber formation. At that point, as he later stated in his combat report, his engine "started missing, after which it started smoking badly." He never saw the Nate that was on his tail.[21]

Second Lieutenant Dumas, flying on Wright's wing, made only two passes. But, with the expected "buck fever" on a first aerial encounter, he fired from too long a range and missed both.[22]

Bob Neale spotted the fighters that Rosbert missed – eight Nates behind and slightly above the bombers. His flight took on the nimble, fixed-gear Nates, with Neale firing a full deflection burst that missed. Swinging around, he took a head-on shot at another, observing smoke coming from the fighter as they passed each other. After another pass on a third Nate, also seen smoking, he found himself on the receiving end from a Nick (which, being unfamiliar with the new twin-engine fighter, he later identified in his combat report as a bomber). Pushing the throttle forward, he dived away in an attempt to elude his pursuer. After a five-mile chase to the west, he had gained enough distance from the Nick to do a 180 for a head-on pass, which again proved ineffective. Turning to the south, he spotted a southwest-bound Nate. In a quick pass, he observed hits on the little fighter but couldn't claim it as confirmed since he didn't see any flame or crash. By the time he reached the field, the remnants of the Japanese force was heading back southeast in retreat. He landed his undamaged Tomahawk No. 7 at 0645, claiming only two "probables." [23]

Bill Bartling, flying Tomahawk No. 16 on Neale's wing, sailed into the Nates with several diving attacks. Like his flight leader, however, he thought he saw his tracers going into several of his targets, but with no apparent effect. Becoming separated from his flight during the twisting, turning air battle and drifting northwest of the field, he spotted a Nick (which he also called a "bomber") diving on a P-40. With a quick wing-over, he dived on the twin-engine fighter. After chasing the Nick south for several minutes, he looked back to see a Nate slightly below but coming up fast in his six-o'clock position. Then, as he reported:

> "*...I swung back and got on the I-97 in a head on run. He burst into flames and went down...I saw a different bomber trying to swing on me. I made a head on run with him and I saw his left engine smoking. By the time I had turned around he was heading for home. I finally caught him from the rear. He made a sharp turn at my first burst from a long range. We then made a head on run, narrowly averting collision. Then I noticed him going down. I hit again in the side of his fuselage and he broke into flames and crashed...*" [24]

Dick Rossi, his radio out, picked out a Nate below and heading away. His first bursts missed and the Japanese fighter turned sharply beneath the Tomahawk's nose. Spotting a second Nate "heading for home," as he put it, Rossi moved in to within fifty yards before pressing his triggers. At that moment all his guns jammed, and he was forced to dive away under the lucky Japanese pilot. After clearing the malfunction, he reported:

> "*I found an I-97 going home (pretty far southeast) smoking slightly. I came in behind firing a burst and set him ablaze...*" [25]

The fourth man in Neale's flight, 2nd Lt. Masters, was flying in Tomahawk No. 15. He made his first pass head-on to a Nate but dived away when he saw another Nate coming up close behind. As the bombers turned to the southeast, he gave chase, finally getting close enough to fire a burst at "tail-end Charlie." Although Masters reported that he "noticed a trail of smoke," the bomber maintained its position in the formation and continued to the southeast.[26]

George Burgard, maintaining top cover at 21,000 feet with his flight of three, waited three minutes after Bond made first contact with the Japanese bombers, then radioed Neale that he was coming down. Five miles southeast of the field, with Lieutenant Carney sticking close on his wing in Tomahawk No. 96, he spotted four Nates several thousand feet below and dived to the attack. Firing long bursts on his first pass, he scored hits before he had to break away, as he put it, "to avoid another Jap who was practicing his deflection shooting on me." After several minutes of the swirling air battle, during which he fired on at least three more Nates, he saw a P-40 2,000 feet below that was smoking badly and being closely pursued by a Nate. Diving on the Nate, he drove it away, then followed the P-40 back to the field. The P-40, of course, was Al Wright's Tomahawk No. 42, which Burgard observed making a crash-landing on the field before turning south and climbing back into the battle. A quick radio call to the field assured him that Wright got out OK. before his ship burned, although he did suffer a wrenched back.[27]

As he climbed out, Lieutenant Masters formed on Burgard and the two cruised south at 13,000 feet. About 50 miles out, they overtook a Lily with a Nate just above. Burgard fired on the Nate, then turned for the bomber which dove for the ground. Overshooting in the dive after the Lily, he climbed out for another run, just as the Nate put a round through his aileron as it dived past. Then, as Burgard reported:

> *"I made a sharp diving turn and poured three long bursts at the fighter on a deflection shot. The fighter skidded badly and fell off on a wing. I pulled up and saw he had run into the side of a sharp peak and blew up."*[28]

Climbing back up, he picked up the bomber hedgehopping through the sharp karsts. Burgard's twin .50s jammed as the two twisted and turned at never more than 150 feet above the ground, banking tightly around the cone-shaped peaks. After five or six such passes, Burgard caught the Lily in a valley, as he described in his combat report:

> *"My speed was too great to stay behind him and when I pulled up to kill speed he headed south again... On a pass from above my right fifty started working and in as brief interval his left engine caught fire. He attempted to maintain his altitude but I throttled back and continued to shoot until he dragged his wing on a small knoll and mushed in..."*[29]

John Dean, in P-40E No. 132, followed Burgard and Carney into the fray. Attacking the bomber formation on the right flank, he hit a Lily, starting its right engine to smoking as it dropped from the formation. Following it down, he continued to pour fire into the crippled bomber as heavy smoke began to pour out. Breaking off the attack, he saw a Lily under attack by two other P-40s and joined the fight, as he reported:

> *"...I fired a short burst into the twin-engine bomber and saw the plane burst into flames and go earthward in a flat glide... At this time I was about 8,000 feet and saw a low flying plane below me. Diving down I discovered it to be an I-97. I closed the range to 800 feet and gave him a burst. Bits flew from the plane and the engine began to smoke. He rolled over to the right and I saw him crash."*[30]

Then, suddenly, it was all over – the sky was empty of Japanese aircraft. One by one, the nine remaining P-40s landed back at Kweilin, the last, Burgard, touching down at 0730. Except for the two aircraft lost to Japanese gunfire, the only other damage was two bullet holes in Bartling's left wing and the one hole in his aileron. Writing their combat reports while the air combat was still fresh on their minds, the pilots agreed that they had destroyed nine Japanese aircraft, half of the raiding force. Rosbert claimed two Lilys, while Bartling, Dean, Burgard and Rossi each downed one Nate. Bartling, Dean and Burgard also claimed a bomber each; but, from the descriptions of their Japanese opponents and their speed and maneuverability, the "bombers" were the new twin-engine Ki-45 Nicks, which they had not seen before.

It was mid-afternoon before Charlie Bond arrived at the Kweilin railway station, a wide, white bandage wrapped around his head. After welcoming him back, Chennault smiled and said, "I hope your bad luck is over."

"So do I," responded Bond.

Third Squadron pilot Bill Reed quipped, "I wonder if old Charlie's got any other hobbies?"[31]

After losing three of Bob Neale's P-40s (including Matt Kuykendall's lost on the way to Kweilin), the next day Chennault ordered Ed Rector to take a flight of six P-40Es to augment the eight remaining ships of the Adam & Eves. Departing Peishiyi at 0920, Rector left John Petach in charge of the Second Squadron. Panda Bear leader Tex Hill and company would arrive in Kunming later in the day from their ferry trip from India. After a 10-day checkout program for the Chinese pilots in Karachi, they had been delayed by a sand storm in Jodphur and monsoon rains in Dinjan, but they were successful in bringing back 11 P-43s for the C.A.F., losing only one aircraft and its Chinese pilot who became disoriented flying through an overcast.[32]

Encouraged by the latest A.V.G. victories, Chennault sent a radiogram to Chungking, making another plea for a bomber force:

> URGENT GEN. STILLWELL CARE GEN BISSELL. WITH REINFORCEMENT OF ONE B-25 AND ONE SQUADRON PURSUIT JAP AIR FORCE INTERIOR CHINA CAN BE DESTROYED BESIDES INFLICTING GREAT DAMAGE OTHER JAP INSTALLATIONS. SITUATION MOST OPPORTUNE NOW BUT IMMEDIATE ACTION NECESSARY BEFORE CHANGE OCCURS.[33]

A series of three photographs featuring the line up of A.V.G. pilots following the Chinese Air Force awards ceremony in Kunming on June 6, 1942. Pictured are Snuffy Smith, Bill Bartling, Charlie Bond, George Burgard, Gil Bright, Dick Rossi, Bob Neale, John Petach, John Dean, Joe Rosbert, Fritz Wolf, Bob Prescott, Bill Reed and Ken Jernstedt. (via J.J. Harrington)

CHAPTER 23

"It Was The Hardest Thing I Have Ever Done..."
– *Robert Neale*

Chennault reluctantly left Kweilin on June 14 for a called meeting in Chungking with Tenth Air Force commander, Maj. Gen. Lewis Brereton, who had flown in from his headquarters in Delhi, India, to establish the final induction board. With barely three weeks remaining before the A.V.G. was formally disbanded, and with a minimal number of new U.S.A.A.F. arrivals in China, he was hoping to get the majority of the A.V.G. to remain in China as members of the U.S.A.A.F. Chennault was also hoping for some positive results from the meeting, gaining control of the five B-25s and one B-24 already in Kunming. Brereton dashed that hope, however, telling Chennault that Army policy allowed only one numbered Air Force in a theater of operations, and that, although he would still have charge of the fighters, the bomber force would be commanded by Col. Caleb Haynes, soon to be promoted to brigadier general. Chennault's new command would be named the China Air Task Force.[1]

Gen. Brereton then appointed a six-man induction board for the recruiting task – Chennault, Col. Homer Sanders, Lt. Col. Walter Urbach, Maj. Edgar J. Ingmire, newly-promoted Maj. Tom Gentry and U.S. Navy Capt. Mark T. Gilkinson to make the appeal to the former Navy personnel.[2] Left off the board was Brig. Gen. Clayton Bissell (who had failed so miserably during his earlier meeting with the A.V.G.). He was also in attendance but said little during the conference.

Although the board wouldn't meet formally until they arrived in Kweilin a week later, Chennault met informally with Tex Hill, who had recently arrived in Chungking from his three-week ferry trip from India. When approached by the "Old Man" to accept induction, Tex simply replied, "General, I'll stay around as long as you need me." [3]

Rain and low clouds kept the A.V.G. on the ground at Kweilin for the next week, with only dawn-to-dusk alerts being maintained and no threats from the I.J.A.A.F. In a replay of the Japanese raids on Kunming six months earlier, and after taking a beating at Kweilin the Japanese Army Air Force didn't return to the Kwangsi capital while the A.V.G. was in the area, except for a small force sent on the 13th to destroy the wreckage of the Ki-45 that Burgard shot down 50 miles south of Kweilin.

On the 19th, the Adam & Eves had the opportunity to meet a survivor from the hapless Nick, when I.J.A.A.F. Corporal Kei Honda, the rear gunner, was brought in. He had held out for three days, armed with a machine gun taken from the wreckage, when, exhausted and hungry, he surrendered to Chinese forces. Under interrogation, he said that he was drafted from his chicken farm and sent to the Air Force where he was trained as an aerial gunner. He revealed that the new twin-engine fighter was armed with a 20-mm cannon and two 12.75-mm machine guns in the nose, which forewarned the group to be cautious of any future head-on passes on the aircraft. After a photo session with the pilots, and being introduced to Burgard, the pilot who shot him down, he was turned back over to the Chinese for final disposition.[4]

Stung by their losses over Kweilin on the 12th, the I.J.A.A.F. turned their attention to the undefended city of Hengyang, 200 miles to the northeast of Kweilin, hitting the city relentlessly. To meet the new threat, Chennault further spread his forces on the 20th, ordering six Second Squadron P-40s to the beleaguered eastern Chinese city. Having just returned from Chungking, Tex Hill led the flight, with newly-arrived U.S.A.A.C. 2nd Lt. M.S. Cluck on his wing. Shortly into the flight, he was forced to turn back with his fresh, young pilot because of weather, leaving Ed Rector in charge of the Panda Bears contingent – Catfish Raines, former C.A.F. instructor Van Shapard, U.S.A.A.C. Lt. Ajax Baumler and First Squadron pilot Charlie Sawyer. The five arrived at Hengyang at 1745, and would be joined by the sixth P-40 the next day, flown in by Frank Schiel.[5]

The monotony of the constant foul weather at Kweilin, unending games of cribbage and acey-deucey and uneventful alert duty was broken on the night of the 20th when the citizens of Kweilin turned out to give their A.V.G. protectors a grand banquet. As it was described by Charlie Bond:

"We were wined and dined and showered with praise and even a few gifts. The hostel club was decorated by the Chinese, and it was without a doubt the most elaborate affair I had attended since being in China. They even had cigars at every place setting. We received beautiful pieces of embroidery – eagles pouncing on the Japanese red sun…"[6]

The Adam & Eves returned the favor the next afternoon, with a nine-plane review for the citizens of Kweilin, with Neale, Bond and Burgard leading the three-plane elements. Commented Burgard, "It was a pretty darned good show and the people ate it up."[7]

Chennault and the induction board arrived at Kweilin on the 21st. Following a short meeting with the men that night, outlining the process of the board, the meeting began at 0800 the following morning in a cave adjacent to the similar cave that served as the A.V.G.'s operations room. Col. Sanders, seated next to Chennault, began the interviews with the ex-U.S. Army personnel. When it came Charlie Bond's turn, Sanders said, "I'm sure you're going to stay with us, aren't you, Charlie?" Bond replied, as he had done before, that he would gladly stay, provided he could be given a regular commission in the U. S. Army. When Sanders replied that he couldn't do anything about that, it ended the conversation.[8]

Bob Neale was one of the last pilots to be interviewed, sitting down across from the board just before noon on the 22nd. As a former U.S. Navy pilot, he, too, was questioned by the board's Navy representative, whom Charlie Bond recalled as being "too abrasive and arrogant." That didn't matter all that much to Neale, however. He had already made up his mind. "It was the hardest thing I have ever done to tell Gen. Chennault that I was going to refuse induction here," he recorded in his diary that night. Immediately after the interview, Neale began to feel sick and went over to see flight surgeon Doc Sam Prevo, who diagnosed his condition as probable dengue fever. At 1500 he asked Charlie Bond to take over temporary command of the Adam & Eves before returning to the hostel and climbing into bed.[9]

At Hengyang the Second Squadron had their hands full. After Ed Rector led a strafing attack on a Japanese gunboat and a few smaller watercraft on the Yangtzi River south of Hankow that morning, the Japanese were quick to retaliate. The Panda Bears were alerted by the Chinese early warning net at 1320 that 14 enemy aircraft were 80 kilometers from Hengyang and Rector requested support from Kweilin.[10] However, the deteriorating weather at Kweilin delayed Charlie Bond from immediately mounting a flight until Chennault cleared Bond to attempt a takeoff into the weather. Bond got into the air with eight P-40s and headed north. Heavy rain and lowering ceilings, however, forced the flight down to tree-top level, and Bond reluctantly turned back to Kweilin.[11]

Meanwhile, Rector led five P-40s into the air to meet a force of 14 Ki-27 Nates (Catfish Raines reported seeing 14, while Charlie Sawyer gave the number as 21-24 in his combat report). Climbing to 20,000 feet, they circled the area before spotting the incoming Japanese aircraft, some 8,000 feet below.[12]

Following Rector and Ajax Baumler as they peeled off for a diving attack, Charlie Sawyer, in P-40E No. 111, made a head-on pass, giving the Nate a quick burst before diving under him. Although he saw smoke streaming from the enemy fighter's engine, he didn't have time to follow him for a "kill" confirmation before he tacked onto another, coming in from the Nate's rear quarter. This one also began smoking as Sawyer dove away. After making several more attacks without a visible result, he finally drove in behind a Nate and, as he described it:

"I came upon him slightly below and directly from the rear. I opened fire at approximately 500 yds and closed to about 250 yds, with pieces flying off and smoke coming out. He turned 90 degrees to the left and I got in a good burst at close range and he seemed to explode, and disintegrate."[13]

Raines, flying P-40E No. 116, was experiencing mechanical problems as they circled Hengyang – his prop governor was cycling off and on for the first 10 minutes after the Japanese aircraft were sighted. Finally it started working and Raines joined the combat, diving on a lone Nate at 13,000 feet. Forced to break off when he was jumped by six more Nates, he made a half-dozen passes on other "targets of opportunity," reporting smoke from one, and the canopy flying off another before both dove into cloud cover.[14]

Frank Schiel joined the battle late. He had flown a reconnaissance mission over the Japanese base at Hankow in his camera-equipped Tomahawk earlier that morning, returning to Hengyang at 1400, a half-hour after the earlier scramble. Refueling, he was in the air by 1430. Climbing to 12,000 feet, he spotted a lone Nate 10-15 miles north of Hengyang. Continuing his climb, he placed the sun at his back and dove on the unsuspecting Japanese fighter. As he stated in his combat report:

> *"Fired a long burst of 50 rounds. Bits of metal flew from the ship as I could see the tracers enter the plane. The Jap pulled up sharply, did a snap roll to the left, and started down in a shallow dive, smoke and gasoline vapor streaming out behind the ship. I turned to the right, made a circle and started back for a second pass, but the Jap ducked into a cloud and I was unable to pick him up again... This plane is a very definite probable which will no doubt be found by the Chinese."*[15]

Van Shapard, flying his second combat sortie since joining the A.V.G., became separated from the flight during the melee. Jumped by five Nates, he dove for home. In desperation, he called Mickey Mihalko, who was manning the radio shack at Hengyang, proposing to fly low over the base, giving ground gunners an opportunity to fire on his pursuers. As Shapard's P-40 came streaking low across the field, the Nates close behind, the A.V.G. ground troops opened fire with all the machine guns they could muster. Their fire was largely ineffective, but it at least broke up the chase and allowed Shapard to land unscathed.[16]

The dismal weather began to break by mid-morning of the 23rd. At 1106, the Hengyang radio net reported four waves of Japanese bombers approaching Hengyang from the southwest.[17] This time, the Adam & Eves were able to respond without delay, with George Burgard leading five P-40s toward Hengyang. When the expected attack proved to be a false alarm, the five returned to Kweilin.[18]

Early that afternoon, satisfied that the Japanese had turned their attention from Kweilin to Hengyang, Chennault directed Charlie Bond to take seven of the First Squadron P-40s to Hengyang to reinforce Ed Rector and his Panda Bears, but to delay takeoff as late as possible, aiming for a landing at Hengyang just before dark.[19] With that, Chennault and the induction board departed for Hengyang in Col. Haynes's DC-3, their mission at Kweilin largely a failure. Although a few ground personnel had signed on, no pilots had been so persuaded.[20] Before leaving, Haynes sent a radiogram to Gen. Stillwell, voicing the concern of the board:

> FROM PRESENT OBSERVATIONS I DEEM IT IMPERATIVE THAT INDUCTION OF A.V.G. BE DEFERRED UNTIL OCTOBER FIRST AND THAT PRESENT CONTRACTS BE CONTINUED OTHERWISE OUR OPERATIONS ARE IN SERIOUS JEOPARDY. INDUCTION BOARD STRONGLY CONCURS IN ABOVE RECOMMENDATIONS. SGD. HAYNES[21]

The plea got no further than Stillwell, however, who promptly dropped it into "File 13."

A light rain was starting to fall at 1730 as Bond's flight taxied out to line up on the runway. As Bond started to take the active, ground control told him to "hold" – a flight of six B-25s from Kunming had been cleared to land. Five minutes later, after the last Mitchell cleared the runway, Bond and company were on their way to Hengyang, with Bond leading a flight of four and George Burgard following with a flight of three.

The seven ships circled Kweilin widely to the north to avoid observers noting that they were leaving the city to the southeast. Weather en route was still marginal, with rain continuing to fall. A headwind didn't help matters either, and it was just getting dark when Bond spotted the field at Hengyang. Angling away from the city to the southeast, they flew at tree-top level as they circled back to the field. Landing as dark closed in, the seven were met by Ed Rector and his pilots who drove them the two miles to their three-story hostel, reminding some of a large college dormitory. As George Burgard commented later, "The hostel here is very nice, but the food stinks!" [22]

It was a very short night. The men had barely gotten to sleep when someone ran down the hall yelling, "The Japs are on their way!" Piling out of bed, the A.V.G. pilots rushed out of the hostel to take shelter along the bank of the nearby river. After waiting for a short while for the bombs to fall, the "all clear" was sounded. Another false alarm. The raid was apparently targeted for Changsha, 100 miles to the northeast.

After going back to bed for a couple hours of sleep, the pilots were roused again by Rector to drive back to the field to position the P-40s for a quick scramble if need be. He then directed a two-plane element into the air, "just in case." Although the early-warning net reported a single ship (probably reconnaissance) some 50 miles distant, the deteriorating weather prevented it from getting any closer.[23]

Landing at Hengyang, Chennault scheduled the first interviews for 1630. One of the first in line during the first two-hour session, former U.S. Navy ensign Jim Howard was interviewed by Capt. Gilkinson. After being commended for his service in the A.V.G., he was offered promotion to major and command of a fighter squadron. Still suffering from the effects of a bout with dengue fever, he politely rejected the offer. After seven months of combat, he just wanted to go home.[24]

Howard's roommate, Frank Schiel, was next in line. After a long back-and-forth discussion in which he, too, was offered a majority, he finally accepted. Perhaps influenced by Second Squadron leader Tex Hill. He and fellow Panda Bears Ed Rector and Gil Bright, had been offered similar promotions to major with the attendant promise of squadron commands, accepted induction into the U.S. Army Air Corps. Charlie Sawyer, offered a captaincy, also signed on. A second two-hour session was held following a quick dinner, which netted a few more ground personnel. It was a small victory for the recruiters – at least better than their reception at Kweilin.[25]

Following a last effort from 0800 to 1000 on the 24th, the board flew to Kunming, where they picked up a few more ground personnel, but no pilots.[26] The pitch had been a losing effort from the beginning. The men had endured seven months of relentless combat against a far larger enemy force; ever-dwindling supplies and equipment, with little chance of resupply; primitive living conditions much of the time; many debilitated from the effects of malaria and dengue fever; and they were just plain tired and homesick after being away from family, friends and familiar American cooking for nearly a year. When all some wanted was a furlough back to the States for a few weeks before returning to combat, that, too, was denied. All the above was exacerbated by the imperious and very undiplomatic approach by Brig. Gen. Clayton Bissell early on.

After making their final plea at Kunming, the board returned to Chungking. Their tally – five pilots and 35 ground personnel (including four who accepted induction into the U.S. Navy).[27] Considering the circumstances, that many was a miracle.

The East China bases were "socked in" from the 24th to the 29th. The pilots at Hengyang passed their time reading, playing cards or listening to the one phonograph record they had for their portable player – "Moon Over Burma." When Bond brought in his Adam & Eves to augment Rector's Second Squadron, most thought that they would be there for only a couple of days and then head back to Kweilin. "All we brought was a toothbrush," observed George Burgard, adding, "These clothes are beginning to stink badly." [28]

With Hengyang established as his forward operating base, on the 29th Chennault sent the six B-25s recently arrived at Kweilin to Hengyang. Escorted by the recently recuperated Bob Neale and Bill Bartling, the flight took off as the weather cleared at 1630 and arrived at Hengyang forty minutes later.[29]

Back in Chungking, a deeply disappointed Chennault assessed the situation. Only 16 Army Air Corps pilots had arrived to add to the five A.V.G. pilots who had signed on to stay – hardly enough to man a squadron, much less an entire fighter group. Also, Chinese intelligence had reported the Japanese were mounting a force to "wipe out the U.S.A.A.C. after the A.V.G. leaves." If only he could get his men to extend just two weeks more to give the Army brass enough time to get the 23rd Fighter Group in place...

With that in mind, and with the subsequent approval of the Generalissimo to extend the contracts, he boarded a C.N.A.C. DC-3 for Hengyang on the 30th. Before leaving, however, he approached Tex Hill, perhaps the most loyal to the "Old Man" of all his men. Recognizing Hill's popularity with the men, he reviewed the grim situation with Hill and asked him to try to persuade as many as he could to extend their contracts for two weeks, emphasizing the importance of maintaining a viable fighter force until the new 23rd Fighter Group could be fully organized and trained.

"Sure, I'll do that." said Hill, nodding in agreement.[30]

Hill had his first opportunity a short while later when he encountered Johnny Petach, travel bag in hand. Exchanging pleasantries, Petach said he was just on his way to say goodbye to Chennault before he boarded the C.N.A.C. transport to India. Explaining the dire situation to Petach, Hill had his first recruit. Petach said, "Sure, Tex, I'll stay. I guess two weeks won't make much difference." [31]

Shortly after arriving at Hengyang, Chennault met with Bob Neale to press the need for his pilots to extend for two weeks to prepare the fledgling 23rd Fighter Group for combat. Neale reluctantly agreed to stay and accompanied Chennault into the alert shack where he called his squadron together to put forth the proposal. Although the first reaction from many was, "Hell, no!" Neale was able to get four Adam & Eves to join him, including Charlie Bond. Bond wasn't very happy about the situation, though. Just earlier in the day Ed Rector had announced that they would be leaving Hengyang that day. "I was mad as hell," Bond wrote later, but added, "I knew my conscience wouldn't let me do anything about it but say yes." [32]

After a short stay at Hengyang, Chennault boarded the DC-3 and, accompanied by Bob Neale, flew to Kweilin to make a pitch to the ground crews.[33] In the final analysis, he was successful in persuading 20 departing A.V.G. pilots and some two dozen ground crew to stay on for the requested two additional weeks.[34] Adding the five who signed on to be inducted earlier, along with the 16 Army Air Corps pilots who had trickled in since the beginning of June, however, he could count on only 41 pilots for the next two-and-one-half weeks. He fared better, number-wise, with the ground personnel. Adding to the 22 who agreed to stay for two more weeks, 24 more had accepted U.S.A.A.C. induction as lieutenants or N.C.O.s. A few of his staff agreed to stay with him as well – Skip Adair and flight surgeons Tom Gentry and Sam Prevo received commissions as majors, while John Williams and Roy Hoffman became captains.[35]

Bob Neale returned to Hengyang on July 1, bringing five crew chiefs from Kweilin with him. Forced to turn back once for low ceilings, they final made it in at 1130 after flying 100 feet off the ground most of the way.[36]

That afternoon the weather broke. Chennault took advantage of "his" new force of six B-25 medium bombers (newly-designated as the 11th Bombardment Squadron, technically under the command of Col. Caleb V. Haynes), and ordered them to strike the Japanese base at the former C.A.F. airfield at Hankow, located on the Yangtze River only 300 miles north-northeast of Hengyang.

The mission was plagued with difficulties from the start. One of the bombers got stuck in the mud, while another had to abort for a broken fuel line.[37] Finally, as the weather lifted, four got airborne at 1630, escorted by Ed Rector and a flight of five P-40s. Heavy cumulus clouds forced the formation to 14,000 feet; and with a solid under-cast, the bombers became lost.

"We milled around for at least a half hour," commented Peter Wright, who was leading Rector's second element. Twice Rector put them back on track, but, unable to find the aerodrome, they attempted to hit their secondary target, shipping on the Yangtze River.

Instead, their 500-lb bombs fell on a small village on the banks of the Yangtze. Turning for home, they inadvertently flew right over Hankow, where Wright observed "about 10 oil tanks or a refinery, eight or ten large ships, and a huge airfield with at least 50 planes all lined up." Luckily, there was no anti-aircraft fire, and no interceptors rose to challenge the bombers and their escorts. "God only knows why," wrote Peter Wright in his diary that night. They landed without incident back at Hengyang at 1830.[38]

That night the citizens of Hengyang hosted a banquet in honor of their A.V.G. defenders. Charlie Bond described it as "almost exactly" as the earlier one in Kweilin, with Bob Neale noting it was "mostly speeches." [39]

The next day, Chennault followed up with another raid on the docks and shipping at Hankow. Although only three B-25s were available, they were escorted by four P-40s. Taking off at 1530, they encountered scattered to overcast clouds for the first 200 miles, then enjoyed CAVU (ceiling and visibility unlimited) conditions to the target area. The formation picked up light anti-aircraft fire 50 miles south of Hankow but proceeded on to destroy a couple of boats and a section of the docks. Again – incredibly – no Japanese fighters were encountered. It was the "lull before the storm." [40]

The air raid alarm rang at Hengyang at 0210 on the 3rd, rousting the pilots from their bunks. Charlie Bond gathered several others, and all piled into their station wagon and rushed to the airfield to find Bob Neale already there. With the sound of heavy engines approaching, they drove down to the riverbank where they took cover as the Japanese bombers passed overhead. Circling once, the bombers dropped their loads, missing the field by several hundred yards.[41]

Ordering an immediate retaliatory strike, Chennault had the armorers out by dawn loading bombs on five Mitchells Their target: the Japanese air base at Nanchang, 250 miles to the northeast. Four P-40s would fly top cover.[42]

Taking off shortly after 0900, the formation flew between a 3,000-foot under-cast and a second cloud layer at 4,500 feet. Just south of Nanchang, the clouds became broken, giving the B-25s a clear view of the Japanese field. Arriving over Nanchang at 1015, they dove down to 2,500 feet and dropped their 500-pounders, which hit along the northwest-southeast runway, causing minor damage to the field. As they turned to the south, they were immediately set upon by six Japanese fighters, a mixture of fixed-gear Ki-27 Nates and Ki-43 Oscars. Catfish Raines, flying above the bombers, turned to engage the attackers. He was "too low and too far behind," as he later reported, "to keep them from hitting the bombers." nevertheless, he caught up with two "stragglers," and fired head on at one Nate, which dived under him. Then, as he continued in his combat report:

> *"Came on another one from behind that was chasing the B-25s. I fired a long burst into him at fairly long range. He rolled over and dove under me trailing smoke. He pulled up in a steep turn off to one side and bailed out of his ship. There were other Japs diving on me so I ran for a cloud and came on home, catching the bombers ten or fifteen minutes from the fight."* [43]

Harry Bolster was also above the Mitchells when they began their bomb run but didn't initially see the defending fighters. As the bombers departed toward the south, he dove on the field in his own attack, strafing a line of hangars and a group of parked Nates, setting one on fire. As he pulled up, he saw three Nates attacking the B-25s. Then, as he later reported:

> *"I opened up on one of the planes and noticed it start into a spin. I then dove out and a shell exploded in my cockpit and a model 'O' zoomed off my tail. I finally lost the Jap, but oil was covering my cockpit and constituted a very dangerous fire hazard. So after flying to friendly territory I bailed out."* [44]

Bolster was subsequently picked up by Chinese civilians and returned to Hengyang. He was given credit for one aircraft destroyed on the ground and a "probable" in the air.

That afternoon the Japanese returned the favor. The Chinese warning net reported a lone aircraft approaching Hengyang. When the "lone aircraft" turned out to be seven Ki-43 Oscars approaching the field from the north, about 25 miles out, the alert flight was scrambled and the five B-25s and a DC-3 took off for Lingling, an auxiliary field 65 miles to the southwest. Bob Neale, leading the P-40s, managed to damage one Oscar in a head-on encounter, even though only three of his six .30 caliber guns were firing. Coming in low over the field, the Japanese strafing attack left one P-40 burning before they climbed out.[45]

Charlie Bond, flying in Kittyhawk No. 114, dropped his centerline tank before engaging the Oscars. Spotting a fighter coming in on Bus Loane's tail, he turned into it, firing short bursts, and forced the Japanese to break off the attack. Climbing out to 17,000 feet, he sighted another aircraft at some distance which, as they approached each other, turned out to be another Oscar. Each pilot banked hard in an attempt to get behind the other, but soon the nimble Japanese fighter had gained the advantage. Diving away, with his airspeed indicator pushing 410 mph, Bond lost his pursuer in the clouds.

Having jettisoned his belly tank earlier, however, he was now running low on fuel, his tachometer was out and his engine had started to run rough. As his fuel low-level light began glowing red, he decided to head for Lingling. Landing with barely enough fuel to taxi in, he was surprised to see most of the other P-40s that were scrambled, as well as the B-25s and DC-3 that escaped the strafing attack.

After the "all clear" at Lingling, Bond returned to Hengyang, only to be told by Bob Neale to follow him, Bus Loane and Dick Rossi back to Lingling. Back again in the Lingling barracks, Neale offered him a shot of bourbon. Bond relayed his experiences of the day, adding that, "At the rate I'm going, maybe I ought to quit." Neale just grinned and said, "Have another drink." [46]

The Fourth of July – Independence Day in America and Dissolution Day for the American Volunteer Group in China – started with a bang. In a follow-on attack on the A.V.G.'s forward fields, the Japanese sent five twin-engine Ki-21 Sallys in the early morning hours to Lingling. Their first salvo awakened Charlie Bond, who saw the flashes from the first explosions through the window of his barracks. The next salvo struck so close to the barracks that "it shook me out of my bunk," as Bond described it.[47]

At 0530, Bob Neale received a report that another wave of bombers was headed for Hengyang. Rushing to their P-40s, Neale and one other pilot took off minutes later, climbed to 13,000 feet, and arrived over Hengyang at 0600. Looking up, Neale saw the Japanese formation 1,000 feet above. The reported "bombers," however, turned out to be 12 Nakajima Ki-27 Nates from the 54th Hiko *Sentai*, based at Hankow. Simultaneously spotting the P-40s, five of the little fixed-gear fighters rolled into a diving attack. Neale dived away, then made a climbing turn to the west. With an altitude advantage and the sun to his back, he opened up on one of the Nates. With smoke streaming from its engine, the damaged fighter started down. As Neale's combat report continued:

> *"Followed fight north and picked out lone enemy fighter about 5,000 feet below. Made diving attack from rear quarter and pulled up. 2nd attack was a front quarter after which fighter started to smoke and was losing altitude... This fighter was joined by 2 more I-97s who maneuvered on my tail... I dove away..."* [48]

As Neale and his wingman were taking off from Lingling, Ed Rector received the same report of the incoming Japanese formation. Leading a flight of six, he climbed west of the field to 18,000 feet, just in time to spot the five Nates diving on Neale's flight, 3,000 feet below. Then, as reported by Rector:

> *"We immediately attacked and had enemy planes outlined against a white cloud layer at 7,000 feet. I fired a 4 second burst at an I-97 in a slight diving turn, and he burst into flames and bits of debris flew from his plane as he dived through the cloud layer. I climbed up to the outside and made several more passes... One I-97 forced me to dive out, but in doing so I got in a head on burst at two fighters coming up at me. I got in a fleeting glimpse of a wingman's engine letting out a large burst of smoke as I passed over them..."* [49]

Flying in P-40E No. 151, Catfish Raines, one of the six in Rector's flight, made several passes, both from head-on and from the rear quarter, with little result. Finally, pulling behind one of the Nates, he fired a short burst before climbing over the Japanese fighter. "I could see him under me going down in a corkscrew dive trailing smoke," he reported. He continued: "I tried to follow him to make sure of a crash, but he went through a small layer of clouds." As three more enemy fighters appeared above him, he had other things to think about. When he landed he claimed one "probable." [50]

As was the case at Lingling, the Japanese attacks resulted in little damage to the parked aircraft or personnel. With no losses to the A.V.G., Neale, Rector, Charlie Sawyer, Van Shapard (scoring his first) and Jim Howard (scoring his last) were credited with the destruction of one plane each.[51]

While the air battles raged above Hengyang, five B-25s lifted off the Hengyang runway in a pre-planned mission to hit the Tien Ho Aerodrome at Canton, 280 miles to the southeast. With an escort of P-40s, the Mitchells encountered only light anti-aircraft fire and no defending enemy aircraft. All five recovered safely at Kweilin as planned.

Charlie Bond and Dick Rossi departed Lingling during the morning to stand alert at Kweilin. They were scrambled once on a reported unknown coming in from the south; the "unknowns" were the B-25s returning from their mission to Canton. No sooner had they landed than they were scrambled on a second alert. Off they went again, only to intercept Bob Neale and Bus Loane coming in from Lingling.[52]

After landing at Kweilin at 1345, Neale was sent up again to escort a DC-3 coming in from Hengyang with a reported Japanese fighter trailing it. Neale found the transport, but no Japanese fighter – another false alarm. Landing at 1845, Neale was through for the day. As he wrote in his diary that night, it was, "Quite a day." Quite a day, indeed.[53]

Meanwhile, in Chungking that night, the Generalissimo and Madame Chiang Kai-shek threw a barbecue in honor of Gen. Chennault and his American Volunteer Group. Held in the mansion of Chinese Republic President Lin Sen, it was attended by Chennault and the few A.V.G. members still in the capital city, along with many Chinese dignitaries including Madame Sun Yat Sen, widow of the Chinese Republic's first president, and Madame H.H. Kung, wife of the Minister of Finance (both sisters of Madame Chiang). The master of ceremonies was General J.L. Huan, director of the War Area Service Corps, responsible for the food and billeting of the A.V.G. in China. Originally scheduled to be an outside party, a rainstorm forced everyone inside where Madame Chiang served non-alcoholic punch and entertained the group with parlor games such as "musical chairs."

As the evening wore on, there were several speeches honoring Chennault and the American Volunteer Group, with Chennault replying in kind. As a finale, Madame Chiang's two sisters unveiled a large oil painting of Chennault with the Generalissimo and Madame, which was duly presented to Chennault. At 2300, Chennault and the other A.V.G. attendees drove back to Peishyi in the drizzling rain. Then, as the clock struck twelve, the American Volunteer Group in China ceased to exist – but will long live on in the history books.[54]

Ed Rector, shown here in the cockpit of his P-40, scored his last of 4.75 victories with the A.V.G. on July 4, 1942, but, as one of the five A.V.G. pilots who remained with Chennault, scored three more with the 23rd Fighter Group. (Fred Hodges photo collection, via Bob Andrade)

Japanese radioman/gunner, Cpl. Kei Honda, survivor of the Ki-45 shot down down by George Bugard on June 12, 1942, stands with 1st Squadron pilots after being brought in on June 20 at Kweilin. Honda is flanked by Snuffy Smith and Bill Bartling, with Joe Rosbert and Dick Rossi above. (via Bob Andrade)

104
Roy Grinnell

First Blood, Last Blood. *Oil on canvas by Roy Grinell. Painting shows Ed Rector downing a Nakajima Ki-27 "Nate" on July 4, 1942, scoring one of the final victories of the A.V.G. Rector scored his first victory during the A.V.G.'s first combat on December 20, 1941, and his last on the day the A.V.G. was disbanded. (Courtesy Roy Grinnell Studio)*

CHAPTER 24

The 23rd Fighter Group Takes Over

The 23rd Fighter Group was officially activated on July 4, 1942, the first U.S. combat unit to be activated on foreign soil. It would continue to be under the overall command of Brig. Gen. Chennault, as head of the newly-activated China Air Task Force (a work-around since the official U.S.A.A.F. policy at that time was to have only one numbered air force per theater of operations, and the Tenth Air Force was already ensconced in India).

To lead the 23rd Fighter Group during the transition, Chennault tapped Bob Neale, who would be leaving China in 14 more days (a reasonable choice, since the bulk of his combat pilots would be fellow A.V.G. members who volunteered to stay for the added two weeks). Col. Robert L. Scott, who would ultimately take command, was in Kunming discharging the administrative duties for organizing the new Group.[1] As promised by the induction board in late June, three of the five pilots accepting U.S.A.A.F. commissions were to be promoted to major and given squadron commands – Frank Schiel would command the 74th Fighter Squadron, Tex Hill would take the 75th, and Ed Rector the 76th. Gil Bright would become a major, while Charlie Sawyer would be inducted as a captain.[2]

On Sunday, July 5, Tex Hill brought nine P-40Es to Kweilin from Peishiyi. The next morning, in the first combat mission for the new fighter group, he led four Kittyhawks – with Johnny Petach and U.S. Army Air Corps Second Lieutenants Minor and Butsch – to escort five of Caleb Haynes's B-25s to Canton to destroy a major oil refinery and fuel storage areas.[3]

As the Kittyhawk top cover held at 11,000 feet, the Mitchells hit their target and turned north for home, with their escort following. A minute later, one of the B-25 pilots radioed that they had "bandits" coming up from below. Johnny Petach, leading an element behind Hill, immediately jettisoned his belly tank and, banking right and left, began searching for the attackers. Five minutes later he spotted a Japanese Nate 6,000 feet below. Making a quick call to his flight leader, with a quick wing-over he dove on the enemy fighter. As he got closer, he saw there were actually three, flying in a string 500 yards apart. Petach continued the narrative in his combat report:

> *"They must have seen me, because the last E.A. fired a burst and all 3 started to turn towards me. I opened fire at 500 yards, but was shooting behind the last man so I pulled the nose of my plane well ahead of the E.A. and gave him a 1 second burst. Then the E.A. pulled into sight right in front of my nose so that my fire raked him. As he passed, I saw large holes in his wings. The other two planes started to make a pass at me so I pulled up and took off…I saw two more I-97s about 3 miles north of our combat. I turned towards them and this time they turned away and headed for a mountain. The first plane was just turning around the mountain top when I overhauled the second plane. I gave him a short burst and he burst into flame and was burning well…"*[4]

Tex Hill also downed a Nate in the ensuing scrap, his first aerial victory as commander of the newly-activated 75th Fighter Squadron. Curving in behind one of the fixed-gear fighters, he let loose a barrage of fire from his six .50-calibers that literally sawed the wing off the little fighter. Running low on gas by that time, he called for a rejoin of his flight and headed for home.[5]

Inclement weather returned to the area on the 7th, but not before Neale sent Hill with the nine Kittyhawks to Hengyang. Later, when a Japanese ship was reported inbound, he dispatched six P-40s to intercept the interloper. Although five made contact, the recce bird eluded them in the clouds. As the ceiling lowered, Neale noted in his diary that night, "looks like we are in for another bad spell of weather." [6]

It rained most of the day on the 8th and 9th, precluding any activity at Kweilin. On Thursday, the 9th, however, Tex Hill took a flight of four on another bomber escort mission to Hankow. After the B-25s completed their bomb run and turned back north, and when it was evident that there would be no aerial opposition, Hill cruised southwest along the Yangtze River looking for "targets of opportunity." Spotting four suspected Japanese gunboats, he took his flight down in a screaming dive, leveling off just above the water, heading straight for the rivercraft. The combined firepower of four Kittyhawks with twenty-four .50-caliber machine guns cut the boats to pieces. After the fourth strafing pass, all four were on the bottom of the Yangtze.[7]

The next day, in response to a Chinese request to hit a Japanese Army headquarters at the village of Linchuan, Hill sent a flight of four on the mission. Reported as a lightly-defended target, it would be a "milk run." Hill selected Johnny Petach to lead the flight, with 2nd Lt. Leonard Butsch on his wing. The second element would be Spanish Civil War veteran Capt. A.J. "Ajax" Baumler, with Arnold "Red" Shamblin as his wingman.[8]

Taking off at 1045, the flight climbed to 6,500 feet and arrived over their target at 1203. Observing no anti-aircraft fire, Petach put the flight into string formation and started his dive. Then, as Ajax Baumler later reported:

> *"Mr. Petack [sic] had just reached terminus of his dive at about 2,300 feet when his airplane burst into flames around the cockpit and main fuel tanks. His airplane went into a violent tumbling spin completely out of control and a portion of his left wing separated from the airplane. His airplane (No. 106) crashed in flames on river edge at North East side of wall around city. The pilot remained in the plane and it is my opinion that his plane received several direct hits from 20 mm anti-aircraft fire immediately after he had released his bombs."* [9]

Following his flight leader, Baumler dropped his bombs and remained low, banking right and left as he followed the twisting river. Reaching their pre-briefed rendezvous point, he noticed that his wingman, Shamblin, wasn't with him. Circling back toward the town at 1,500 feet, he observed that their bomb run had been successful – numerous fires had been set and it appeared that all bombs had fallen within the city. But there was no sign of a parachute or any sign of Shamblin's P-40. With still no enemy aircraft sighted, he and Butsch joined up and returned to Hengyang at 1315, reporting the death of Petach and the "missing in action" of Red Shamblin.[10]

The news hit Tex Hill badly, since he had personally persuaded both men to remain for the additional two weeks. The fact that they had readily volunteered, recognizing the risks that all the other two-week extenders faced, did little to assuage Hill's feelings of personal guilt.

Later that day, Peter Wright and new arrival 2nd Lt. Lester Hall took off from Hengyang to intercept another Japanese reconnaissance aircraft over the field. After climbing to 16,000 feet in P-40E No. 112, Wright left his wingman at altitude while he descended to search for the enemy aircraft. He finally spotted his quarry, a twin-engine plane (probably a Ki-51 Sonia) at 10,000 feet and "very far away." As he reported the action:

> *"I kept chasing him until we ran into bad weather and he started to turn. I caught up rapidly then from behind and below...I opened up at about 150 yds., directly behind him. At the first burst part of his tail came off and his left hand motor caught fire. I gave him two more bursts and he rolled over and went down. I did not see him crash as oil from his engine covered my windshield. I returned to the field and heard that the Chinese had found the wreck an hour later...about 100-200 kilometers north east of Hengyang."* [11]

Even though technically credited to the 23rd Fighter Group, Wright's aerial victory would be the last recorded by an A.V.G. pilot (but not the last A.V.G. shoot-down).

Also that morning, Maj. Johnny Alison, new commander of the 16th Fighter Squadron, brought nine more

P-40Es to Kweilin from Kunming. After refueling, Bob Neale sent them right along to Lingling and designated Charlie Bond to lead the way. Shortly after they landed at their new base of operations, an air raid alert was sounded and Alison ordered the nine back into the air. Another false alarm. [12]

Alison wasted no time in getting his pilots oriented, sending them up on practice flights throughout the day on the 11th. That evening, however, many of the men suddenly became violently ill, throwing up their just-finished dinner. Before the siege of stomach upset was over, 18 of the 25 men were affected. A hurried investigation by the Chinese revealed that the culprit was the cooking oil used by the Chinese cooks. Chinese helpers had been sent into town for more cooking oil. The helpers – an enterprising lot – purchased the cheaper tung oil, used for producing paint and lacquer, and pocketed the difference. The fish on the menu that night had been fried in tung oil! [13]

That did it for Charlie Bond, who was already disgruntled with staying in China for the additional two weeks. "I am unhappy, disgusted, feeling ill and just about at the end of my rope," he wrote. "These are awful conditions," he continued. "The barracks are overcrowded and when I get in my bunk I have a fight with the bedbugs. Lingling is the worst base we have occupied in China." [14]

After hearing of the food poisoning, Bob Neale flew to Lingling the next morning, bringing castor oil, Epsom salts, aspirin and cans of "Flit" insect repellent to combat Bond's bedbugs. With Bond still suffering from the effects of the tung oil-laced fish, he sent Bond back to Kweilin to check in with Doc Sam Prevo and called Jim Howard in to oversee the 16th Fighter Squadron's transition. Later in the day, two more B-25s arrived, bringing the 11th's strength at Kweilin to six.[15]

Accompanied by Col. Caleb Haynes, Chennault left Chungking on the 13th, flew to Kweilin, with short visits en route to Hengyang and Lingling, bringing additional belly tanks for the Kittyhawks, food for the troops and sacks of long-awaited mail from the States. That night the Chinese threw another party for the Americans, with Bob Neale commenting that he "saw the best bunch of Chinese girls since he'd been in China." Despite a planned 0200 wake-up for the next day's planned escort mission to Hankow, Neale couldn't get his crew back to the barracks much before midnight.[16]

Rain and a low ceiling at Kweilin on Tuesday the 14th postponed the bombing mission, but Chennault was able to take off for Kunming with two transports to bring back Army replacements for the departing A.V.G. Before leaving, however, Chennault asked Neale once more if he hadn't changed his mind about accepting induction into the U.S.A.A.C. The letter that he had just received from his sweetheart back in the States sealed that answer.[17]

Bad weather continued over the next two days. Col. Haynes, however, elected to get five of his six B-25s off on Thursday, the 16th, with an escort of five P-40Es led by Tex Hill. Leading the B-25s, Haynes landed at Hengyang to refuel. There he picked up his escort of five P-40Es led by Hill before pressing on for Hankow. A few minutes after departing Hengyang, one of the B-25s developed engine trouble and had to turn back and two P-40s also turned around, leaving four B-25s and escorted by four P-40s.

The mission proved to be a true "milk run" – the bombs hit in the targeted dock areas, there were no enemy fighters encountered and the bombers and their escort turned south for Hengyang. Fifteen minutes after they landed, however, the early warning net reported "unknown ships north of the field." Tex Hill, in the operations shack, ordered that no planes take off, surmising that if it was a Japanese attack, they would have no time to get airborne and would be "sitting ducks" for the attackers. The word obviously didn't get out to those on the field, however, since pilots and bomber crews raced to their planes, and soon were taxiing rapidly to the runway. In the words of Maj. Alison, "From that point on, everything was confusion." In the rush to get airborne, P-40s and B-25s were "scrambling like a bunch of geese," as Alison put it, barely missing one another in the melee.[18]

Freeman Ricketts, one of the remaining A.V.G. pilots, circled the field, awaiting further orders. Conflicting reports of "enemy" aircraft, first to the north and then to the south, were being reported by the warning net. Ordered to return to the field, Ricketts then:

> *"... saw a ship to the west, very low and circling. I called WC-5 and reported ship and said I was going to investigate... Estimated the ship to be three miles away and as I started towards it, it turned in the opposite direction. I overtook ship and reported to WC-3...that it appeared to be a twin engine Jap reconnaissance plane. I overtook*

this ship from directly behind and a very little below…I opened fire at approximately 350-400 yards. The airplane then turned away from me…I overtook and fired another burst. I then pulled off to the right and recognized the insignia of the B-25. I watched the B-25 to see if it could continue flight and about one or two miles later saw five parachutes…" [19]

Miraculously, all the B-25 crew members survived, including the pilot, who rode the stricken aircraft in for a belly landing. Maj. Alison, in the air at the time, reported that he heard the ground radio warn pilots that two B-25s were in the air nearby. Also, he heard another P-40 pilot near Ricketts tell him that the "unknown" aircraft was a B-25, but apparently Ricketts's radio was operating intermittently and he didn't receive the transmissions. The B-25 crew later reported that they thought they had been shot down by "two Zeros." It would be the last "victory" for the A.V.G.[20]

That afternoon, A U.S.A.A.F. C-47 arrived at Kweilin with ten fresh pilots and four ground crew to be assigned to the 16th Fighter Squadron. Just three more days and all the "Christmas help" would be gone.

Charlie Bond was up at 0430 on Friday, 18 July – "going home day," as he recorded – and rousted out Van Shapard, who was still asleep on his "agony bench." After a good breakfast, they boarded the Army C-47 with 19 other A.V.G. personnel – all but Bob Neale – for the first leg of their journey back to the States. Neale had elected to remain at Kweilin for a few hours to join the A.V.G. men coming from Hengyang.

Skip Adair met the transports at Peishiyi. Taking Neale and Bond aside, he asked them if they would accompany Emma Jane "Red" (Foster) Petach on the trip back home. Accepting readily, they would care for the young, pregnant widow as if she were their sister. Boarding a C.N.A.C. DC-3 the next morning, the three made a brief stop at Kunming before proceeding on to Calcutta, via Dinjan. [21]

As the last remnants of the A.V.G. were winging their way to India, eleven of those who stayed were receiving their formal induction into the U.S. Army Air Corps. In a muggy tent at Peishiyi, Col. Homer Sanders, head of the induction board, had each man raise his right hand, read the printed oath and sign the document. Tex Hill, Ed Rector, Frank Schiel and Gil Bright were sworn in as majors, while Charlie Sawyer became a captain. Two communications men, Rolland Richardson and A.A. Miller, were made first and second lieutenants, respectively; and armorer E.R. McKinney and crew chiefs George Bailey, J.H. Musick and R.P. Rasmussen became master sergeants.[22]

In addition to the 11 mentioned in the preceding paragraph, Skip Adair and flight surgeons Thomas Gentry and Sam Prevo were later inducted as majors; Roy Hoffman and John Williams were commissioned as captains; W.H.S. "Daffy" Davis, Donald Bell, Ernest O. Bonham, Carson Roberts, John Engler, Joseph Lussier, J.J. Harrington, W. A. Sykes and Don Rodewald became first lieutenants, while Edward Harris, Elton Loomis, Bill Towery, Arvold Miller and Marlin Hubler were inducted as second lieutenants. Robert Gallagher, Gerhard Neuman and Chinese-American Francis Yee enlisted as sergeants while Chinese mechanics Kee Pon, Pak Lee, William Mah, George Lum, Chun Yuen Gee and Lem Wu enlisted as privates. Alex Mihalko and Edgar McClure received commissions as ensigns, while Richard Ernst and Ralph Sasser became chief petty officers in the U.S. Navy.[23]

Although only 39 members of the A.V.G. who accepted immediate induction into the U.S. Army Air Corps or U.S. Navy remained in China with the Old Man, it would be unfair to consider those returning to the States as trying to avoid further combat. All had endured nearly a year of harsh, sometimes primitive, living conditions; flying (and repairing) aircraft with a minimum of spare parts; and always on the move from one location to another as the Japanese advanced. In short, the men were just worn out. Several would have stayed if their request for a furlough back to the States had been granted – which was not, due to the pressing need for manpower in the theater.

Indeed, after a brief rest, most would return to their parent organizations, or join civil aviation as airline pilots, mechanics or aviation company test pilots. Whatever their service, in uniform or not, they would always be proud to wear the enameled pin signifying they were once a member of the legendary "Flying Tigers."

EPILOGUE

"The China Air Task Force was patched together in the midst of combat with whatever happened to be available in China during the gloomy summer of 1942." [1]

So wrote Brig. Gen. Claire Chennault as he assumed command of the newly-named China Air Task Force (C.A.T.F.) on July 5, 1942, consisting of the 23rd Fighter Group (with the 74th, 75th and 76th Fighter Squadrons replacing the A.V.G.'s First, Second and Third Pursuit Squadrons), augmented by the Tenth Air Force's 16th Fighter Squadron and Col. Caleb Haynes's 11th Bombardment Squadron of B-25s. His fighter force numbered 46 P-40s (28 Tomahawks and 18 P-40Es transferred from the A.V.G.), of which 35 were deemed combat-ready. He barely had enough pilots to fly the planes – five ex-A.V.G. who had accepted U.S.A.A.F. commissions and 12 new U.S.A.A.F. arrivals, augmented by the 19 ex-A.V.G. pilots who agreed to remain for an extra 14 days. The ground support was scarcely more favorable – 35 ex-A.V.G. inducted into the U.S.A.A.F., plus 20 newly-arrived from the States, assisted by 22 two-week volunteers. Perhaps to boost the morale of the new arrivals, Chennault continued to term his force as the "Flying Tigers" (much to the consternation of most of the original A.V.G., including R.T. Smith, who would years later denounce the Fourteenth Air Force Association's use of the name.) [2]

Despite the emergency assistance offered to the 39 ex-A.V.G. members staying the extra two weeks, Washington officialdom's continued hostility toward the recalcitrant "mercenaries" reared its ugly head. The departing members of the A.V.G. found that, to the U.S. Army bureaucracy, they were now nothing more than unemployed "soldiers of fortune," civilians with no priority for either military accommodations or air transportation back to the States. Bus Keeton spoke for the frustration of many in his diary entries from July 5 through July 11:

> *"...After all the promises the General and all the Army generals gave us they are not living up to one of them...(July 5); "...Arrived in Karachi (via C.N.A.C. DC-3) about noon... Talked with Colonel Logg in charge of the Ferry Command and it looks as though the Army will not help us any further from now on, we are on our own. A hell of a place to get stranded, nothing going out but boats...(July 8); "...Got back to the airport about 10:30 P.M. The corporal at the desk said they received orders from Gen. Bradley that no A.V.G. would be allowed to stay at the Airport Hotel. We would either have to stay in town or in a tent on the field... (July 9); "...The Army is out to get us and it looks as though they will..."* (July 11).

Keeton and approximately 80 other A.V.G. members departed Karachi on the SS *Mariposa*, an American cruise ship converted to an Army troop carrier, on July 27.

The two-week volunteers fared little better, despite a "letter of authorization" Chennault gave each of those staying over requesting transportation priority. In late July, Frank Schiel wrote to Chennault that, according to Erik Shilling:

> *"...Two or three days ago, all Air Transport Command control officers in India and Africa received a radio from Washington stating that absolutely no priority would be given to former A.V.G. members. This leaves the last bunch (those who volunteered for extra combat duty) including Red (Mrs. John) Petach (then pregnant) stranded in India. They are trying hard to get boat passage."* [3]

Chennault's letter finally had some effect for Bob Neale, Charlie Bond and Red Petach; they were able to get a flight to Khartoum, North Africa, and from there to Miami, arriving on August 7.[4]

With the China Air Task Force now under the command of Brig. Gen. Earl L. Naiden's Tenth Air Force, headquartered in New Delhi, India, Chennault faced multiple challenges of communication and supply. His worst nightmare came to pass on August 18, 1942, however, when Brig. Gen. Clayton Bissell replaced Naiden as commander of Tenth Air Force.[5] Sandwiched between Lt. Gen. Stilwell, commander of the China-Burma-India Theater, and Bissell – both hostile to Chennault's tactics and free-flowing command procedures – Chennault fought continuing battles, both with the Japanese and his designated superiors.

At one point, Bissell sent Brig. Gen. Francis Brady to inspect Chennault's operations. Brady concluded that "any similarity between the C.A.T.F. and a military organization is purely accidental." Brady was apparently appalled that strict adherence to military customs – like saluting – was not required.[6]

Chennault endured the situation, with continuing difficulties with supply and disagreements on tactics, until March 10, 1943, when the C.A.T.F. was reorganized as the Fourteenth Air Force, with newly-promoted Major General Chennault as its commander. Unfortunately, his problems continued, with Bissell and Stilwell sniping from the sidelines to the sympathetic ears of Gen. George Marshall (a close friend of Stilwell) and Gen. "Hap" Arnold in Washington. That there was a "clash of personalities" is putting it mildly. When Chiang succeeded in getting Stilwell replaced in October 1944, Marshall credited Chennault with influencing the Generalissimo.[7] Then there was the issue of personal correspondence between President Roosevelt and Chennault (heartily suggested by Roosevelt), bypassing Marshall.

In late 1944, reports began to surface of alleged smuggling of gold, drugs and other valuables by Fourteenth Air Force personnel, along with sexual "improprieties" and prostitution among some of Chennault's staff. Gen. Albert C. Wedemeyer, Chiang Kai-shek's chief of staff and commander of all American forces in China, found that complaints about prostitution and other illegal activities in the theater were often gross exaggerations and told Chennault to "forget it." However, four months following the Washington-directed investigation in early 1945 – which found "not a shred of evidence" against Chennault – Gen. Arnold, at Marshall's direction, strongly suggested that Chennault retire for "health reasons." [8] With a date of retirement set for October 10, 1945, on August 1 he was officially relieved of command, just two weeks from the formal surrender of Japan. He was not invited to attend the surrender ceremonies aboard the U.S.S. *Missouri*. He later wrote of his bitter feelings: "On V-J Day it was my fondest hope to be aboard the battleship *Missouri* in Tokyo Bay and watch the Japanese formally acknowledge their defeat." [9]

Chennault returned to Louisiana to the adulation of an admiring public. Following a huge parade in New Orleans, complete with a flyover of B-25s and P-51s, the city honored him with a banquet at the Roosevelt Hotel.[10] Later he was approached by a delegation led by Louisiana Governor James Noe to run for governor in the upcoming election. He declined, saying he would prefer the post of state game director.[11]

Things were not so happy at home, however; over the eight years that Chennault had been away in China, with few interludes back in Louisiana, he and Nell had grown further apart. Letters between the two had become increasingly infrequent. When he arrived at their home in Waterproof, he found her "cold and distant." Both agreed to a divorce. Home for less than three weeks, he left for Washington on August 20 to reunite with old friends, including Whiting Willauer, whom he had known since in China before the A.V.G. days.[12] He would subsequently marry Anna Chan, a former news reporter whom he had first met when she interviewed him in 1943. Married on December 21, 1947, she would bear him two daughters, Cynthia Louise and Claire Anna, in 1950 and 1952 respectively.

For some time, the still-restless Chennault had been quietly considering returning to China to form an airline. He was quite pleased, then, to learn that Willauer had entertained the same idea, and in January 1946 they were back in China. It proved to be propitious timing. Since November 1945 supplies provided to China by the United Nations Relief and Rehabilitation Administration (U.N.R.R.A.) had been piling up on the docks in Shanghai with no means of moving them inland. So it was that in February 1946 Chennault and Willauer were invited by the U.N.R.R.A. director of operations to draft a proposal for a relief airline. Working through the Chinese National Relief and Rehabilitation Administration (C.N.R.R.A.), U.N.R.R.A.'s counterpart in China, the two spent several frustrating months negotiating with various competing political and business interests before finally signing the contract with C.N.R.A.A. for the Chinese National Relief and Rehabilitation Air Transport, or "C.A.T." as it became known, on October 25, 1946.[13] With $2 million advanced by U.N.R.R.A., the company was able to purchase five surplus C-47s in the Philippines and seventeen C-46s in Hawaii.[14]

Several returning A.V.G. personnel later joined Chennault in the venture included radio specialist John Williams, who set up a radio network; Dr. Tom Gentry, who signed on as flight surgeon; and former A.V.G. pilots Erik Shilling, Dick Rossi and Joe Rosbert, who would eventually become director of operations for the airline. Based in Canton, C.A.T. began operations on January 31, 1947 as a commercial supply line for bringing food, medicine and building material into devastated areas of China.[15]

In September 1946, the U.S. government, in an attempt to placate the "agrarian reformer" Communists, had stopped all export of military hardware to the Nationalist government.[16] Supplied by Soviet Russia, the Communists began to turn the tide. In late 1947, as fighting between Chiang's Nationalist forces and Mao Tse-tung's Communists increased, C.A.T. morphed into Civil Air Transport, a paramilitary airline operating under the direct control of the Chinese Ministry of Communications to assist in the military effort of hauling food, supplies and personnel for the Nationalist Chinese in their long battle with the Chinese Communists.[17]

C.A.T. operations began on January 28, 1948 with a military airlift in support of besieged Mukden, bringing in food and supplies and evacuating people, all while under fire from Communist antiaircraft batteries. Similar rescue operations continued in March, with flights into Linfen in Shansi Province, and in support of Weihsien in April. Continuing to move south, the Communists threatened the Shantung Province, and on October 29 C.A.T. abandoned their base at Mukden. By the end of November, they were forced out of Hsuchow. Nanking and Shanghai fell to the Communists in April 1949 as Chiang retreated to Canton.[18] On May 1 Mao declared victory, proclaiming the People's Republic of China, with Peking as its capital.

With the fall of mainland China to the Communists October 1949, Chiang and the remnants of his army retreated a month later to the island province of Formosa (later renamed Taiwan), 140 miles off the east coast of China. C.A.T., reorganized under the name Civil Air Transport, Inc. (maintaining the acronym C.A.T.) remained in operation in Chungking until November 29, when they left for Taiwan. The evacuation of Kunming in early January 1950 marked the end of C.A.T.'s operations on the mainland and Chennault and C.A.T. joined Chiang on Taiwan.[19]

Unfortunately, with its now-restricted area of operations, C.A.T. had lost $671,000 by March 1950. At that point, with the instability in the Far East, including the restive Communist government in Korea, the U.S. Central Intelligence Agency (CIA) on March 24 agreed to underwrite the airline with a "loan" of $350,000. C.A.T. thus became a virtual CIA-owned entity for covert activities in the Far East, albeit retaining its commercial activities.[20]

On June 25, 1950 the North Korean army invaded South Korea across the 38th parallel dividing the two countries. Three days later, the CIA made the airline venture official with the purchase of Civil Air Transport, Inc. for $1 million. Chennault was retained as chairman of the board, and Joe Rosbert continued as director of operations. Although nominally holding his high-sounding title, Chennault ceased his daily involvement with C.A.T. operations, becoming the primary conduit between the Nationalist government and the CIA. With his reduced influence in the operations Chennault divided more of his time between his Taiwan residence and the house he purchased in Monroe, Louisiana. During the time when he was back in the States, he continued to lobby for more support for Chiang's Nationalist government through speeches, visits to Washington and his memoir, *Way of a Fighter*, published in 1949.[21]

During a routine physical exam in early 1956, Chennault was diagnosed with lung cancer, undoubtedly the result of his lifelong, two-packs-a-day smoking habit. An operation in 1957, which removed most of his left lung, gave him enough time to visit Taiwan one last time. Hospitalized once again in late 1957 for radiation treatment, he grew steadily weaker and passed away on July 27, 1958, just twelve days after Congress gave him his third star as a lieutenant general. He was buried in the National Cemetery in Arlington, Virginia. On November 14, 1958, Lake Charles Air Force Base was renamed Chennault Air Force Base in honor of the Old Man.[22] On October 8, 1976, in Baton Rouge, Louisiana, a life-size bronze statue of Chennault, a gift to Louisiana from the Republic of China, was unveiled by Anna Chennault as A-7s of the 23rd Tactical Fighter Wing from England Air Force Base flew over, followed by a lone P-40 from the Confederate Air

Force; and on September 6, 1990, a U.S. commemorative postage stamp was issued in his honor.[23]

Of the 220 men still with the A.V.G. when it was disbanded on July 4, 1942, approximately 187 returned to active duty with the U.S. Army, Navy or Marines, including those who remained in China with the newly-formed China Air Task Force, such as the ex-pat German national Gerhard Neumann (a.k.a. "Herman the German"), who enlisted in the U.S. Army as a staff sergeant and remained with the 23rd Fighter Group.[24] Some returned to fly for Chennault in the 23rd Fighter Group, including Chuck Older and Don "Rode" Rodewald who returned to China after completing pilot training back in the States. Tex Hill and Ed Rector, two of the five pilots who remained with Chennault in 1942, returned to China after a tour in the States and both became commanders of the 23rd Fighter Group, in 1943 and 1945 respectively. Communications specialist Robert M. Smith re-enlisted upon his return to the States, entered Officer Candidate School, and returned to China to command the 159th Army Airways Communications System (A.A.C.S.) Squadron in Chungking. Many of Chennault's clerical and support staff also made their way back to China to serve on Chennault's Fourteenth Air Force staff – Joe Alsop, Tom Trumble and Doreen Davis as civilians, John Williams, who had been in charge of the A.V.G.'s early-warning net, and former chaplain Paul Frillman returned as an Army intelligence officer.[25]

Other former A.V.G. pilots returning to China included Bill Reed, who commanded the 3rd Fighter Group of the Chinese-American Composite Wing (C.A.C.W.), adding six victories to his three with the A.V.G., and George McMillan, who joined the Fourteenth Air Force's 51st Fighter Group. Oley Olson and R.T. Smith returned to the C.B.I. with the 1st Air Commando Group, Smith as commander of the Group's squadron of P-51s and B-25s, and Olson commanding the CG-4 gliders that supported British General Ord Wingate's operations behind the Japanese lines in Burma.[26]

Jim Howard rejoined the U.S.A.A.F. in January 1943, flew P-51s as commander of the 354th Fighter Group in Europe, added six victories to his 2.33 with the A.V.G. and became the only U.S.A.A.F. fighter pilot in the European Theater to receive the Medal of Honor; Charlie Bond was given command of a Stateside fighter squadron, later going to Russia as Ambassador Averill Harriman's aide and serving as chief of the Air Division of the U.S. Military Mission to the U.S.S.R.; P.J. Greene commanded the 97th Fighter Squadron, flying P-38s in Italy. Greg "Pappy" Boyington rejoined the U.S. Marine Corps in November 1942, eventually commanded VMF-214 and destroyed 22 more Japanese aircraft before being shot down himself. Surviving two years as a P.O.W., he also received the Medal of Honor. Percy Bartelt, Keith Christensen, Frank Lawlor and Fritz Wolf returned to the Navy flying Hellcats in the Pacific Theater. As commander of VBF-9, Lawlor added two aerial victories to his A.V.G. score, and Wolf, commanding VFB-3, became an ace, adding one more to his four with the A.V.G.[27]

Among the ground crew who reenlisted back in the States was Charlie Bond's crew chief, Walt Dolan, who returned to China; Chuck Baisden, who joined R.T. Smith in the 1st Air Commando Group as Smith's crew chief and top turret gunner on "Barbie," Smith's B-25; and, as previously mentioned, Don Rodewald who was accepted for flight training and returned to China to fly with the 23rd Fighter Group. Including Rodewald, at least nine A.V.G. ground personnel entered pilot training with the Air Force or Navy – Robert Rasmussen, Jesse Crookshanks, Bill Sykes, Carson Roberts, Marlin Hubler, E. B. McClure, Rolland Richardson and Ed Stiles. Richardson was reunited with Tex Hill at the Air Proving Ground Test Facility at Eglin Field, Florida, while Stiles became a B-25 pilot. Crookshanks was shot down over Germany and spent 18 months as a P.O.W.[28]

Additionally, 20 pilots either remained in China or returned to fly cargo over the "Hump" for C.N.A.C. – Frank Adkins, Bill Bartling, Carl Brown, John Dean, Clifford Groh, Lester Hall, Duke Hedman, Freddy Hodges, Link Laughlin, Bus Loane, Einar Mickelson, Moose Moss, Bob Prescott, Catfish Raines, Joe Rosbert, Van Shapard, Erik Shilling, Allen Wright, Peter Wright, and Dick Rossi, who held the record of 753 "Hump" flights. Three would die as their cargo ships crashed into the Himalayas – Dean on November 17, 1942, Allen Wright on December 10, 1943 and Mickelson on February 20,1944. Also joining C.N.A.C. were Doc Richards, crew chiefs Frank Losonsky and Melvin Woodward, and communication specialists Robert King, Loy Seamster and Roger Shreffler.[29]

Others became test pilots – Ken Jernstedt, Erik Shilling and Parker Dupouy for Republic Aviation, and Ed Overend, who flew for Douglas Aircraft before rejoining the Marines and subsequently adding three victories as commander of VMF-321. George Burgard and Bob Neale signed on as pilots for American Overseas Airlines; Peter Wright and Matt

Kuykendall joined American Export to ferry aircraft across the Atlantic, while Tommy Haywood joined Consairway, ferrying B-24s between the United States and Australia; Edgar Goyette joined Eastern Airlines, while Bob Layher, after rejoining the Navy, flew with American Export Airlines in the Naval diplomatic run. After a short stint with Trans World Airlines and later flying the "Hump" with C.N.A.C., in June 1945 Bob Prescott formed Flying Tiger Lines as a cargo hauler, hiring former A.V.G. pilots such as Dick Rossi and Bus Loane, each of whom had flown earlier with both C.N.A.C. and C.A.T., and Duke Hedman, Bill Bartling, Link Laughlin and Bob Raines who had all flown with C.N.A.C.[30]

Sadly, several Tigers would not survive the war. Frank Schiel, while commanding the 74th Fighter Squadron, died in the crash of his P-38 in 1943; George McMillan died in June 1944 when his P-38 was hit by anti-aircraft fire and Bill Reed died bailing out of his P-40N in December 1944. Former armorer Roy Hoffman died as a crewman on a B-24 in Europe. Harry Bolster and Bob Brouk rejoined the U.S.A.A.F., only to be killed later in flying accidents, as did Ralph Gunvordahl and Maurice McGuire, who became test pilots. Bill Sykes was killed in a flying accident at Williams Field, Arizona in 1943, while both Robert Rasmussen and Carson Roberts were later killed in action in North Africa in 1943.[31] On a happier note, the three A.V.G. pilots who became prisoners of the Japanese did survive the war. Lew Bishop escaped from a moving train in May, 1945, returning to the U.S. to rejoin the Navy; Charlie Mott, who was forced to work on the infamous River Kwai railroad, and Billy McGarry were repatriated in May, 1945, with McGarry later receiving a law degree from Loyola University.[32]

Most who remained on active duty or joined the Reserves or National Guard retired as Navy commanders/captains or Air Force lieutenant colonels/colonels. Three reached star rank – Jim Howard with the Air Force Reserve and Tex Hill with the Texas Air National Guard became brigadier generals, while Charlie Bond rose to major general as commander of the Twelfth Air Force.[33]

Several achieved prominence in the civilian world following the war. Gerhard Neumann immigrated to the United States and found work as an engineer for General Electric, rising in a 32-year career to become one of the vice-presidents of GE. Ken Jernstedt was mayor of his hometown of Hood River, Oregon, Oregon State Representative in 1967 and State Senator from 1969 until he retired in 1991. Chuck Older obtained a law degree from the University of Southern California, and later became a superior court judge for Los Angeles where he presided over the infamous Charles Manson mass murder trial in 1970.[34] "Rode" Rodewald had a combat tour in Korea, where, as an F-86 pilot, he was credited with destroying a MiG-15. Back in the States in 1954, he suffered a near-fatal crash in a flamed-out T-33 which broke his back and left him a paraplegic. Medically discharged from the Air Force, he returned to civilian flying in 1969, re-qualifying with a specially modified, hand-controlled aircraft. In 1984, flying a single-engine Piper Comanche 260, he set a world record as the first paraplegic to fly solo around the world.[35]

As early as 1944, A.V.G. veterans were talking about organizing a Flying Tigers Association. Pappy Paxton spearheaded the effort, writing several letters to Chennault soliciting his support. There was one problem, however; Bill Pawley, at that time U.S. ambassador to Peru, had offered a $10,000 "donation" for seed money to start the organization in which, of course, he would become a member. He had also produced a small, 75-page, soft-bound book, entitled "Americans Valiant and Glorious," honoring the 22 A.V.G. members who died while serving, and the five additional former members who died later in the war. When told of Pawley's request, Chennault flatly refused, telegramming Paxton, "Will not accept membership in any organization with Pawley who was never a member of the A.V.G. and who failed to serve us in the most critical hour."[36]

Although incorporated in New York on January 15, 1945, it was not until 1952, however, before the Association got off the ground, meeting for the first time in Hollywood, California. Dick Rossi was elected president, a post he held until his death on April 17, 2008. They met annually in Ojai, California, for the first time in 1957, and continued in that location for several years, branching out later with meetings across the United States and even meeting twice in Taiwan, in 1964 and 1978.[37]

The A.V.G. veterans had long been disappointed, and somewhat bitter, that their year-long combat service fighting the Japanese in the early months of World War II had not been recognized as wartime service by U.S. officialdom. Indeed, one of the first efforts to gain such recognition was initiated by the fledgling American Fighter Aces Association in a "Bill of Rights Requesting Veteran Recognition by Congress for The Original Flying Tigers, American Volunteer Group, Chinese Air Force, China 1941-1942," drafted by Pappy Paxton, first Secretary-Treasurer of the Association. After 45 years of bureaucratic intransigence, this was finally corrected in September 1992 in Washington, DC, at the annual Air Force Association convention. There, the American Volunteer Group was presented with the Presidential Unit Citation, signed by Secretary of the Air Force Donald B. Rice, which gave all A.V.G. members "veteran" status and hence eligible for veterans' benefits and U.S. military awards and decorations. Four years later, on December 8, 1996, at the Flying Tigers Association's meeting in Dallas, Texas, U.S. Air Force Chief of Staff, Gen. Ronald R. Fogleman, presented each attending A.V.G. veteran or family member of those deceased with an appropriate decoration – each pilot with the Distinguished Flying Cross and each ground personnel with the Bronze Star for their service in China. Those not attending and the families of those deceased would all be eligible to receive the decorations.[38]

Despite R. T. Smith's long-ago lament, the Flying Tiger name lives on today in America's military. The A-10 "Warthogs" of the 23rd Fighter Group, stationed at Moody Air Force Base, Georgia, now composed of the 75th and 76th Fighter Squadrons (successors to the A.V.G.'s 2nd and 3rd Squadrons), still proudly wear the leering sharks teeth and baleful eye on their nose made famous by the A.V.G.'s P-40s – the only U.S.A.F. unit authorized to do so. The unit emblem of their parent 23rd Wing is the familiar winged tiger handed down by the Fourteenth Air Force, beneath which are the words "Flying Tigers."

In 1988, the Flying Tigers Association formally granted the use of their name and logo to the U.S. Army's 2nd Battalion, 229th Aviation Regiment, stationed at Fort Rucker, Alabama. Adorned with the familiar sharks teeth and eye on the nose of their AH-64 Apache attack helicopters, the unit made U.S. Army history by capturing 467 Iraqi troops during Operation Desert Storm in 1991.[39]

On February 6, 2020, Frank Losonsky, crew chief with the 3rd Squadron, died in Columbus, Georgia, at the age of 99. He was the last of the breed, the end of an era and the likes of which we shall never see again – the Flying Tigers!

ACKNOWLEDGMENTS

Dozens of people contributed to this book – A.V.G. veterans and family members, former enemy airmen and many other personal friends and historians dedicated to the history of the American Volunteer Group.

First and foremost of the A.V.G. members, I must recognize David L. "Tex" Hill for starting me on this journey. I spent hours as a guest in his San Antonio home, listening to (and recording) his many first-hand stories of those days in China during 1941-1942. He put me in touch with George L. "Pappy" Paxton, who shared his scrapbooks, photo albums and personal documents with me. Robert T. "R.T." Smith invited me to my first Flying Tigers reunion. Later I met Jasper J. Harrington, who gave me a delightful recorded interview; Merlin D. Kemph, Irving Stolet, Clarence Riffer, Paul J. Perry, William Towery and John Williams provided me with additional photos and documents. I was honored to also develop personal friendships and long correspondence with Chuck Baisden, Dick Rossi, Charlie Bond, Ed Rector, Erik Shilling, Robert "Buster" Keeton, Joe Rosbert, Peter Wright, and Jim Cross, each of whom added to my growing files on the A.V.G.. A.V.G. family members, too, gave me much support and provided valuable resource material. Lydia Rossi, widow of Dick Rossi, gave me access with permission to use, her vast files of A.V.G. photos. Lee Burgard gave me a copy of his dad's diary; Brad Smith, son of R.T., shared A.V.G. documents with me and gave me permission to quote liberally from his dad's diary. Ed Reed, nephew of Bill Reed, provided me with a copy of Bill's diary. As a professional school teacher and academic administrator, Ed gave me years of encouragement for my project and, finally, volunteered to edit the final manuscript, going through, line-by-line, my manuscript, offering punctuation corrections and suggestions for better writing.

Aside from the A.V.G. family, I corresponded with Robert W. Fausel, former Curtiss-Wright demonstration pilot in pre-war China and later author, who gave me his personal insights into the personalities of Claire Chennault and Bill Pawley, and aviation author and historian Dr. William Leary, who graciously reviewed my earlier drafts and provided valuable, professional critiques.

Several friends and fellow historians must be recognized: aviation artist Roy Grinnell introduced me to Dick Rossi and Jim Cross and gave permission to use his A.V.G.-based artwork in the production of this book. Bob Andrade, long-time friend of Dick Rossi and other A.V.G. members residing in California, introduced me to William "Black Mac" McGarry and has continued to provide me with photos and rare documents, including a copy of the signed letter of the 24 protesting pilots during the "Loiwing Revolt." Additionally, Bob and his wife, Jan, provided continuing editorial support right up until the manuscript was published. Jim Lansdale, military historian and authority the Japanese Army Air Force, provided me with valuable insight, and photographs, on Japanese aviation during World War II. The late Mark Burken, official historian of the Flying Tigers Association for many years, became a particularly close friend and generously shared his voluminous A.V.G. archives with me, including over 25 A.V.G. diaries, combat reports and other documents acquired over 20 years of research. Bill Hess, long-time friend, aviation historian and author of over 30 books, proof-read my early drafts and gave me continuing encouragement along the way. My good friends and accomplished authors, aviation historians Carl Molesworth and Jon Guttman made final reviews of the manuscript, and Carl guided me through an effective book proposal to prospective publishers. Finally, aviation historian Bob Baldwin, whom I met at the 2019 Flying Tigers Association reunion in San Diego, introduced me to Mick Prodger, head of San Antonio-based Elm Grove Publishing, who graciously accepted my manuscript for publication, and who has guided me through to the final product.

From the "other side," I was able to establish correspondence with several Japanese aviation historians and former Japanese airmen, who provided me with information concerning the Japanese air units who opposed the A.V.G.: Yohei

Hinoki, surviving fighter ace of the 64th *Sentai;* Masahiro Nakayama, who provided me with photographs of the wreckage of Jack Newkirk's Tomahawk and related articles; Toshiki Kudo, who put me in touch with retired I.J.A.A.F. General Yuruzu Kochiyama, with whom I developed a lengthy correspondence; Hiroshi Ichimura, who gave me a detailed report on the April 10, 1942, Japanese attack on Loiwing; Fumio Watanabe, grandson of 82nd *Chutai* pilot Gouichi Suzuki, provided me a translation of his grandfather's account of the December 20, 1941, raid on Kunming, complete with the names of all the pilots of the 10 attacking Ki-48 "Lily" bombers, as well as valuable information on I.J.A.A.F. units operating in Burma and China, 1941-1942; and retired Major Kouki Kawamoto, who provided me with a history of the 50th *Chutai* in Burma. Finally, retired I.J.A.A.F. Colonel Masakazu Shimizu, former Mitsubishi Ki-51 reconnaissance officer with the 71st *Chutai* in China, with whom I have maintained a 26-year correspondence and who put me in touch with 71st *Chutai* squadron mate, retired Captain Toriyuki Ieda. His assistance has been invaluable.

I have saved the best for last. This book could not have been completed without the encouragement, patience, prodding ("You're going to research this book to death!"), computer expertise and typing assistance of Mary, my wife and life-partner of 59 years. She has truly been "the wind beneath my wings."

Appendix I.
Passenger Lists

S.S. *President Pierce*, H.M.S. *Van Houetz*
June 9-July 28,1941

Baisden, Charles N.
Blaylock, Glen O.
Bugler, Carl F.
Cross, Harvey G.
Dolan, Walter J.
Engle, Charles R.
Fauth, John E.
Frillman, Paul W.
Gasdick, Joseph
Hoffman, Roy G.
Hubler, Marlin R.
Jordan, Joe T.
Keller, Daniel W.
Kenner, Charles D.
McHenry, Sharon L.
Metasavage, Frank G.
Misenheimer,Charles
Olson, Henry L.
Perry, Paul J.
Regis, James E.
Riffer, Clarence
Rogers, Robert W.
Schramm, Leo J.
Schur, Carl E.
Smith, Robert A.
Stiles, Edward L.
Sweeney, Joseph W.
Uebele, John. J.
Wyke, William R.
Wylie, Harold G.

M. S. *Jaegersfontein*
July 10-August 15,1941

Atkinson, Peter
Armstrong,John D.
Bacon, Noel R.
Bailey, George R.
Baughman, Edmund C.
Bell, Donald
Blackburn,William J.
Blackwell, Harold
Bonham, Ernest O.
Bright, John G.
Brouk, Robert R.
Bruce, Everett, DDS
Buxton, Richard H.
Carter, John B.
Chaney, Charles
Cook, Elmer J.
Cribbs, Charles D.
Criz, Albert
Crookshanks,Jesse R.
Crotty, John D.
Curran,George F.
Cushing, Albert D.
Daube, Otto W.
Dorris, Carl E.
Dudzik, Francis P.
Dupouy, Parker S.
Ernst, Richard J.
Fobes, Edwin L.
Foster, Emma J.
Fox, Henry E.
Francisco, Charles H.
Gallagher, Robert
Gentry,Thomas C.,MD
Geselbracht,H.M.,Jr.
Gorham, Lloyd L.
Gove, Irving P.
Graham, Richard E.
Harris, David H.
Hastey, Raymond L.
Hauser, John B.
Heller, John E.
Henson, Thomas M.
Hodges, Fred S.
Hoyle, Daniel J.
Jacobson, Frank A.
Johnston, Leon H.
Jones, Jack D.
Kemph, Merlin D.
Lindstedt, Robert K.
Little, Robert L.
Losonsky, Frank S.
McAllister, Gale E.
McClure, Edgar B.
McDowell, Mark H.
McGuire, Maurice G.
McKinney, Eugene R.
Mangleburg, Lacey F.
Martin, Neil G.
Merritt, Kenneth T.
Miller, Arvold A.
Moss, Robert C.
Mott, Charles D.
Musgrove, Willard L.
Musick, James H.
Neal, Robert J.
Neale, Robert H.
Newkirk, John V.
Olson, Arvid E., Jr.
Overly, John L.
Peret, Richard C.
Petach, John E.,Jr.
Pietsker, Joseph H.
Pistole, Herbert
Poshefko, Joseph A.
Power, John D.

Prevo, Samuel, MD
Probst, Albert E.
Quick, Carl
Rasmussen, Robert P.
Regis, Stanley J.
Richards, Lewis J.,MD
Richardson, R.L.
Ricks, Wayne
Rodewald, Donald L.
Rumen, John N.
Rushton, Edwin H.
Sandell, Robert J.
Sasser, Ralph W.
Sawyer, Charles W.
Schaper, William E.
Schiller, Ralph F.
Seamster, Loy F.
Seavey, Edward H.
Shaw, John E.
Shilling, Eriksen E.
Shreffler, Roger
Smith, Robert H.
Smith, Robert M.
Sommers, John T.
Stewart, Jo B.
Swindle, Estes T. Jr.
Sykes, William A.
Towery, William H.
Tulley, Chester A.
Tyrrell, George
Unger, William H.
VanTimmeron,Frank E.
Vaux, Morgan H.
Viverette, Hugh J.
Walker, Harold H.
Wallace, Stanley H.
Walroth, Robert H.
Wiggin, Edwin D.
Wirta, Harvey C.
Wolf, Fritz E.
Woodward, Melvin H.
Wright, Peter
Wyat, L.G.
Yarberry, Glen L.
Young, John P.

M.S. *Bloemfontein*, Penang Trader

July 24-September 15,1941

Christensen, Keith J.
Christman, Allen B.
Colguette, Leon P.
Cornelius, Jack
Durbin, Estill E.
Dyson, James P.
Foshee, Ben C.
Goyette, Edgar T.
Greene, Paul J.
Hammer, Maax C.
Harrington, Jasper J.
Hedman, Robert P.
Hill, David L.
Jones, Thomas A.
King, Robert J.
Kuykendall, Matthew
Leibolt, Edward J.
McGarry, William D.
McMillan, George B.
Mihalko, Alex
Osborne, Harold L.
Paull, Preston B.
Rector, Edward F.
Reed, William N.
Schiel, Frank, Jr.
Smith, Corbet J.
Smith, Robert T.
Watson, Eugene W.

M.S. *Zaandam*, Pancrius

August 26-October 11,1941

Allard, James
Bartling, William
Bent, Morton
Clouthier, Paul
Farrell, John
Gallagher, Edward
Haywood, Thomas
Jernstedt, Kenneth
Locke, Robert
Kustay, Stephen
Older, Charles
Regis, James
Swartz, Frank
Wagner, Earl
White, John

M.V. *Klipfontein*, Tarakan

September 4–October 29,1941

Bohman, Morris P.
Brady, James E.
Breeden, Kenneth V.
Brown, Carl
Cole, Thomas J.
Conant, Edwin S.
Gilbert, Henry G.
Harpold, Clayton M.
Houle, Leo J.
Jaeger, George B.
Jourdan, Walter C.
Sutherland, William
Kaelin, Albert V.
Kiner, Melvin W.
Lancaster, George R.
Linton, Jack R.
Moore, Lawrence C.
Moss, Kenneth R.
Paxton, George L.
Raines, Robert J.
Rasbury, James D.
Ringey, Joseph E.
Rosbert, Camille J.
Sanger, Kenneth C.
Shapiro, William D.
Walsh, Andrew A.
White, Richard J.
Wilson, Clifford H.

M.S. *Boschfontein*

September 24-November 12,1941

Adkins, Frank W.
Bernsdorf, Donald R.
Bishop, Lewis S.
Bond, Charlie R. Jr.
Boyington,Gregory
Burgard, George T.
Croft, John S.
Cross, James D.
Donavon, John T.
Fish, W. H.
Groh, Clifford G.
Gunvordahl, Robert W.
Hennessy, John J.
Hodges, Fred S.
Hurst, Lynn A.
Janski, Edwin A.
Keeton, Robert B.
Kepka, George B.
Knapp, Donald R.
Laughlin, Chauncey H.
Layher, Robert F.
Mickelson, Einar I.
Moss, Robert C.
Overend, Edmund F.
Prescott, Robert W.
Ricketts, Freeman I.
Rossi, John R.
Smith, Curtis E. Jr.
Wright, A. M.

M.S. *Zaandam*

October-November 25,1941

(Instructors for the CAF)

Baugh, Marion F.
Bolster, Harry R.
Clothier, Paul
Fuller, Henry W.
Hall, Lester J.
Loane, E.W.
Peeden, Joseph
Seiple, Wilfred
Shapard, Van
Shamblin, Arnold
Stolet, Irving
Stubbs, Gail L.

Appendix II.
General Orders

HEADQUARTERS FIRST AMERICAN VOLUNTEER GROUP CLC/wrw
Office of the Commanding Officer

Toungoo, Burma,
August 23, 1941.

GENERAL ORDERS)
:
NUMBER 1)

1. In compliance with Order No. 05987, dated August 1, 1941, issued by H. E., Generalissimo Chiang Kai-Shek, the preliminary organization of the First American Volunteer Group is announced as follows:

Col. C. L. CHENNAULT	Group Commander.
B. R. CARNEY	Chief of Staff & Executive Officer.
C. D. MOTT	Chief of 1st Section, (G-1).
J. M. WILLIAMS (DS)	Chief of 2nd Section, (G-2).
J. H. HOWARD	Chief of 3rd Section, (G-3).
R. W. WHITEHEAD (DS)	Chief of 4th Section, (G-4).
P. S. DUPOUY	Assistant Chief, 2nd Section, (G-2).
C. W. SAWYER	Assistant Chief, 3rd Section, (G-3).
E. S. RUSHTON	Assistant Chief, 3rd Section, (G-3).
E. E. SHILLING	Assistant Chief, 3rd Section, (G-3).
R. L. EASTEY	Assistant Chief, 4th Section, (G-4).
S. H. WALLACE	Assistant Chief, 4th Section, (G-4).
Dr. T. C. GENTRY	Chief Surgeon.
Dr. E. W. BRUCE	Dental Surgeon.
Dr. JOS. LEE	Assistant Surgeon.
Dr. S. B. PREVO	Assistant Surgeon.
Dr. L. J. RICHARDS	Assistant Surgeon.
Miss J. B. STEWART	Chief Nurse.
Miss E. J. FOSTER	Nurse.
R. GALLAGHER	Nurse (Male).
Rev. P. W. FRILLMANN	Chaplain & Recreation Officer.
N. G. MARTIN	Assistant Recreation Officer.
W. R. WYKE	Sergeant Major.
R. C. PERET	Chief Mechanic.
R. W. ROGERS	Hangar Chief.
W. W. RICKS	Propellor Specialist.
D. J. HOYLE	Clerk, Administrative.
F. P. DUDZIK	Clerk, Administrative.
J. T. JORDAN	Clerk, Finance.
E. G. WYLIE	Clerk, Finance.
L. H. JOHNSTON	Clerk, Intelligence.
E. H. SEAVEY	Clerk, Meteorology.
J. E. REGIS	Photographer.
J. H. PIETSKER	Photographer.
J. D. CROTTY	Clerk, Operations.
M. R. KUHLER	Clerk, Operations.
L. S. JOHNSTON	Clerk, Operations.
A. D. CUSHING	Clerk, Engineering.

- 1 -

General orders No. 1, Hq. 1st American Volunteer Gp (8-23-41) contd.

J. P. YOUNG	Clerk, Engineering.
S. L. McHENRY	Clerk, Supply & Transportation.
J. E. HELLER	Clerk, Supply & Transportation.
D. BELL	Clerk, Supply & Transportation.
R. H. BUXTON	Medical Technician.
C. D. CRIBBS	Medical Technician.
T. M. BENSON	Medical Technician.
E. E. VIVERETTE	Medical Technician.
J. E. SHAW	Medical Technician.
J. A. SWEENEY	Radio Operator.
R. SHREFFLER	Radio Operator.
R. G. HOFFMAN	Chief Armorer.
G. F. CURRAN	Parachute Rigger & Inspector.
C. C. LEAGHTY	Parachute Rigger & Inspector.

2. a. The 1st Pursuit Squadron is organized as follows:

SQUADRON LEADER
Sandell, R. J.

WING MEN
Atkinson, P. W.
Little, R. L.
McGuire, M. G.
Mangleburg, L.F.
Merritt, K. T.
Neale, R. H.
Probst, A.
Sawyer, C.W. (*)
Shilling, E. B. (*)
Smith, R. H.
Wolf, F. E.

CREW CHIEF
Blackburn, W. J.
Dolan, W. J.
Gasdick, J.
Graham, R.E.
Jacobson, F. A.
Kemph, M. D.
Kenner, C.
McClure, E. B.
Meisenheimer, C. V.
Metasavage, F. G.
Musgrove, W. L.
Schaper, W. E.
Uebele, J.
Yarbery, G. L.
Gove, I. P.
Regis, Sat

CHIEF, ADMINISTRATION
Wyke, W. R. (*)

CLERK, ADMINISTRATIVE RECORDS
Dudzik, F. P. (*)

CLERK, OPERATIONS
Dorris, C. E.
Seavey, E. H. (*)
Hubler, M. H. (*)

CLERK, ENGINEERING
Jones, J. D.

CLERK, SUPPLY & TRANSPORTATION
Heller, J. E. (*)

COMMUNICATIONS
Miller, A. A.
Smith, R. M.
Vaux, M. H.
Wyatt, L. G.

ARMORER
McKinney, E. R.
Neal, R. J.
Rodewald, D. L.
Unger, W. H.

- 2 -

General Orders No. 1, Hq. 1st American Volunteer Gp (8-23-41) contd.

b. The 2nd Pursuit Squadron is organized as follows:

SQUADRON LEADER
Newkirk, J. V. K.

FLIGHT LEADER
Howard, J. H. (*)

WING MEN
Armstrong, J. D.
Bacon, N. R.
Bright, J. G.
Geselbracht, H. M. Jr.
Moss, R. C.
Petach, J. E. Jr.
Power, R. H.
Rushton, E. H. (*)
Wright, P.
Mott, C. D. (*)
Hastey, R. L. (*)

LINE CHIEF
Peret, R. C. (*)
Rogers, R. W. (*)

CREW CHIEF
Bailey, G. R.
Blackwell, E. J.
Carter, J. B.
Chaney, C.
Daube, O. W.
Fox, H. E.
Gorham, L. L.
Hauser, J. B.
Overley, J. L.
Quick, C.
Tulley, C. A.
Tyrrell, G. J.
Walker, H. H.
Woodward, M. H.

CLERK, ADMINISTRATIVE RECORDS
Fobes, E. L.

CLERK, OPERATIONS
Bugler, C. F. (*)
Crotty, J. D. (*)

CLERK, ENGINEERING
Cushing, A. D. (*)

CLERK, SUPPLY & TRANSPORTATION
Bell, D. (*)

MESS
Towery, W. H.

ARMORER
Musick, J. H.
Pistole, H.
Rumen, J. N.
Wirta, H. C.
Schiller, R. F.

COMMUNICATIONS
Baughman, E. G.
Ernst, R. J.
Sasser, R. W.
Seamster, L. F.
Shreffler, R. P. (*)
Sykes, W. A.

- 3 -

Appendix III.
American Volunteer Group Roster

Volunteers Serving Until Disbandment

Adair, C.B., HQ staff
Adkins, Frank, pilot
Allard, James, mechanic
Bailey, George, crew chief
Baisden, Charles, armorer
Bartling, William, pilot*
Baughman, Edmund, communications
Beaupre, Leo, clerk
Bell, Donald, clerk
Bent, Morton, clerk*
Blackburn, John, pilot*
Blackburn, William, crew chief*
Blackwell, Harold, crew chief*
Bolster, Harry, pilot*
Bond, Charles, pilot*
Bonham, Ernest, communications
Brady, James, clerk
Breeden, Kenneth, clerk
Brice, George, crew chief*
Bright, J. Gilpin, pilot
Brown, Carl, pilot*
Brouk, Robert R., pilot
Bruce, Everett, DDS
Bugler, Carl, admin chief
Burgard, George, pilot
Callan, Michael, ground crew
Carney, Boatner, HQ staff
Carter, John, line chief*
Cavanah, Herbert, pilot
Ceder, Melvin, HQ staff
Chaney, Charles, crew chief
Chennault, Claire Lee, Commander
Christensen, Keith, armorer
Chun, Yuen Gee, engineering
Conant, Edwin, pilot
Cornelius, Jack, crew chief
Cribbs, Charles, medical staff
Croft, John, pilot
Crookshanks, Jesse, crew chief
Cross, Henry, communications
Cross, James, pilot
Curran, George, crew chief
Cushing, Albert, clerk
Daube, Otto, crew chief*
Davis, Doreen, HQ staff*
Davis, William, HQ staff*
Dean, John, pilot
DeWolfe, Robert, HQ staff
Dolan, Walter, crew chief
Doran, Francis, clerk
Dupouy, Parker, pilot
Engle, Charles, ground crew
Engler, John, communications
Ernst, Richard, communications
Farrell, John, pilot
Fobes, Edwin, clerk
Fox, Henry, line chief*
Francisco, Charles, communications
Frillman, Paul, HQ staff
Fritzke, Allen, armorer*
Gallagher, Robert, medical staff
Gasdick, Joseph L., crew chief
Gentry, Thomas, MD
Geselbracht, Henry, pilot*
Gorham, Lloyd, crew chief
Gove, Irving, crew chief
Goyette, Edgar, pilot
Greene, Paul J., pilot
Greenlaw, Harvey, HQ staff
Greenlaw, Olga, HQ staff
Groh, Clifford, pilot
Hall, Lester, pilot*
Harrington, Jasper J., line chief
Harris, David, pilot/HQ staff
Harris, Edward, admin chief
Haywood, Thomas, pilot
Hedman, Robert, pilot
Hennessy, John, pilot
Henson, Thomas, medical staff
Hill, David Lee, pilot
Hodges, Fred, pilot
Hoffman, Roy, HQ staff
Hooker, Burton, parachute rigger*
Howard, James, pilot*
Hoyle, David, admin chief
Hubler, Marlin, clerk
Jacobson, Frank, crew chief
Janski, Edwin, prop. specialist
Jernstedt, Kenneth, pilot
Jordan, Joe, clerk
Kaelin, Albert, clerk*
Keeton, Robert, pilot
Keller, Daniel, crew chief
Kelly, Thomas, communications
Kemph, Merlin, crew chief*
Kenner, Charles, crew chief
Kepka, George, crew chief
Kiner, Melvin, communications
King, Robert, communications
Kustay, Stephen, armorer
Kuykendall, Matthew, pilot
Kwong, Lawrence, HQ staff
Laughlin, C.H., pilot
Lawlor, Frank, pilot
Layher, Robert, pilot*
Leighty, Charles, parachute rigger
Lee, Joseph, medical staff
Lee, Pak On, engineering
Lindstedt, Robert, communications
Linton, Jack, armorer*
Loane, Ernest, pilot*
Locke, Robert, prop. specialist
Loomis, Elton, communications
Losonsky, Frank, crew chief
Lum, George, engineering
Lussier, Joseph, communications
McAllister, Gale, crew chief
McClure, Edgar, crew chief
McHenry, Sharon, supply
McKinney, Eugene, armorer*
McMillan, George, pilot
Mickelson, Einar, pilot
Mihalko, Alex, communications
Miller, Arvold, communications*
Misenheimer, Charles, crew chief
Moss, Kenneth, weather*
Moss, Robert, pilot
Musgrove, Willard, crew chief
Neal, Robert, armorer*
Neale, Robert, pilot*
Neumann, Gerhard, mechanic
Older, Charles, pilot

Olson, Arvid, pilot
Olson, Henry, crew chief
Osborne, Harold, crew chief
Overend, Edmund, pilot
Overley, John, crew chief*
Paull, Preston, crew chief
Paxton, George, pilot*
Peeden, Joseph, crew chief
Peret, Richard, HQ staff
Perry, Paul, armorer
Petach, Emma Jane (Foster), nurse*
Petach, John, pilot*
Pietsker, Joseph, photographer
Pistole, Herbert, armorer
Pon, Kee Jeung, engineering
Poshefko, Joseph, armorer*
Prescott, Robert, pilot
Prevo, Samuel, MD
Probst, Albert, pilot*
Quick, Carl, crew chief*
Raines, Robert, pilot*
Rasmussen, Robert, crew chief
Rector, Edward, pilot
Reed, William, pilot
Regis, James, photographer
Regis, Stanley, crew chief
Richards, Lewis, MD
Richardson, Rolland, communications.
Ricketts, Freeman, pilot*
Ricks, Wayne, prop. specialist
Riffer, Clarence, armorer
Roberts, Carson, communications
Rodewald, Donald, armorer
Rogers, Robert, crew chief
Rosbert, C. Joseph, pilot*
Rossi, J. Richard, pilot
Rumen, John, armorer*
Sasser, Ralph, communications
Sawyer, Charles, pilot
Schaper, Wilfred, crew chief*
Schiel, Frank, pilot
Schields, Milan, prop specialist*
Schiller, Ralph, armorer
Schramm, Leo, crew chief
Seamster, Loy, communications
Seavey, Edward, clerk
Seiple, Wilfred, crew chief
Shamblin, Arnold, pilot*
Shapard, Van, pilot*
Shaw, John, medical staff
Shilling, Eriksen, pilot
Shreffler, Roger, communications
Smith, Curtis, pilot
Smith, Robert A., crew chief
Smith, Robert M., communications
Smith, Robert T., pilot
Stewart, Jo, nurse
Stiles, Edward, crew chief
Stolet, Irving, crew chief
Sutherland, William, mechanic
Sweeney, Joseph, communications
Sykes, William, communications
Terry, Julian, clerk
Towery, William, food service
Trumble, Thomas, HQ staff*
Tuley, Chester, crew chief
Tyrell, George, crew chief
Uebele, John, crew chief
Van Timmeran, Frank, line chief
Vaux, Morgan, communications
Viverette, Hugh, med. staff
Wagner, Earl, armorer*
Wakefield, Manning, crew chief*
Walters, George, clerk
Whelpley, Donald, weather
Whitwer, Eloise, clerk
Williams, John, communications
Wilson, Clifford, mechanic
Wing Shee, George, engineering
Wirta, Harvey, armorer
Wolf, Fritz, pilot
Woodward, Melvin
Wright, Allen, pilot*
Wright, Peter, pilot*
Wu, Lem Fong, engineering
Wyatt, Louis, communications*
Yee, Francis, engineering
Young, John, clerk

Died While Serving

Armstrong, John, flying accident, September 8, 1941
Hammer, Maax, flying accident, September 22, 1941
Atkinson, Peter, flying accident, October 25, 1941
Martin, Neil, KIA, December 23, 1941
Gilbert, Henry, KIA, December 23, 1941
Mangleburg, Lacy, flying accident, December 23, 1941
Baugh, Marion, flying accident, January 3, 1942
Merritt, Kenneth, ground accident, January 8, 1942
Christman, Allen, KIA, January 23, 1942
Hoffman, Louis, KIA, January 26, 1942
Cole, Thomas, KIA, January 30, 1942
Sandell, Robert, flying accident, February 7, 1942
Leibolt, Edward, presumed KIA, February 25, 1942
Fauth, John, bombing attack, March 22, 1942
Newkirk, John, KIA, March 24, 1942
Swartz, Frank, bombing attack, April 24, 1942
Blackburn, John, flying accident, April 28, 1942
Foshee, Ben, bombing attack, May 4, 1942
Donovan, John, KIA, May 12, 1942
Jones, Thomas, flying accident, May 16, 1942
Little, Robert, KIA, May 22, 1942
Petach, John, KIA July 10, 1942

* *Volunteered for additional two weeks.*
** *Returned to States to enter Navy pilot training.*
*** *Approved for leave back to the States to marry fiancée, but never returned to A.V.G..*

MIA/Prisoners of War

Alsop, Joseph, HQ staff, interned December 25, 1941
Mott, Charles, January 8, 1942
McGarry, William, March 24, 1942
Bishop, Lewis, May 17, 1942
Shamblin, Arnold, MIA/assumed KIA, July 10, 1942

Resigned Prior to December 7, 1941

Durbin, Estill, September 1941 (en route)
Dyson, James, September 1941 (en route)
Cook, Elmer, pilot, September 24, 1941
Kelleher, John, pilot, September 24, 1941
McGuire, Maurice, pilot, Sept 24, 1941
Power, Robert, pilot, September 24, 1941
Schur, Carl, crew chief, September 24, 1941
Wallace, Stanley, pilot, September 24, 1941
Watson, Eugene, pilot, September 30, 1941
Bryant, Alfred, clerk, October 1, 1941
Wiggin, Edwin, admin. chief, Oct 15, 1941
Rushton, Edwin, admin. chief, Oct. 24, 1941
Johnston, Leon, clerk, October 31, 1941
Walroth, Robert, pilot, October 31, 1941
Bohman, Moriss, pilot, November 5, 1941
White, Richard, pilot, November 5, 1941
Ringey, Joseph, mechanic, Nov 8, 1941
Swindle, Estes, pilot, November 12, 1941
Smith, Corbet, food service, Nov 15, 1941
Smith, George, clerk, December 1941
Criz, Albert, pilot, December 5, 1941

Discharged Prior to Disbandment

Fuller, Henry, pilot, December 31, 1941
Stubbs, Gail, pilot, December 31, 1941
Moore, Lawrence, January 13, 1942
Sanger, Kenneth, January 13, 1942
Gunvordahl, Ralph, pilot, January 14, 1942
Houle, Leo, pilot, January 14, 1942
Knapp, Donald, pilot, January 15, 1942
Bernsdorf, Donald, pilot, January 15, 1942
Shapiro, William, clerk, January 19, 1942
Swindle, Estes, pilot, January 20, 1942
Walsh, Andrew, clerk, January 21, 1942
Henderson, George, clerk, January 21, 1942
Buxton, Richard, medical staff, Feb 7, 1942
Dudzik, Frank, clerk, February 10, 1942
Hurst, Lynn, pilot, February 1942
Lancaster, George, clerk, February 11, 1942
Newell, Ferris, clerk, February 11, 1942
Rasberry, James, food serv., Feb 11, 1942
Richardson, Randall, weather, Feb 11, 1942**
Hanley, Leo, armorer, March 11, 1942
Richardson, Charles, crew chief, March 11, 1942
Hardesty, Martin, crew chief, March 14, 1942
Mundelin, Charles, crew chief, , Mar 14, 1942
Reynolds, George, crew chief, Mar 14, 1942
Heller, John, clerk, March 17, 1942
Bartelt, Percy, pilot, March 21, 1942
Power, John, crew chief, March 27, 1942
Colquette, Leon, crew chief, March 1942
Gallagher, Edward, crew chief, March 1942
White, John, crew chief, March 23, 1942
Sommers, John, clerk, March 26, 1942
Wyke, William, HQ staff, March 26, 1942
Unger, William, armorer, March 27, 1942
Bacon, Noel, March, 1942 ***
Jaeger, George, mechanic, April 8, 1942
Harpold, Clayton, food service, Apr 16, 1942
Metasavage, Frank, crew chief, April16, 1942
Yarberry, Glen, crew chief, April 16, 1942
McDowell, Mark, crew chief, April 18, 1942
Boyington, Gregory, pilot, April 21, 1942
Walker, Harold, crew chief, April 26, 1942
Crotty, John, clerk, April 27, 1942
Hauser, John, crew chief, April 27, 1942
Graham, Richard, crew chief, April 28, 1942
Blaylock, Glenn, crew chief, April 28, 1942
Fish, William, pilot, May 1942
Hastey, Raymond, pilot, May 14, 1942
Whitehead, R.G., HQ staff, May 31, 1942
Wylie, Harold, clerk, June 1942
Jones, Jack, armorer, April 28, 1942

Group Headquarters Section
Group Commader:
Claire Chennault

C.B. SKIP ADAIR

JOHN ALLARD

JOSEPH ALSOP

EDMOND BOUGHMAN

LEO BEAUPRE

DONALD BELL

ERNEST BONHAM

JAMES BRADY

KENNETH BREEDON

EVERETT BRUCE, M.D.

BOATNER CARNEY

MELVIN CEDER

CHARLES CRIBBS

HARVEY CROSS

ALBERT CUSHING

DOREEN DAVIS

WILLIAM DAVIS

FRANCIS DORAN

EUGENE DURALL

JOHN ENGLER

RICHARD ERNST

EDWIN FOBES

E.J. RED FOSTER

PAUL FRILLMANN

ROBERT GALLAGHER

JOSEPH GASDICK

CHUN YUEN GEE

THOMAS GENTRY, M.D.

HARVEY
GREENLAW

OLGA
GREENLAW

THOMAS
HENSON

ROY
HOFFMAN

DAVID
HARRIS

BURTON
HOOKER

MARLIN
HUBLER

JOSEPH
JORDAN

WALTER
JOURDAN

THOMAS
KELLY

MELVIN
KINER

ROBERT
KING

CHARLES
LEAGHTY

JOSEPH
LEE

PAK ON
LEE

ROBERT
LINDSTEDT

ROBERT
LOCKE

GEORGE
LUM

SHARON
McHENRY

ARVOLD
MILLER

KENNETH
MOSS

GERHARD
NEUMANN

RICHARD
PERET

JOSEPH
PIETSKER

KEE JEUNG
PON

SAMUEL
PROVO, M.D.

JAMES
REGIS

LEWIS
RICHARDS, M.D.

ROLLAND
RICHARDSON

WAYNE
RICKS

ROBERT ROGERS

RALPH SASSER

JOHN SHAW

LOY SEAMSTER

EDWARD SEAVEY

MILAN SHIELDS

ROGER SHREFFLER

CURTIS SMITH

ROBERT M. SMITH

JO STEWART

WILLIAM SUTHERLAND

WILLIAM SYKES

JULIAN TERRY

WILLIAM TOWERY

THOMAS TRUMBLE

MORGAN VAUX

HUGH VIVERETTE

MANNING WAKEFIELD

GEORGE WALTERS

DON WELPLEY

ELOISE WHITWER

JOHN WILLIAMS

CLIFFORD WILSON

GEORGE WINGSHEE

HARVEY WIRTA

MELVIN WOODWARD

LEM FONG WU

HAROLD WYLIE

FRANCIS YEE

JOHN YOUNG

1st Pursuit Sqdn. *Adam & Eves* Commanders: Sandell & Neale

PETER ATKINSON

WILLIAM BARTLING

JOHN BLACKBURN

WILLIAM BLACKBURN

CHARLES BOND

CARL BROWN

GEORGE BURGARD

MICHAEL CALLAN

JOHN CARTER

EDWIN CONANT

JACK CORNELIUS

JOHN CROFT

JIM CROSS

GEORGE CURRAN

JOHN DEAN

WALTER DOLAN

CARL DORRIS

JOHN FARRELL

IRVING GOVE

EDGAR GOYETTE

JASPER HARRINGTON

JOHN HENNESSEY

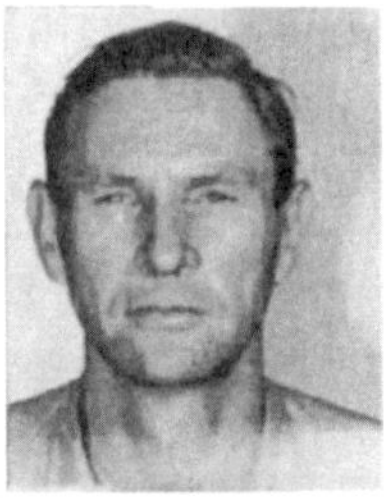
LOUIS HOFFMAN

FRANK JACOBSON

AL KAELIN

MERLIN KEMPH

CHARLES KENNER

STEPHEN KUSTAY

MATTHEW KUYKENDALL

EDWARD LEIBOLT

JACK LINTON

ROBERT LITTLE

ERNEST LOANE

JOSEPH LUSSIER

EDGAR McCLURE

WILLIAM McGARRY

EUGENE McKINNEY

EINAR MICKELSON

CHARLES MISENHEIMER

WILLARD MUSGROVE

ROBERT NEAL

ROBERT NEALE

ROBERT PRESCOTT

ALBERT PROBST

ROBERT RASMUSSEN

DONALD RODEWALD

JOSEPH ROSBERT

J. RICHARD ROSSI

CHARLES SAWYER

ROBERT SANDELL

FRANK SCHIEL

WILFRED SCHAPER

ROBERT H. SMITH

JOHN UEBELE

FRITZ WOLF

ALLEN WRIGHT

LOUIS WYATT

2nd Pursuit Sqdn. *Panda Bears* Commanders: Newkirk & Hill

JOHN ARMSTRONG

GEORGE BAILEY

MORTON TWISTY BENT

HAROLD BLACKWELL

HARRY BOLSTER

GEORGE BRICE

JOHN BRIGHT

CARL BUGLER

CHARLES CHANEY

ALLEN CHRISTMAN

JOHN CARTER

THOMAS COLE

OTTO DAUBE

WILLIAM FISH

HARRY FOX

ALLEN FRITZKE

HENRY GESELBRACHT

LLOYD GORHAN

LESTER HALL

MAAX HAMMER

DAVID LEE HILL

JAMES HOWARD

LYNN HURST

THOMAS JONES

ROBERT KEETON

ROBERT KING

FRANK LAWLOR

ROBERT LAYHER

LACY MANGLEBURG

KENNETH MERRITT

ALEX MIHALKO

ROBERT MOSS

CHARLES MOTT

JAMES MUSICK

JOHN NEWKIRK

JOHN OVERLEY

PRESTON PAULL

GEORGE PAXTON

JOHN PETACH

HERBERT PISTOLE

CARL QUICK

EDWARD RECTOR

FREEMAN RICKETTS

CARSON ROBERTS

JOHN RUMEN

ARNOLD SHAMBLIN

VAN SHAPARD

RALPH SCHILLER

FRANK SWARTZ

CHESTER TULEY

GEORGE TYRELL

EARL WAGNER

PETER WRIGHT

3rd Pursuit Sqdn.
Hell's Angels
Commander:
Arvid "Oley" Olson

FRANK ADKINS

CHARLES BAISDEN

LEWIS BISHOP

ROBERT BROUK

HERBERT CAVANAH

KEITH CHRISTENSEN

L. PAUL CLOUTHIER

LEON COLQUETTE

JESSE CROOKSHANKS

JOHN DONOVAN

PARKER DUPOUY

CHARLES ENGLE

JOHN FAUTH

BEN FOSHEE

CHARLES FRANCISCO

ROBERT GALLAGHER

HENRY GILBERT

PAUL GREENE

CLIFFORD GROH

THOMAS HAYWOOD

ROBERT HEDMAN

FRED HODGES

DANIEL HOYLE

KENNETH JERNSTEDT

THOMAS JONES

DANIEL KELLER

GEORGE KEPKA

C.H. LINK LAUGHLIN

ELTON LOOMIS

FRANK LOSONSKY

LACY MANGLEBURG

NEIL MARTIN

GALE McALLISTER

GEORGE McMILLAN

CHARLES OLDER

ARVID OLSON

HENRY OLSON

HAROLD OSBORNE

EDMUND OVEREND

PAUL PERRY

JOSEPH POSHEFKO

ROBERT RAINE

WILLIAM REED

STANLEY REGIS

CLARENCE RIFFER

LEO SCHRAMM

WILFRED SEIPLE

ERIK SHILLING

ROBERT A. SMITH

ROBERT T. SMITH

EDWARD STILES

IRVING STOLET

JOSEPH SWEENEY

JULIAN TERRY

FRANK VAN TIMMEREN

Appendix IV.
Enemy Aircraft Confirmed Destroyed*

Pilot	Air	Ground
Adkins, F.W.	1.00	-
Bacon, N.R.	3.00	0.50
Bartelt, P.R.	5.00	2.00
Bartling, W.E.	5.00	3.00
Bishop, L.S.	2.20	3.00
Blackburn, J.E.	2.00	-
Bolster, H.R.	2.00	-
Bond, C.H.	7.00	2.50
Boyington, G.	2.00	1.50
Bright, J.G.	3.00	3.00
Brouk, R.R.	3.50	-
Burgard, G.T.	10.00	0.78
Cole, T.J.	1.00	-
Dean, J.J.	3.00	-
Donovan, J.T.	1.00	3.00
Dupouy, P.S.	3.50	-
Farrell, J.W.	1.00	-
Geselbracht, H.M.	1.50	-
Greene, P.J.	2.00	-
Groh, C.G.	2.00	-
Gunvordahl, Ralph	1.00	-
Hastey, R.L.	1.00	-
Haywood, T.C.	4.25	0.83
Hedman, R.P.	6.00	-
Hill, D.L.	10.25	2.00
Hodges, F.S.	1.00	-
Howard, J.H.	2.33	4.00
Jernstedt, K.A.	5.00**	7.00
Jones, T.A.	1.00	3.00
Keeton, R.B.	1.00	-
Kuykendall, M.W.	1.00	-
Laughlin, C.H.	2.20	3.00
Lawlor, F.L.	7.00	1.50
Layher, R.F.	0.33	0.50
Little, R.L.	10.00	0.55
McGarry, W.D.	8.00	3.68
McMillan, C.B.	4.25	-
Merritt, K.T.	1.00	-
Mickelson, E.I.	1.00	-
Moss, R.C.	2.00	2.00

Pilot	Air	Ground
Mott, R.D.	-	3.00
Neale, R.H.	13.00	3.68
Newkirk, J.V.	7.00	2.00
Older, C.H.	9.25	-
Olson, A.E.	1.00	-
Overend, E.F.	5.00	-
Petach, J.E.	3.98	-
Prescott, R.W.	5.50	0.78
Raines, E.J.	3.20	-
Rector, E.F.	4.00	3.50
Reed, W.N.	3.00	8.00
Ricketts, F.I.	1.20	-
Rosbert, C.J.	6.00	-
Rossi, J.R.	6.00	0.28
Sandell, R.J.	5.00	-
Sawyer, C.W.	2.00	-
Schiel, F.	4.00	3.00
Shapard, V.	1.00	-
Shilling, E.E.	0.75	-
Smith, R.H.	5.00	0.50
Smith, R.T.	8.78	-
Wolf, F.E.	4.00	-
Wright, P.	2.53	1.00
Total	**231.5**	**73.08**

* *The author determined these figures from A.V.G. combat reports, diary entries, and confirmation reports, some changed after the fact by A.V.G. boards. Most were confirmed by the pilot's wingman or discovery of crashed aircraft.*

** *The American Fighter Aces Association Victory Confirmation Board upgraded Ken Jernsted's aerial victories to five on March 27, 1997, based on Parker Dupoy's testimony that he observed Jernstedt shooting down two Japanese bombers during the devastating raid on Magwe on March 21, 1942. Rushed to the hospital upon landing, Jernstedt never filed a combat report.*

Appendix V.
Map of China, 1941

Notes and Sources

Preface

1. The Robert Short episode was taken from *Robert Short, the Making of a Hero,* by Martin Cole in the *American Aviation Historical Society Journal,* Vol. 15, No. 1, pages 19-22.
2. William M. Leary, Jr., *The Dragon's Wings,* p. 60.
3. Crozier, *The Man Who Lost China,* p.114
4. Gordon Keith Pickler, *The United States Aid to the Chinese Air Force, 1931-1949,* unpublished PhD dissertation, Florida State University, 1971, p.4.
5. Ibid., p. 5.
6. Rosholt, *Flight in the China Air Space, 1910-1950,* p. 31.
7. Ibid., p. 35.

Chapter 1: From Jouett to Chennault

1. Rosholt, *Flight in the China Air Space, 1910-1950,* pp. 35-36
2. Ibid., p.36.
3. John Jouett letter to W.A. Spencer, May 26, 1932. (Author's collection.)
4. Rosholt, *Flight in the China Air Space, 1910-1950,* p. 35.
5. Russell Whelan, *The Flying Tigers,* pp. 16-17.
6. Pickler, *United States Aid to the Chinese Air Force, 1931-1949,* p. 7.
7. Rosholt, *Flight in the China Air Space, 1910-1950,* p. 36.
8. Ray Wagner, *The Chinese Air Force, 1931-1940,* in the *American Aviation Historical Society Journal,* Fall, 1974, p. 163.
9. Rosholt, *Flight in the China Air Space, 1910-1950,* p 36.
10. Ibid., p. 37
11. Ibid., p. 40.
12. Wagner, *The Chinese Air Force, 1931-1940,* in the *American Aviation Historical Society Journal,* Fall, 1974, p. 163.
13. Pickler, *United States Aid to the Chinese Air Force, 1931-1949,* p. 15.
14. Claire L. Chennault, *Way of a Fighter,* p. 36.
15. Gerry Beauchamp, *Hawks of the Eastern Skies,* in *Air Classics* Magazine, October 2003, p. 24.
16. Wagner, *The Chinese Air Force, 1931-1940,* in the *American Aviation Historical Society Journal,* Fall, 1974, p. 163.
17. Beauchamp, *Hawks of the Eastern Skies,* p. 24.
18. Pickler, *United States Aid to the Chinese Air Force, 1931-1949,* p. 15.
19. Rosholt, *Flight in the China Air Space, 1910-1950,* p. 39.
20. Chennault, *Way of a Fighter,* p. 29.
21. Pickler, *United States Aid to the Chinese Air Force, 1931-1949,* p. 18.
22. Chennault, *Way of a Fighter, p.* 29
23. Ibid., p. 28.
24. Pickler, *United States Aid to the Chinese Air Force, 1931-1949,* p.19.
25. Martha Byrd, *Chennault: Giving Wings to the Tiger,* pp. 61-62. (hereafter, Chennault.)
26. Chennault, *Way of a Fighter,* p.10.
27. Ibid., pp. 12-16.
28. Ibid., pp. 18-19.
29. Ibid., p. 22.
30. Ibid., p. 22.
31. Ibid., pp. 28-30.
32. Byrd, *Chennault,* p. 62.
33. Chennault, *Way of a Fighter,* p. 32.
34. Ibid., p. 33.
35. Crozier, *The Man Who Lost China,* p. 114.
36. Chennault, *Way of a Fighter,* p. 35.
37. Ibid., p. 38.

38. Toland, The Rising Sun, p.46.
39. Ibid., p. 56.
40. Ibid., p. 57.
41. Chennault, *Way of a Fighter,* p. 39.

Chapter 2: Italians, Russians and the 14th International Squadron

1. Chennault, *Way of a Fighter,* p. 39.
2. Ibid., p. 40.
3. Ibid., pp. 40-41.
4. Toland, *The Rising Sun,* p. 58.
5. Chennault, *Way of a Fighter,* p. 45.
6. Ibid., p. 45.
7. Wagner, *The Chinese Air Force, 1931-1940,* in the *American Aviation Society Journal,* Vol. 19, No. 3, pp. 167-168.
8. Holcomb, *Flying Behind the Twelve-Pointed Star,* p. 39.
9. Ibid., pp. 39,41.
10. Rosholt, *Flight in the China Air Space, 1910-1950,* p. 78.
11. Crozier, *The Man Who Lost China,* p. 204.
12. Ibid., p. 204.
13. Ibid., p. 205.
14. Chennault, *Way of a Fighter,* p. 61.
15. Ibid., p. 23.
16. Pickler, *United States Aid to China, 1931-1949,* pp. 23-23.
17. Chennault, *Way of a Fighter,* p. 87.
18. Ibid., pp. 57-58
19. Holcomb, *Flying Behind the Twelve-Pointed Star,* p. 41.
20. Crozier, *The Man Who Lost China,* pp. 202-204.
21. Ibid., p. 204, and Toland, *The Rising Sun,* p. 63.
22. Holcomb, *Flying Behind the Twelve-Pointed Star,* p. 42.
23. Pickler, *United States Aid to the Chinese Air Force, 1931-1949,* p. 30.
24. Holcomb, *Flying Behind the Twelve-Pointed Star,* p. 42.
25. Pickler, *United States Aid to the Chinese Air Force, 1931-1949,* p. 46.
26. Chennault, *Way of a Fighter,* p. 70.
27. Ibid., p. 70.
28. Ibid., p. 70.
29. Ibid., p. 70.
30. Rosholt, *Flight in the China Air Space, 1910-1950,* p. 103.
31. R.E. Baldwin and Thomas William McGarry, *Last Hope, The Blood Chit Story,* p. 35.
32. Pickler, *United States Aid to the Chinese Air Force, 1931-1949,* pp. 40-41.
33. Holcomb, *Flying Behind the Twelve-Pointed Star,* p. 42.
34. Wagner, *The Chinese Air Force, 1931-1940,* pp. 168-169.
35. Chennault, *Way of a Fighter,* p.71.
36. Ibid., p. 71.
37. William Labussiere, *The Truth About the 14th Squadron,* in the *American Aviation Society Journal,* Winter, 1987, p. 286.
38. Chennault, *Way of a Fighter,* pp. 64-65.
39. Rosholt, *Flight in the China Air Space, 1910-1950,* p. 94.
40. Ibid., p. 94.
41. Ibid., p. 94.
42. Chennault, *Way of a Fighter,* p. 65, and Rosholt, *Flight in the China Air Space, 1910-1950,* p. 95.
43. Ibid., p. 65.
44. Pickler, *United States Aid to the Chinese Air Force, 1931-1949,* p. 45.
45. Ibid., p. 45.
46. Immanuel C.Y. Hsu, *The Rise of Modern China,* p. 703.

47. Rosholt, *Flight in the China Air Space, 1910-1950,* pp. 85-86.
48. Ibid., p. 96.
49. Chennault, *Way of a Fighter,* p. 73.
50. Ibid., p. 73.
51. Ibid., p. 74.
52. Ibid., p. 80.

Chapter 3: Chiang Gains American Support

1. Crozier, *The Man Who Lost China,* p. 209.
2. Wanda Cornelius and Thayne Short, *Ding Hao: America's Air War in China, 1937-1945*, pp. 9-26.
3. Hsu, *The Rise of Modern China,* p. 704.
4. Crozier, *The Man Who Lost China,* p. 209.
5. Ibid., p. 210.
6. Chennault, *Way of a Fighter,* p. 87.
7. Ibid., p. 88.
8. Rosholt, *Flight in the China Air Space, 1910-1950,* p. 99.
9. Pickler, *United States Aid to the Chinese Air Force, 1931-1949,* p. 63.
10. Ibid., p. 65.
11. Ibid., p. 69.
12. Toland, *The Rising Sun,* p. 75.
13. Ibid., p. 80.
14. Ibid., pp. 80-81.
15. Rosholt, *Flight in the China Air Space, 1910-1950,* p. 99.
16. Wagner, *The Chinese Air Force, 1931-1940,* in the *American Aviation Historical Society Journal,* Vol. 19, No.3, p. 169.
17. Rosholt, *Flight in the China Air Space, 1910-1950,* p.99.
18. Crozier, *The Man Who Lost China,* p. 225.
19. Pickler, *United States Aid to the Chinese Air Force, 1931-1949.* pp. 69-72.
20. Ibid., p. 66.
21. William D. Pawley, *Americans Valiant and Glorious, self-published, 1945,* p. 64.
22. Byrd, Chennault, p. 106.
23. Chennault, *Way of a Fighter,* p. 90.
24. Ibid., p. 90.
25. Pawley, *Americans Valiant and Glorious,* p. 56.
26. Edward L. Leiser, *Memoirs of Pilot Elwyn H. Gibbon, the Mad Irishman,* in the *American Aviation Historical Society Journal,* Vol. 23, No.1, p. 14.
27. Pawley, *Americans Valiant and Glorious,* p. 62, and Byrd, *Chennault,* p. 61.
28. Chennault, *Way of a Fighter,* p. 90.
29. Ibid., p. 92.
30. Byrd, *Chennault,* p. 108.
31. Jack Samson, *Chennault: A Definitive Biography of the Legendary Leader of the Flying Tigers and the U.S. Air Force in China,* p. 62. (hereafter, Chennault.)
32. Byrd, *Chennault,* p. 108.
33. Daniel Ford, *Flying Tigers,* p. 53.
34. Over the past several years, there has been some controversy regarding whether the H-81-A2 flown by the A.V.G. was the export model of the P-40B or P-40C. Dimensionally identical, the major differences between the "B" and "C" were an increase of 22 pounds for the "C" (with a concurrent drop in speed from 352 to 345 mph), the addition of belly shackles for a 52-gallon drop tank, and the addition of two .30-caliber wing guns. The aircraft the A.V.G. received had no external tank shackles and had only four wing guns. Also, Chennault himself, in *Way of a Fighter*, continually spoke of them as P-40Bs. The author is satisfied they were P-40Bs.
35. Chennault, *Way of a Fighter,* p. 100.
36. Byrd, *Chennault,* pp. 111-112.
37. Ibid., p. 112.

38. Anthony R. Carrozza, *William D. Pawley: The Extraordinary Life of the Adventurer, Entrepreneur, and Diplomat Who Cofounded the Flying Tigers,* p. 77. (hereafter, William D. Pawley.)
39. Chennault, *Way of a Fighter,* p. 98.
40. Pickler, *United States Aid to the Chinese Air Force, 1931-1949,* p. 73.
41. Chennault, *Way of a Fighter,* p. 99.
42. Byrd, *Chennault,* p. 119, and Chennault, *Way of a Fighter,* p. 101.
43. Fairbanks memo, p. 11.
44. Ibid., p. 14
45. Ford, *Flying Tigers,* p. 86.
46. Byrd, *Chennault,* p. 115.
47. Ibid., p. 116.
48. Chennault, *Way of a Fighter,* p. 101.
49. Ibid., p. 101.
50. Pickler, *United States Aid to the Chinese Air Force, 1931-1949,* p 75.
51. Byrd, *Chennault,* p. 115.
52. Toland, *The Rising Sun,* pp. 58-59.
53. Contrary to the popular account of President Roosevelt signing "an unpublished executive order," no such order was ever signed – it was all verbal, no doubt to protect the secrecy of the operation. (See pages 116-117of *Chennault: Giving Wings to the Tiger,* by Martha Byrd.)

Chapter 4: Recruiting

1. Fairbank memo, pp. 18-19.
2. Chennault, *Way of a Fighter,* pp. 102-103.
3. Ibid., p. 103.
4. Pickler, *United States Aid to the Chinese Air Force, 1931-1949,* p. 105.
5. C. Joseph Rosbert, *Flying Tiger Joe's Adventure Story Cookbook,* p. 34.
6. Ibid., p. 34.
7. Ibid., p. 34.
8. Ibid., p. 38.
9. Charles R. Bond, Jr. and Terry Anderson, *A Flying Tiger's Diary,* p. 20.
10. Ibid., p. 20
11. R.T. Smith, *Tale of a Tiger,* p. 13
12. Ibid., p. 18.
13. Ibid., p. 22.
14. Ibid., p. 22.
15. Cornelius and Short, *Ding Hao: America's Air War in China, 1937-1945,* p. 104.
16. Ibid., p. 104.
17. Robert B. Hotz., *With Chennault: The Story of the Flying Tigers,* p. 97.
18. Ibid., pp. 96-97.
19. Ibid., p. 97.
20. Pickler, *United States Aid to the Chinese Air Force, 1931-1949,* p. 20.
21. Fairbank memo, p. 20
22. Pickler, *United States Aid to the Chinese Air Force, 1931-1949,* p. 107.
23. Fairbank memo, pp. 21-25.
24. Walter E. Pentecost and J.J. Sloan, *Advance of the Flying Tigers,* in the *American Aviation Historical Society Journal,* Vol. 15., No.2, pp. 138-139.
25. Ibid., pp. 139-143.
26. Ibid., p. 143.
27. Pickler, *United States Aid to the Chinese Air Force, 1931-1949,* p. 113.
28. Chennault, *Way of a Fighter,* p. 103.
29. Byrd, *Chennault,* pp. 92 and 118.
30. Ibid., p. 120

31. Chennault, *Way of a Fighter,* p. 107
32. Byd, *Chennault,* pp. 92 and 118
33. Paul Frillman and Graham Peck, China: *The Remembered Life,* p.57.
34. Ibid., p. 55.
35. Charles N. Baisden , letter to author, April 25, 1988.
36. Frillman and Peck, China: *The Remembered Life,* p. 58.
37. Ibid., p. 55.
38. Clarence Riffer diary, July 3, 1941.
39. Frillman and Peck, *China: The Remembered Life,* p. 61
40. Clarence Riffer diary, July 22, 1941.
41. Frillman and Peck, China: *The Remembered Life,* pp. 62-63.
42. Ibid., p. 65.
43. Baisden, letter to author, May, 1988.
44. R.T. Smith, *Tale of a Tiger,* p. 58.
45. Paul J. Perry, letter to author, July 15, 1988.
46. Olga Greenlaw, *Lady and the Tigers,* p. 32.
47. Baisden, *Flying Tiger to Air Commando,* p. 20.

Chapter 5: Training

1. Ford, *Flying Tigers,* p. 69.
2. Pentecost and Sloan, *Advance of the Flying Tigers,* in the *Journal of the American Aviation Historical Society,* Summer, 1970, p. 139.
3. Paul J. Perry, letter to author, August 8, 1988.
4. Carrozza, *William D. Pawley,* p. 89.
5. Fairbank memo, p. 23.
6. Robert M. Smith, *With Chennault in China: A Flying Tiger's Diary,* p. 17.
7. Ibid., p. 19.
8. Ibid., pp. 19-20.
9. Ibid., p. 20.
10. Ibid., p. 21.
11. Ibid., p. 21.
12. Pentecost and Sloan, *Advance of the Flying Tigers,* in the *Journal of the American Aviation Historical Society,* p. 139.
13. Ibid., p. 139.
14. Chennault, *Way of a Fighter,* pp. 107-108.
15. Ibid., p. 109.
16. William E. Schaper diary, August 22, 1941.
17. Ibid., August 22, 1941.
18. Byrd, *Chennaultr,* p. 125.
19. Greenlaw, *Lady and the Tigers,* pp. 11-15.
20. Ford, *Flying Tigers,* p. 76, and Greenlaw, *Lady and the Tigers,* p. 37.
21. Chennault, *Way of a Fighter,* p. 111, and Frillman and Peck, *China: The Remembered Life,* p. 71.
22. Ibid., p.71
23. Author's interview with George "Pappy" Paxton, 1977.
24. Chennault, *Way of a Fighter,* p. 112.
25. Ibid., pp.112-113.
26. Ibid., p. 113.
27. Ibid., p. 113.
28. Ibid., p. 114.
29. Jack Samson, *Chennault,* pp. 74 and 78.
30. Duane Schultz, *The Maverick War: Chennault and the Flying Tigers,* p. 117. (hereafter, The Maverick War.)
31. Ibid., pp. 117-118.
32. Cornelius and Short, *Ding Hao: America's Air War in China, 1937-1945,* p. 123.

33. Schultz, *The Maverick War,* p. 118.
34. Robert T. Smith, *Tale of a Tiger,* p. 47.
35. Ibid., p. 46.
36. Ibid,. p. 48.
37. Ibid., p. 50.
38. Ibid., p. 50.
39. Ibid., p. 58.
40. Ibid., p. 67.
41. Ibid., p. 68.
42. Schultz, *The Maverick War,* p.118.
43. Ibid., p.118.
44. Ford, *Flying Tigers,* pp. 389-396.

Chapter 6: Accidents, Another Fatality and Resignations

1. R.M Smith, *With Chennault in China: A Flying Tiger's Diary,* pp. 138-141.
2. Erik Shilling, *Destiny: A Flying Tiger's Rendezvous With Fate,* pp. 109-110. (hereafter, Destiny.)
3. R.T. Smith, *Tale of a Tiger,* p. 78.
4. Rosbert, *Flying Tiger Joe's Adventure Story Cookbook,* p.41.
5. Schultz, *The Maverick War,* pp. 103-104, and Byrd, *Chennault,* pp. 128-129.
6. R.T. Smith, *Tale of a Tiger,* p. 98.
7. Ibid., p. 98.
8. Ibid., p. 104.
9. Ibid., pp. 105-107.
10. Ibid., p. 108
11. Chennault, *Way of a Fighter,* p. 111, and Shilling, *Destiny,* p. 95.
12. R.T. Smith, *Tale of a Tiger,* p. 109.
13. Ford, *Flying Tigers,* p. 85.
14. Chennault, *Way of a Fighter,* p. 111.
15. Gregory Boyington, *Baa, Baa, Black Sheep,* p. 25.
16. Bond and Anderson, *A Flying Tiger's Diary,* p. 25.
17. Ibid., p. 24.
18. George Burgard diary, October 18, 1941.
19. Bond and Anderson, *A Flying Tiger's Diary,* p. 35.
20. Pickler, *United States Aid to the Chinese Air Force, 1939-1941,* p. 110.
21. Letter, November 17, 1985, from Harold E. Klein to Terry Smith.
22. Bond and Anderson, *A Flying Tiger's Diary,* pp. 43-44.
23. Ibid., p. 44.
24. Shilling, *Destiny,* pp. 107-108.
25. Ibid., p. 108.
26. Bond and Anderson, *A Flying Tiger's Diary,* p. 43.
27. Ibid., p. 44.
28. Burgard diary, November 17, 1941.
29. Ibid., November 18, 1941.
30. Bond and Anderson, *A Flying Tiger's Diary,* p. 51.

Chapter 7: North to Kunming

1. Chennault, *w*p. 121.
2. Bond and Anderson, *A Flying Tiger's Diary,* p. 46.
3. Gordon W. Prange, *At Dawn We Slept,* pp. 416-419.
4. R.T. Smith, *Tale of a Tiger,* p.128
5. Burgard diary, November 30, 1941.

6. R.T. Smith, *Tale of a Tiger,* p. 134.
7. Robert Keeton diary, December 8, 1941.
8. Byrd, *Chennault,* p. 134.
9. Samson, *Chennault,* p.87.
10. Burgard diary, December 8, 1941.
11. R.T. Smith, *Tale of a Tiger,* p. 145.
12. Shilling, interview with author, February 1, 1990.
13. Schultz, *The Maverick War,* p. 131.
14. Burgard diary, December 10, 1941.
15. R.T. Smith, *Tale of a Tiger,* p. 146.
16. Chennault, *Way of a Fighter,* p. 125.
17. Ibid., p. 126.
18. Mike Minnich, *Tiger in the Sky, the Saga of the A.V.G., Air Enthusiast Quarterly*/Four, p. 121.
19. Claire L. Chennault papers, Hoover Institute.
20. R.T. Smith, *Tale of a Tiger,* p. 153.
21. Burgard diary, December 17, 1941.
22. Keeton diary, December 18, 1941.
23. Rosbert, *Flying Tiger Joe's Adventure Story Cookbook,* p. 73.
24. Burgard diary, December 18, 1941.
25. Ray Wagner, *The Curtiss P-40 Tomahawk,* Profile Publications, No. 35, p. 9.
26. Bond and Anderson, *A Flying Tiger's Diary,* p. 58, and Ford, Flying Tigers, p. 109.
27. Ibid., p. 58.
28. Rosbert, *Flying Tiger Joe's Adventure Story Cookbook,* p. 73.
29. Burgard diary, December 19, 1841.
30. Rosbert, *Flying Tiger Joe's Adventure Story Cookbook,* p. 75.
31. Burgard diary, December 19, 1941.
32. Bond and Anderson, *A Flying Tiger's Diary,* p. 60.

Chapter 8: "Next Time, Get 'em All!"

1. Fumio Watanabe, letter to author, April 15, 1991.
2. Chennault, *Way of a Fighter,* p. 128.
3. Ford, *Flying Tigers,* p. 116.
4. Chennault, *Way of a Fighter,* p. 129.
5. Ibid., p. 129
6. Bond and Anderson, *A Flying Tiger's Diary,* p. 61.
7. Schultz, *The Maverick War,* p. 141.
8. Bond and Anderson, *A Flying Tiger's Diary,* P. 61.
9. Hotz, *With General Chennault,* p. 20.
10. Ibid., p. 20.
11. Ibid., p. 21.
12. Stuart Leutherner and Oliver Jensen, *High Honor: Recollections by Men and Women of World War II Aviation,* pp. 29-30. (hereafter, High Honor.)
13. Bond and Anderson, *A Flying Tiger's Diary,* p. 62.
14. Chennault, *Way of a Fighter,* p. 130.
15. Watanabe, letter to author, April 15, 1991.
16. Leuthner and Jensen, *High Honor,* pp. 30-31.
17. Rosbert, *Flying Tiger Joe's Adventure Story Cookbook,* p. 77.
18. Ford, *Flying Tigers,* p. 119.

Chapter 9: Christmas Over Rangoon

1. R.T. Smith, *Tale of a Tiger,* p. 157.

2. Ibid., p. 150.
3. Ibid., p. 151.
4. Ibid., p. 158.
5. Shilling, *Destiny*, p. 124.
6. Arvid Olson diary, December 23, 1941.
7. Interview, Frank Rounds with Charles Older for Columbia University, August 1962.
8. Ibid.
9. R.T. Smith combat report, A.V.G. archives.
10. Charles Older combat report, A.V.G. archives.
11. Ibid.
12. R.T. Smith combat report.
13. Paul J. Greene combat report.
14. R.T. Smith combat report.
15. Ed Overend combat report.
16. Parker Dupouy combat report.
17. Thomas Haywood combat report.
18. Ford, *Flying Tigers,* p. 129.
19. Dupouy combat report.
20. Ralph Gunvordahl combat report.
21. William Reed combat report.
22. Schultz, *The Maverick War,* p. 149.
23. Kenneth Jernstedt combat report.
24. Robert Hedman combat report.
25. Third Squadron Diary, December 23, 1941.
26. Ron Heiferman, *Flying Tigers,* p. 48.
27. R.T. Smith, *Tale of a Tiger,* p. 162.
28. Schultz, *The Maverick War,* p. 151.
29. R.T. Smith, *Tale of a Tiger,* p. 162.
30. Ibid., p. 162.
31. Hotz, *With General Chennault,* p. 41.
32. Combined Combat Reports for December 25, 1941.
33. George McMillan combat report.
34. Ford, *Flying Tigers,* p. 140.
35. Dupouy combat report.
36. Reed combat report.
37. Jernstedt combat report.
38. Dupouy combat report.
39. McMillan combat report.
40. R.T. Smith combat report.
41. Hedman combat report.
42. Older combat report.
43. Overend combat report.
44. Robert Brouk combat report.
45. Lewis Bishop combat report.
46. Heiferman, *Flying Tigers,* p. 48.
47. R.T. Smith, *Tale of a Tiger,* p. 165.
48. Ibid., p. 164.

Chapter 10: The Pandas Get Their Turn

1. Olga Greenlaw, *Official Diary of Operations, The First American Volunteer Group, December 1941-July 1942,* December 27, 1941. (hereafter, *Official Diary of Operations.*)
2. Ibid., December 29, 1941.

3. Keeton diary, December 29, 1941.
4. R.T.Smith, *Tale of a Tiger,* p. 170.
5. Keeton diary, December 30, 1941,
6. Ford, *Flying Tigers,* p. 160.
7. Bob Cunningham interview with Tex Hill, 1980.
8. Ibid.
9. Hotz, *With General Chennault,* p. 128.
10. George Paxton diary, January 4, 1941.
11. Kenneth Merritt combat report.
12. Hotz, *With Chennault,* p. 132, and Paxton diary, January 5, 1942.
13. Paxton diary, January 5, 1942.
14. Peter Wright, *I Learned About Flying From That, Flying* magazine, May, 1944.
15. Ibid.
16. Ibid.
17. Charles Mott, *Three Years on the River Kwai: A P.O.W.'s Story, Wings of Gold* magazine, Fall 1991.
18. Ford, *Flying Tigers,* p. 171.
19. Robert Neale diary, January 12, 1942.
20. George Burgard diary, January 12, 1942.
21. Neale diary, January 12-13, 1942.
22. Ibid., January 13, 1942.

Chapter 11: The A.V.G. Strikes Back

1. Bond and Anderson, *A Flying Tiger's Diary,* p. 73.
2. Ibid., January 7, 1942.
3. Ford, *Flying Tigers,* p 198.
4. Bond and Anderson, *A Flying Tiger's Diary,* p. 75.
5. Daniel Hoyle, *Diary of the Third Pursuit Squadron,* January 17, 1942.
6. Correspondence with Gen. Y. Kochiyama.
7. Thomas Haywood Combat Report, January 17, 1942.
8. Minich, *Tiger in the Sky, Air Enthusiast Quarterly*/Four, p.123.
9. Greenlaw, *Official Diary of Operations,* January 19, 1942.
10. Howard combat report.
11. Percy Bartelt combat report.
12. Howard combat report.
13. Greenlaw, *Official Diary of Operations,* January 20, 1942.
14. Heifermamn, *Flying Tigers,* p. 51.
15. Kochiyama correspondence.
16. David L. Hill and Reagan Schaupp, *Tex Hill: Flying Tiger,* p. 122.
17. Neale diary, January 23, 1942.
18. Ibid.
19. Paxton diary, January 23, 1942.
20. Bacon diary, January 23, 1942.
21. Newkirk combat report.
22. Lawlor combat report.
23. Schultz, *The Maverick War,* p. 178.
24. Noel Bacon combat report.
25. Kochiyama correspondence.
26. Greenlaw, *Official Diary of Operations,* January 24, 1942.
27. Neale diary, January 24, 1942.
28. Bacon diary, January 24, 1942.
29. Ibid.
30. Hotz, *With Chennault,* p. 147.

31. Bond and Anderson, *A Flying Tiger's Diary,* p. 78.
32. Ibid., p. 81.
33. Greenlaw, *Official Diary of Operations,* January 25, 1942, and Bond and Anderson, *A Flying Tiger's Diary*, p. 81.
34. Bond and Anderson, *A Flying Tiger's Diary,* p. 82.
35. Newkirk combat report, January 26, 1942.
36. Albert Probst combat report, January 26, 1942.
37. Neale diary, January 26, 1942.
38. Bond combat report, January 26, 1942.
39. Schultz, *A Maverick War,* p. 179.
40. Bond and Anderson, *A Flying Tiger's Diary,* p. 83.
41. Howard, combat report, January 27, 1942.
42. Bond and Anderson, *A Flying Tiger's Diary,* p. 85.
43. Sandell combat report, January 28, 1942.
44. Frank Scheil combat report, January 28, 1942.
45. Bond and Anderson, *A Flying Tiger's Diary,* p. 86.
46. Bacon diary, 29 January, 1942.
47. Bond and Anderson, *A Flying Tiger's Diary,* p. 86.
48. Sandell combat report, January 29, 1942.
49. Newkirk combat report, January 29, 1942.
50. Bond and Anderson, *A Flying Tiger's Diary,* p. 87.
51. Ibid.
52. Ibid.
53. Ibid., p. 88.
54. Howard combat report, January 30, 1942.
55. Minich, *Tiger in the Sky, Air Enthusiast Quarterly*/Four, p. 123.
56. Keeton diary, February 3, 1942.
57. Ibid.
58. Greenlaw, *Official Diary of Operations,* February 4, 1942.
59. Keeton diary, February 4, 1942.
60. Greenlaw, *Lady and the Tigers,* p. 137.
61. Ibid.
62. Bond and Anderson, *A Flying Tiger's Diary,* p. 95, and Greenlaw, *Official Diary of Operations, February 7, 1942.*
63. Greenlaw, *Official Diary of Operations,* February 7, 1942.
64. Ibid., February 10, 1942.

Chapter 12: Evacuation of Rangoon

1. Bond and Anderson, *A Flying Tiger's Diary,* pp. 98-99.
2. R.T. Smith, *Tale of a Tiger,* p. 206.
3. Frank Losonsky, *Flying Tiger: A Crew Chief's Story,* p. 82.
4. R.T. Smith, *Tale of a Tiger,* p. 219.
5. For a detailed account of the first A.V.G. ferry trip, read R.T. Smith's *Tale of a Tiger,* published by Tiger Originals, Van Nuys, CA, 1986, pages 245-260. (needs to move to endnotes)
6. Bond and Anderson, *A Flying Tiger's Diary,* p. 102.
7. Frillman and Peck, *China,* p. 114.
8. Bond and Anderson, *A Flying Tiger's Diary,* p. 103.
9. Burgard diary, February 18, 1942.
10. Greenlaw, *Official Diary of Operations,* February 18, 1942.
11. Bond and Anderson, *A Flying Tiger's Diary,* p. 107.
12. Robert H. Smith combat report, February 21, 1942.
13. Burgard diary, February 21, 1942.
14. Bond and Anderson, *A Flying Tiger's Diary,* pp. 104-106.
15. Ibid.

16. Greenlaw, *Official Diary of Operations,* February 23, 1942.
17. Ibid., February 24, 1942.
18. Bond and Anderson, *A Flying Tiger's Diary,* p. 109.
19. Ibid.
20. Kochiyama correspondence.
21. Robert Prescott combat report, February 25, 1942.
22. Neale diary, February 25, 1942.
23. Bond and Anderson, *A Flying Tiger's Diary,* pp. 109-112.
24. Prescott diary, February 25, 1942.
25. Neale diary, February 25, 1942.
26. Bond and Anderson, *A Flying Tiger's Diary,* p. 113.
27. Prescott diary, February 26, 1942.
28. Burgard diary, February 26, 1942.
29. J.J. Harrington, interview with author, March 28, 1981.
30. Ibid.
31. Greenlaw, *Official Diary of Operations,* February 26, 1942.
32. Kochiyama correspondence.
33. Bond and Anderson, *A Flying Tiger's Diary,* p. 113.
34. Ibid.
35. Ibid., p. 114.
36. Harrington, interview with author, March 28, 1981.
37. Bond and Anderson, *A Flying Tiger's Diary,* p. 114.
38. Ibid., pp. 114-115.
39. Frillman and Peck, *China,* p. 116.
40. Ibid., p. 117.
41. Bond and Anderson, *A Flying Tiger's Diary.* P. 115.
42. Ford, *Flying Tigers,* p. 242.

Chapter 13: "Chennault Was Plenty Steamed"

1. Ford, *Flying Tigers,* p. 242.
2. Bond and Anderson, *A Flying Tiger's Diary,* p. 120.
3. Neale diary, March 1, 1942.
4. Ford, *Flying Tigers,* p. 249.
5. Ibid.
6. Ibid.
7. Keeton diary, March 5, 1942.
8. Ford, *Flying Tigers,* p. 250.
9. Reed diary, March 6, 1942.
10. Losonsky, *Flying Tiger: A Crew Chief's Story,* pp. 85-86.
11. Reed diary, March 10, 1942.
12. Jennifer Holik-Urban and Robert Brouk, *To Soar with the Tigers: The Life and War Diary of Robert Brouk,* p. 62. (hereafter, *To Soar with the Tigers.*)
13. Ibid.
14. 14. Reed diary, March 10, 1942.
15. Burgard diary, March 11, 1942.
16. Holik-Urban and Brouk, *To Soar with the Tigers,* p. 64.
17. Losonsky, *Flying Tiger: A Crew Chief's Story,* p. 86.
18. Ibid., p. 87.
19. Ibid.
20. Frillman and Peck, *China,* pp. 118-119.
21. Ibid., pp. 119-120.
22. Ibid., p. 121.

23. Ibid., p. 124.
24. Barbara W. Tuchman, *Stilwell and the American Experience in China,* pp. 313-314.
25. Ibid., p. 311.
26. Ibid., p. 314.
27. Ibid.
28. Ibid., p. 345.
29. Frillman and Peck, *China,* pp. 124-125.
30. Ibid., p. 126.
31. Ibid., p. 128.
32. Ibid.
33. Ibid., pp. 129-132.
34. Ibid., pp. 132-133.
35. Ibid., p. 134.
36. Ibid. For a more detailed rendition of Frillman's saga, see *China, the Remembered Life*, by Paul Frillman and Graham Peck, Houghton-Mifflin, pages 117-134.

Chapter 14: Attack on Magwe

1. Ford, *Flying Tigers,* pp. 252-253.
2. Hotz, *With General Chennault,* p. 191.
3. Olson diary, March 14, 1942.
4. Hoyle, *Diary of the 3rd Pursuit Squadron,* March 16, 1942.
5. Ford, *Flying Tigers,* p. 256.
6. Reed diary, March 16, 1942.
7. Olson diary, March 17, 1942.
8. Baisden, *Flying Tiger to Air Commando,* p. 50.
9. Reed diary, March 18, 1942.
10. Ibid.
11. Jernstedt combat report, March 18, 1942.
12. Reed diary, March 18, 1942.
13. Ford, Flying Tigers, p. 260.
14. Ibid., p. 259.
15. Ibid., p. 261.
16. Dupouy combat report, March 21, 1942.
17. Ford, *Flying Tigers,* p. 262.
18. Ibid.
19. Clifford Groh combat report, March 21, 1942.
20. Ford, *Flying Tigers,* p. 265.
21. Baisden, *Flying Tiger to Air Commando,* pp. 48-50.
22. Hotz, *With General Chennault,* p. 195.
23. Hoyle, *Diary of the 3rd Pursuit Squadron,* March 21, 1942.
24. Hotz, *With General Chennault,* p. 195.
25. Greenlaw, *Official Diary of Operations,* March 21, 1942.
26. Ford, *Flying Tigers,* p. 267.
27. Ibid., pp. 268-269.
28. Greenlaw, *Official Diary of Operations,* March 22, 1942.
29. Ibid.
30. Baisden, *Flying Tiger to Air Commando,* p. 51.
31. Greenlaw, *Official Diary of Operations,* March 22, 1942.
32. Olson diary, March 23, 1942.
33. George Paxton diary, May 7, 1942.
34. Olson diary, March 23, 1942.

Chapter 15: "Hell of a Nice Mission..."

1. Ford, *Flying Tigers,* p. 269.
2. Greenlaw, *Official Diary of Operations,* March 22, 1942.
3. Bond and Anderson, *A Flying Tiger's Diary,* p. 130.
4. Ibid.
5. Ibid. p. 131.
6. Neale diary, March 23, 1942.
7. Bond and Anderson, *A Flying Tiger's Diary,* p. 131.
8. Ibid.
9. Ibid., p. 132.
10. Neale diary, March 23, 1942.
11. Greenlaw, *Official Diary of Operations,* Robert Neale's mission report, March 24, 1942.
12. Bond and Anderson, *A Flying Tiger's Diary,* p. 132.
13. Neale diary, March 23, 1942.
14. Bond and Anderson, *A Flying Tiger's Diary,* p. 132.
15. Neale mission report, March 24, 1942.
16. Keeton diary, March 24, 1942.
17. Frank Lawlor combat report, March 24, 1942.
18. Keeton combat report. March 24, 1942.
19. Bond and Anderson, *A Flying Tiger's Diary,* p. 134.
20. William Bartling combat report, March 24, 1942.
21. Bond and Anderson, *A Flying Tiger's Diary,* p. 134, and Neale combat report, March 24, 1942.
22. Ibid.
23. Edward Rector combat report, March 24, 1942.
24. Neale combat report, March 24, 1942.
25. Ibid.
26. Bond and Anderson, *A Flying Tiger's Diary,* p. 135.
27. Lawlor combat report, March 24, 1942.
28. Bond and Anderson, *A Flying Tiger's Diary,* pp. 135-136.
29. Schultz, *The Maverick War,* p. 223.
30. Neale combat report, March 24, 1942.
31. Bond and Anderson, *A Flying Tiger's Diary,* p. 137, and Ford, *Flying Tigers,* pp 275-276.
32. Olson diary, March 25, 1942.

Chapter 16: Action at Loiwing

1. Greenlaw, *Official Diary of Operations,* March 26, 1942.
2. Diary, March 26, 1942.
3. Schultz, *The Maverick War,* p. 227.
4. R.T. Smith, *A Tale of a Tiger,* p. 259.
5. Ibid., p. 260.
6. Ibid.
7. Ibid.
8. Older combat report, March 29, 1942.
9. R.T. Smith, *A Tale of a Tiger,* p. 263.
10. Olson diary, March 31, 1942.
11. Gregory Boyington, *Baa, Baa, Black Sheep,* pp. 107-108.
12. R.T. Smith, *A Tale of a Tiger,* p. 263.
13. Olson diary, April 1, 1942.
14. R.T. Smith, *A Tale of a Tiger,* p. 264.
15. Ibid.
16. Brouk, combat report, April 1, 1942.

17. Hoyle, *Diary of the 3rd Pursuit Squadron,* April 1, 1942.
18. Greenlaw, *Official Diary of Operations,* April 2, 1942.
19. Schultz, *A Maverick War,* pp. 230-231.
20. Hotz, *With General Chennault,* p. 204.
21. Ford, *Flying Tigers,* p. 288.
22. Smith, *Tale of a Tiger,* p. 266.
23. Ibid., p. 268.
24. Ibid., p. 271
25. Smith, *Tale of a Tiger,* p. 271, and Olson diary, April 8, 1942.
26. Ford, *Flying Tigers,* p. 167.
27. R.T. Smith combat report, April 8, 1942.
28. C.H. Laughlin combat report, April 8, 1942.
29. R.T. Smith combat report, April 8, 1942.
30. Ibid.
31. Fred Hodges combat report, April 8, 1942.
32. Clifford Groh combat report, April 8, 1942.
33. Fritz Wolf combat report, April 8, 1942.
34. John Donovan combat report, April 8, 1942.
35. Olson combat report, April 8, 1942.
36. Wolf combat report, April 8, 1042.
37. Emund Overend combat report, April 8, 1942.
38. Ford, *Flying Tigers,* p. 284.
39. Greenlaw, *Official Diary of Operations,* April 8, 1942.
40. Smith, *Tale of a Tiger,* p. 271.
41. Greenlaw, *Official Diary of Operations,* April 8, 1942.

Chapter 17: Loiwing Under Attack

1. Keeton diary, March 30, 1942.
2. Ford, *Flying Tigers,* p. 287.
3. Bond and Anderson, *A Flying Tiger's Diary,* p. 148.
4. Greenlaw, *Official Diary of Operations,* April 8, 1942.
5. Keeton diary, April 9, 1942.
6. Greenlaw, *Official Diary of Operations,* April 9, 1942.
7. Smith, *Tale of a Tiger,* p. 273.
8. Keeton diary, April 9, 1942.
9. Smith, *Tale of a Tiger,* pp. 274-275.
10. Ford, *Flying Tigers,* p. 290.
11. Greenlaw, *Official Diary of Operations,* April 10, 1942.
12. Ibid.
13. Ford, *Flying Tigers,* p. 290.
14. Smith combat report, April 10, 1942.
15. Ibid.
16. Reed combat report, April 10, 1942.,
17. Smith combat report, April 10, 1942.
18. Brouk combat report, April 10, 1942.
19. Older combat report, April 10, 1942.
20. Keeton combat report, April 10, 1942.
21. Ibid.
22. Ford, *Flying Tigers,* p. 294.
23. Reed, combat report, April 10, 1942.
24. Smith, *Tale of a Tiger,* p. 276.
25. Olson diary, April 11, 1942.

26. Keeton diary, April 11, 1942.
27. Greenlaw, *Official Diary of Operations,* April 11, 1942.
28. Olson diary, April 11, 1942.
29. David L. Hill combat report, April 12, 1942.
30. Peter Wright combat report, April 11, 1942.
31. Ibid.
32. Smith, *Tale of a Tiger,* p. 277.
33. Reed diary, April 12, 1942.
34. Ibid., April 13, 1942.
35. Smith, *Tale of a Tiger,* p. 278.
36. Reed diary, April 13, 1942.
37. Greenlaw, *Official Diary of Operations,* April 13, 1942.
38. Smith, *Tale of a Tiger,* p. 278.

Chapter 18: "Our Morale Won't Be Good Fast"

1. Keeton diary, April 14, 1942.
2. Ibid.
3. David L. Hill and Reagan Schaupp, *Tex Hill: Flying Tiger,* p. 145.
4. Smith, *Tale of a Tiger,* p. 281.
5. Keeton diary, April 16, 1942.
6. Smith, *Tale of a Tiger,* p. 282.
7. Brouk combat report, April 18, 1942.
8. Dennis R. Okerstrom, *Dick Cole's War: Doolittle Raider, Hump Pilot, Air Commando,* p. 79.
9. Smith, *Tale of a Tiger,* p. 283.
10. Ibid., p. 284.
11. Ibid.
12. Ibid.
13. Hill and Schaupp, *Tex" Hill: Flying Tiger,* p. 146.
14. Smith, *Tale of a Tiger,* p. 284.
15. Copy of original letter of resignation, provided by Robert Andrade.
16. Smith, *Tale of a Tiger,* p. 285.
17. Ibid. p. 284.
18. Rector combat report, April 20, 1942.
19. Holik-Urban and Brouk, *To Soar with the Tigers,* p. 72.
20. Ibid.
21. Ibid., p. 73.
22. Ibid.
23. Greenlaw, *Official Diary of Operations,* April 21, 1942.
24. Reed diary, April 21-22, 1942.
25. Greenlaw, *Official Diary of Operations,* April 21, 1942.
26. Ford, *Flying Tigers,* p. 305.
27. Ibid., pp. 303-304.
28. Ibid.
29. Ibid., p. 307.
30. Greenlaw, *Official Diary of Operations,* April 24, 1942.
31. Hill combat report, April 24, 1942.
32. Ibid.
33. Older combat report, April 24, 1942.
34. Ford, *Flying Tigers,* pp. 308-309.

Chapter 19: Retreat from Loiwing

1. Greenlaw, *Official Diary of Operations,* April 25, 1942
2. Rector combat report, April 25, 1942.
3. Ibid.
4. Raines combat report, April 25, 1942.
5. Greenlaw, *Official Diary of Operations,* April 25, 1942.
6. Smith, *Tale of a Tiger,* p. 291.
7. Laughlin combat report, April 25, 1942.
8. Smith combat report, April 25, 1942.
9. Raines combat report, April 25, 1942.
10. Hoyle, *Diary of 3rd Pursuit Squadron,* April 25, 1942.
11. Donovan combat report, April 16, 1942.
12. Ibid.
13. Greenlaw, *Official Diary of Operations,* April 26, 1942.
14. Smith, *Tale of a Tiger,* p. 293.
15. Greenlaw, *Official Diary of Operations,* April 28, 1942.
16. Ford, *Flying Tigers,* p. 312.
17. Bishop combat report, April 28, 1942.
18. Hill combat report, April 28, 1942.
19. Ibid.
20. Bishop combat report, April 28, 1942.
21. Fred Adkins combat report, April 28, 1942.
22. Bishop combat report, April 28, 1942.
23. Smith combat report, April 28, 1942.
24. Greene combat report, April 28, 1942.
25. Older combat report, April 28, 1942.
26. Overend combat report, April 18, 1942.
27. Olson combat report, April 28, 1942.
28. Jernstedt, combat report, April 28, 1942.
29. Thomas Haywood combat report, April 28, 1942.
30. Ford, *Flying Tigers,* p. 316.
31. Ibid.
32. Olson combat report, April 28, 1942.
33. Smith combat report, April 28, 1942.
34. Smith, *Tale of a Tiger,* p. 295.
35. Ibid.
36. Greenlaw, *Official Diary of Operations,* April 28, 1942.
37. Neale diary, April 18, 1942.
38. Ford, *Flying Tigers,* p. 320.

Chapter 20: The Salween Gorge Missions

1. Ford, *Flying Tigers,* p. 323.
2. Greenlaw, *Official Diary of Operations,* May 3, 1942.
3. Neale, combat report, May 3, 1942.
4. Ford, *Flying Tigers,* p. 324.
5. Bond combat report, May 4, 1942.
6. Bond and Anderson, *A Flying Tiger's Diary,* pp. 164-166.
7. Ford, *Flying Tigers,* p. 325.
8. Ibid.
9. Greenlaw, *Official Diary of Operations,* May 5, 1942.
10. Ibid.

11. Hill combat report, May 5, 1942.
12. Lawlor combat report, May 5, 1942.Frank Schiel, combat report, May 5, 1942.
13. Frank Schiel combat report, May 5, 1942.
14. J. Gilpen Bright combat report, May 5, 1942
15. Raymond Hastey combat report, May 5, 1942.
16. Freeman Ricketts combat report, May 5, 1942.
17. Matthew Kuykendall combat report, May 5, 1942.
18. Hill combat report, May 5, 1942.
19. Ibid.
20. Schiel combat report, May 5, 1942.
21. Ford, *Flying Tigers,* p. 326.
22. Schultz, *The Maverick War,* p. 261.
23. Chennault, *Way of a Fighter,* p. 163, and Losonsky, *Flying Tiger,* p. 93.
24. Greenlaw, *Official Diary of Operations,* May 6, 1942.
25. Ibid., May 7, 1942.
26. Chennault, *Way of a Fighter,* p. 164.
27. Greenlaw, Official Diary of Operations, May 7, 1942.
28. Chennault, *Way of a Fighter,* p. 166.
29. Hill and Schaupp, *"Tex" Hill: Flying Tiger,* pp. 169-170.
30. Greenlaw, *Official Diary of Operations,* May 8, 1942.
31. Lewis Bishop combat report, May 8, 1942.
32. Greenlaw, *Official Diary of Operations,* May 9, 1942.
33. Losonsky, *Flying Tiger,* p. 93.

Chapter 21: "The Train Wasn't Worth Bishop"

1. Neale combat report, May 9, 1942.
2. Greenlaw, *Official Diary Operations,* May 10, 1942.
3. Rector Combat report, May 10, 1942.
4. Ford, *Flying Tigers,* p. 331.
5. James Howard combat report, May 10, 1942.
6. Hill and Schaupp, *Tex Hill: Flying Tiger,* p. 163.
7. Greenlaw, *Official Diary of Operations,* May 11, 1942.
8. Olson diary, May 11, 1942.
9. Hotz, *With General Chennault,* pp. 230-231.
10. Ford, *Flying Tigers,* p. 334.
11. Bishop combat report, May 12, 1942.
12. Ibid.
13. Ford, *Flying Tigers,* p. 334.
14. Greenlaw, *Official Diary of Operations,* May 12, 1942.
15. R.T. Smith, *Tale of a Tiger,* p. 309.
16. Ibid.
17. Greenlaw, *Official Diary of Operations,* May 15, 1942.
18. Ibid.
19. R.T. Smith, *Tale of a Tiger,* p. 312.
20. Baisden, *Flying Tiger to Air Commando,* pp. 68-70.
21. Greenlaw, *Official Diary of Operations,* May 17, 1942.
22. Keeton diary, May 127, 1942.
23. Greenlaw, *Official Diary of Operations,* May 17, 1942.
24. R.T. Smith, *Tale of a Tiger,* p. 314.
25. Peter Wright, combat report, May 17, 1942.
26. Lewis Sherman Bishop and Sheila Bishop Irwin, *Escape from Hell,* pp. 194-195.
27. Robert. L. Scott, *God is My Co-Pilot,* p. 106.

28. R.T. Smith, *Tale of a Tiger,* p. 315.
29. Ibid., p. 318.
30. Ibid.
31. Keeton diary, May 17, 1942.
32. Ibid.
33. Bond and Anderson, *A Flying Tiger's Diary,* p. 178.
34. R.T. Smith, *Tale of a Tiger,* p. 318.
35. Greenlaw, *Official Diary of Operations,* May 22, 1942.
36. Einar Michelson combat report. May 22, 1942.
37. Robert H. Smith combat report, May 22, 1942.
38. Neale diary, May 22, 1942.
39. Keeton diary, May 23, 1942.
40. Burgard diary, May 11-May 23, 1942.
41. Greenlaw, *Official Diary of Operations,* May 24, 1942.
42. Bond and Anderson, *A Flying Tiger's Diary,* p. 180.
43. Keeton diary, May 24, 1942.
44. Neale diary, May 24, 1942.
45. Schiel combat report, May 24, 1942.
46. Ford, *Flying Tigers,* p. 344.
47. Greenlaw, *Official Diary of Operations,* May 29, 1942.
48. Schiel combat report, May 29, 1942.
49. Greenlaw, *Official Diary of Operations,* May 30, 1942.
50. Keeton diary, May 30, 1942.
51. Ibid., May 31, 1942.
52. Ibid.

Chapter 22: The Move to Kweilin

1. Greenlaw, *Official Diary of Operations,* June 1, 1942.
2. Smith, *Tale of a Tiger,* p. 331.
3. Bond and Anderson, *A Flying Tiger's Diary,* p. 183.
4. Burgard diary, June 5, 1942.
5. Ford, *Flying Tigers,* p. 345, and Bond and Anderson, *A Flying Tiger's Diary,* p. 184.
6. Bond and Anderson, *A Flying Tiger's Diary,* p. 184.
7. Ibid., p. 185.
8. Ford, *Flying Tigers,* pp. 349-350.
9. Bond and Anderson, *A Flying Tiger's Diary,* p. 186.
10. Burgard diary, June 11, 1942.
11. Bond and Anderson, *A Flying Tiger's Diary,* p. 187.
12. Ibid.
13. Hotz, *With General Chennault,* p. 244.
14. Neale diary, June 11, 1942.
15. Ford, *Flying Tigers,* p. 351.
16. Various combat reports, June 12, 1942.
17. Bond and Anderson, *A Flying Tiger's Diary,* p. 187.
18. Ibid., p. 188.
19. Ibid., pp. 188-189.
20. Rosbert combat report, June 12, 1942.
21. A. M. Wright combat report, June 12, 1942.
22. James Dumas combat report, June 12, 1942.
23. Neale combat report, June 12, 1942.
24. Bartling combat report, June 12, 1942.
25. Rossi combat report, June 12, 1942.

26. R.W. Masters combat report, June 12, 1942.
27. Burgard combat report, June 12, 1942.
28. Ibid.
29. Ibid.
30. John Dean combat report, June 12, 1942.
31. Bond and Anderson, *A Flying Tiger's Diary,* p. 191.
32. Hill and Schaupp, *Tex Hill: Flying Tiger,* pp. 166-167.
33. Greenlaw, *Official Diary of Operations,* June 13, 1942.

Chapter 23: It Was The Hardest Thing I Have Ever Done..."

1. Ford, *Flying Tigers,* p. 356.
2. Special Order No. 64, dated June 14, 1942, HQ American Army Forces in China, Burma and India.
3. Hill and Schaupp, *Tex Hill: Flying Tiger,* p. 170.
4. Ford, *Flying Tigers,* pp. 355-356, and Bond and Anderson, *A Flying Tiger's Diary,* p. 196.
5. Greenlaw, *Official Diary of Operations,* June 20, 1942.
6. Bond and Anderson, *A Flying Tiger's Diary,* p. 196.
7. Burgard diary, June 21, 1942.
8. Bond and Anderson, *A Flying Tiger's Diary,* p. 197.
9. Neale diary, June 22, 1942.
10. Greenlaw, *Official Diary of Operations,* June 22, 1942.
11. Bond and Anderson, *A Flying Tiger's Diary,* p. 198
12. Charles Sawyer combat report, June 22, 1942.
13. Ibid.
14. Robert Raines combat report, June 22, 1942.
15. Schiel combat report, June 22, 1942.
16. Hotz, *With General Chennault,* p. 255.
17. Greenlaw, *Official Diary of Operations,* June 23, 1942.
18. Bond and Anderson, *A Flying Tiger's Diary,* pp. 198-199.
19. Ibid., p. 199.
20. Ford, *Flying Tigers,* p. 360.
21. Greenlaw, *Official Diary of Operations,* June 23, 1942.
22. Bond and Anderson, *A Flying Tiger's Diary,* p. 199.
23. Ibid., pp. 199-200.
24. James H. Howard, *Roar of the Tiger,* p. 164
25. Ibid., pp. 164-165.
26. Ford, *Flying Tigers,* p. 360.
27. Chennault, *Way of a Fighter,* p. 172.
28. Burgard diary, June 25, 1942.
29. Ford, *Flying Tigers,* p. 362.
30. Hill and Schaupp, *Tex Hill: Flying Tiger,* p. 170.
31. Ibid., p. 171.
32. Bond and Anderson, *A Flying Tiger's Diary,* p. 200.
33. Ibid.
34. Schultz, *The Maverick War,* p. 277. The pilots remaining for the two-week extension were Bill Bartling, Harry Bolster, Charlie Bond, Carl Brown, Hank Geselbracht, Lester Hall, Jim Howard, Bob Layer, Bus Loane, Bob Neale, Georghe Paxten, Al Probst, Johnny Petach, Bob Raines, Freeman Rickets, Joe Rosbert, Arnold Shamblin, Van Shapard and Peter Wright.
35. Ford, *Flying Tigers,* p. 361.
36. Neale diary, July 1, 1942.
37. Bond and Anderson, *A Flying Tiger's Diary,* p. 201.
38. Wright diary, July 1, 1942.
39. Neale diary, July 1, 1942.
40. Greenlaw, *Official Diary of Operations,* July 1, 1942.

41. Bond and Anderson, *A Flying Tiger's Diary,* p. 201.
42. Ford, *Flying Tigers,* p. 363.
43. Raines, combat report, July 3, 1942.
44. Harry Bolster combat report, July 3, 1942.
45. Neale diary, July 3, 1942.
46. Bond and Anderson, *A Flying Tiger's Diary,* July 3, 1942.
47. Ibid., p. 203.
48. Neale combat report, July 4, 1942.
49. Rector combat report, July 4, 1942.
50. Raines combat report, July 4, 1942.
51. Ford, *Flying Tigers,* p. 356.
52. Bond and Anderson, *A Flying Tiger's Diary,* p. 203.
53. Neale diary, July 4, 1942.
54. Greenlaw, *Official Diary of Operations,* July 4, 1942
55. Chennault, *Way of a Fighter,* pp. 173-174.

Chapter 24: The 23rd Fighter Group Takes Over

1. Ford, *Flying Tigers,* p.367.
2. Hill and Schaupp, *Tex Hill: Flying Tiger,* p. 177.
3. Greenlaw, *Official Diary of Operations,* July 5. 1942.
4. John Petach, combat report, July 6, 1942.
5. Hill and Schaupp, *Tex Hill: Flying Tiger,* p. 178.
6. Neale diary, July 7, 1942.
7. Hill and Schaupp, *Tex Hill: Flying Tiger,* p. 179.
8. Ibid., p. 180.
9. Albert Baumler combat report, July 10, 1942.
10. Ibid.
11. Wright combat report, July 10, 1942.
12. Neale diary, July 10, 1942.
13. Bond and Anderson, *A Flying Tiger's Diary,* p. 205.
14. Ibid.
15. Ibid.
16. Neale diary, July 13, 1942.
17. Ibid., July 14, 1942.
18. John Alison action report, July 16, 1942.
19. Ricketts combat report, July 16, 1942.
20. Alison action report, July 16, 1942.
21. Bond and Anderson, *A Flying Tiger's Diary,* p. 207.
22. Greenlaw, *Official Diary of Operations,* July 19, 1942.
23. Ibid., July 4, 1942; Ford, *Flying Tigers,* pp. 360-361 and Cornelius and Short, *Ding Hao,* pp. 260-261.

Epilogue

1. Chennault, *Way of a Fighter,* p. 176.
2. Ford, *Flying Tigers,* p. 373.
3. Chennault, *Way of a Fighter,* p. 175.
4. Bond and Anderson, *A Flying Tiger's Diary,* pp. 211-213.
5. Samson, *Chennault,* p. 165.
6. Byrd, *Chennault,* p. 162.
7. Ibid., p. 201.
8. Ford, *Flying Tigers,* p. 378.
9. Chennault, *Way of a Fighter,* p. 355.

10. Byrd, *Chennault,* p. 286.
11. Samson, *Chennault,* p. 269.
12. Ibid., pp. 270-271.
13. Byrd, *Chennault,* pp. 290-291.
14. Samson, *Chennault,* pp. 282-283.
15. Byrd, *Chennault,* p. 206.
16. Samson, *Chennault,* p. 286.
17. Ibid., p., 287.
18. Ibid., pp. 291-294.
19. Ibid., pp. 307-308.
20. Ibid., p. 311.
21. Ibid., p. 313, and Byrd, *Chennault,* p. 349.
22. Samson, *Chennault,* pp. 330-335.
23. Ford, *Flying Tigers,* p. 382.
24. Ibid., p. 378.
25. Ibid., p. 377.
26. Ibid., p. 379.
27. Ibid., and biographies from Flying Tigers Association website.
28. Ford, *Flying Tigers,* p. 380.
29. Schultz, *The Maverick War,* p. 285, and biographies from the Flying Tigers Association website.
30. Ibid.
31. Ford, *Flying Tigers,* pp. 379-380, and Schultz, *The Maverick War,* pp. 285-286.
32. Ford, *Flying Tigers,* pp. 370 and 380.
33. Biographies from the Flying Tigers Association website.
34. Ibid.
35. Biography from the Flying Tigers Association website.
36. Samson, *Chennault,* p. 254.
37. Flying Tigers Association reunion programs, 1952-2014.
38. Flying Tigers Association reunion program, December 5-9, 1996, *Persistence Brings Rewards,* p.29.
39. *The Tiger Rag,* newsletter of the Flying Tigers Association, October 1989.

Bibliography

Books

Archibald, Joe. *Commander of the Flying Tigers: Claire Lee Chennault.* New York: Julian Meisner, 1966.

Armstrong, Alan. *Preemptive Strike: The Secret Plan That Would Have Prevented The Attack On Pearl Harbor.* Guilford, CT; The Lyons Press, 2006.

Ayling, Keith. *Old Leatherface of the Flying Tigers.* New York: Bobbs-Merrill, 1945.

Baldwin, R.E. and Thomas McGarry. *Last Hope: The Blood Chit Story.* Atglen, PA: Schiffer Publishing, 1997.

Bishop, Lewis Sherman, and Irwin, Shiela Bishop. *Escape From Hell.* Bloomington, IL: Tiger Eye Press, 2004.

Bond, Charles R., Jr. and Terry Anderson. *A Flying Tigers Diary*. College Station, TX; Texas A&M University Press, 1984.

Boyington, Gregory. *Baa Baa Black Sheep.* NY; G.P. Putnam's Sons, 1958.

Byrd, Martha. *Chennault: Giving Wings to the Tiger.* Tuscaloosa; University of Alabama Press, 1987.

Caidin, Martin. *The Ragged, Rugged Warriors,* New York; E.P. Dutton, 1966.

Carrozza, Anthony. *William D. Pawley: The Extraordinary Life of the Adventurer, Entrepreneur and Diplomat Who Cofounded the Flying Tigers.* Dulles, VA: Potomac Books, 2012.

Chennault, Anna. *A Thousand Springs.* New York: Paul Erikson, 1962.

Chennault, Claire Lee. *Way of a Fighter: The Memoirs of Claire Lee Chennault.* New York: G.P. Putnam's Sons, 1949.

Chivers, Sydney P. *The Flying Tigers: A Pictorial History of the American Volunteer Group.* Canoga Park, CA: Challenge Publications.

Clemens, Terrill. *American Volunteer Group Colours and Markings,* Oxford. UK: Osprey Publications, 2001.

Cornelius, Wanda and Thayne Short. *Ding Hao!: America's Air War in China. 1937-1945.* Gretna, LA: Pelican Publishing Company, 1980.

Crozier, Bryan, *The Man Who Lost China.* New York: Scribner's, 1976.

Ford, Daniel. *Flying Tigers: Claire Chennault and the American Volunteer Group.* Washington, DC: Smithsonian Institute Press, 1991.

Frillman, Paul and Graham Peck. *China: The Remembered Life.* Boston: Houghton Mifflin, 1968.

Greenlaw, Olga S. *The Lady and the Tigers.* New York: E.P. Dutton, 1943. Gretna, LA; Pelican, 1980.

Gurney, Gene. *Five Down and Glory.* New York: Putnam's Sons, 1958.

Heiferman, Ron. *Flying Tigers: Chennault in China.* New York: Ballantine Books, 1971.

Hill, David L. and Reagan Schaupp. *Tex Hill, Flying Tiger.* Spartanburg, SC: Honoribus Aviation History Series, 2007.

Holik-Urban, Jennifer and Robert R. Brouk. *To Soar With The Tigers: The Life and War Diary of Robert Brouk.* Woodbridge, IL: Generations, 2011

Hotz, Robert B. *With General Chennault: The Story of the Flying Tigers.* New York: Howard-McCann, 1943.

Howard, James H. *Roar of the Tiger.* New York: Orion Books, 1991

Hsu, Immanuel C.Y. *The Rise of Modern China.* New York: Oxford University Press, 1970.

Jimison, Susan Clotfelter. *Through the Eyes of a Tiger: The John Donavan Story.* Privately published.

Johnson, Wayne and Don Van Cleve. *Chennault's Flying Tigers, 1941-1945.* Dallas, TX: Taylor Publishing, 1983.

Kleiner, Sam. *The Flying Tigers.* Viking, 2018.

Leary, William M., Jr. *The Dragon's Wings: The China National Aviation Corporation and the Development of Commercial Aviation in China.* Athens, GA: University of Georgia Press, 1981.

Lee, Keith. *A Chinese in the A.V.G. – The Story of NG Bak Lim in the American Volunteer Group, the China Air Task Force, the US Army Fourteenth Air Force and the United States Air Force.* Portland, Oregon: Privately published, 2016.

Leonard, Royal. *I Flew for China,* Garden City, NY: Doubleday, 1942.

Leuthner, Stuart and Oliver Jensen. *High Honor: Recollections by Men and Women of World War II Aviation.* Washington: Smithsonian Institution Press, 1989

Losonsky, Frank and Terry M. Losonsky. *Flying Tiger: A Crew Chief's Story.* Atglen, PA: Schiffer, 1996.

McDonald, William C. III and Barbara L. Evenson. *The Shadow Tiger.* Birmingham, AL: Shadow Tiger Press, 2016

McDowell, Ernest. *The P-40 Kittyhawk.* New York: ARCO Publishing, 1968
Mims, Sam. *Chennault of the Flying Tigers.* Philadelphia: Macrae-Smith, 1943.
Mitter, Rana. *Forgotten Ally: China's World War II, 1937-1945.* Boston, MA: Houghton Mifflin Harcourt, 2013.
Molesworth, Carl, *Flying Tiger Ace.* Osprey Publishing, 2020.
Molesworth, Carl. *Wing to Wing: Air Combat in China, 1943-45,* New York: Orion Books, 1990.
Moser, Don. *China-Burma-India.* Alexandria, VA: Time-Life Books, 1978
Nalty, Bernard C. *Tigers Over Asia.* New York: Elsevier-Dutton, 1978.
Neumann, Gerhard. *Herman the German.* New York: Morrow, 1984
Pawley, William D. *Americans Valiant and Glorious.* New York: Caleb Printing, 1945.
Pistole, Larry M. *The Pictorial History of the Flying Tigers.* Orange, VA: Moss, 1981.
Rodewald, Donald and Kenneth T. Meredityh. *Tiger Tenacity.* Lake Cuty, CO: Granite Falls Books, 1991.
Rosbert, C. Joseph. *Flying Tiger Joe's Adventure Story Cookbook.* Lakehurst, GA, 1998.
Rosholt, Malcolm. *Days of Ching Pao: A Photographic Record of the Flying Tigers – 14th Air Force in China in World War II.* Amherst, Wisconsin: Palmer Publications, 1978.
Samson, Jack. *Chennault.* New York: Doubleday, 1987.
Schramm, Leo J. *Leo the Tiger: True Stories about the Flying Tigers from WWII.* Camp Hill, PA: Green Shield Services, 1998
Schultz, Duane. *The Maverick War.* New York: St. Martin's Press, 1987.
Scott, Robert L. Jr. *God is My Co-Pilot.* Garden City, NY: Blue Ribbon Books, 1944.
Scott, Robert L. Jr. *Flying Tiger: Chennault of China.* Garden City, NY: Doubleday, 1959.
Seagrave, Sterling. *Soldiers of Fortune.* Alexandria, VA: Time-Life Books. 1981
Shilling, Erik. *Destiny: A Flying Tiger's Rendezvous with Fate.* Privately published, 1993.
Smith, Robert M. *With Chennault in China.* Blue Ridge Summit: TAB, 1984.
Smith, Robert T. *Tale of a Tiger.* Van Nuys: Tiger Originals, 1986.
Szuscikiewicz, Paul. *Flying Tigers.* New York: Gallery Books, 1990.
Toland, John. *The Flying Tigers.* New York: Random House, 1963.
Tuchman, Barbara W. *Stilwell and the American Experience in China, 1911-45.* New York: MacMillan, 1970.
Tullis, Thomas A. *Tigers Over China: The Aircraft of the A.V.G.* Hamilton, MT: Eagle Editions, 2001
Whelan, Russell. *The Flying Tigers: The Story of the American Volunteer Group.* New York: Viking, 1942.
White, Theodore and Annalee Jacoby. *Thunder Out of China.* New York: William Sloan Associates, 1946
Yenne, Bill. *When Tigers Ruled the Sky – The Flying Tigers: American Outlaw Pilots over China in World War II.* New York: Berkley Caliber, 2016.

Periodicals

Bauer, Daniel. "Legend of the China Skies," *Air Classics*, October 1988.
Bergin, Bob. "Tiger Attack at Moulmein," *Air Classics.*
Bergin, Bob. "The Last Flight of Scarsdale Jack," *Air Classics,* January 1995.
Bergin, Bob. "Pearl Harbor Payback," *Flight Journal*, April 2005.
Bergin, Bob. "Flying Tiger, Burning Bright," *Aviation History,* July 2008.
Byrd, Martha. "Crash Course in Combat," *The Retired Officer,* January 1991.
Carrol, John Scott. "John Ed's Legacy," *The Retired Officer,* May 1986.
Cole, Martin. "Robert Short: The Making of a Hero," *American Aviation Historical Society Journal,* Spring 1970.
Cleaver, Thomas McKelvey. "A Tiger's Tale," *Flight Journal,* December 2002.
Correll, John. T. "The Flying Tigers," *Air Force,* December 2006.
Cox, Clifford. "Last Dare of Jack Newkirk," *Stag,* December 1960.
Ford, Daniel. "One Hundred Hawks for China," *Air & Space,* April/May 1988.

Grant, Rebecca. "Flying Tiger, Hidden Dragon," *Air Force,* March 2002.

Grinnell, Roy. "The Christmas Day Battle That Began a Legend," *Popular Mechanics,* December 1975.

Hasbrouk, Robert W., Jr. "Scarsdale Jack and the Flying Tigers," *Officer Review,* January/February 2016.

Hill, David Lee. "Shooting Fish in a Barrel," *Flight Journal,* October 2005.

Holcomb, Mal. "Flying Behind the Twelve-Pointed Star," *Wings,* October 1993.

Johnston, Russ. "Japs Are Their Specialty," *Flying,* August 1942.

Labussiere, William. "The Truth About the 14th Squadron Foreign Volunteers in China," *American Aviation Historical Society Journal,* Winter 1987.

Leiser, Edward. "Memoirs of Pilot Elwyn H. Gibbon, The Mad Irishman," *American Aviation Historical Society Journal,* Spring 1978.

McLeon, Don. "Tigers of a Different Stripe," *Soldier of Fortune,* January 1989.

Minich, Mike. "Tiger in the Sky: The Saga of the A.V.G.," *Air Enthusiast,* Quarterly/Four.

Ohlinger, John F. "From Africa to China – Via P-40," *Air Combat,* Winter 1973.

Pentecost, Walter. "Here Come the Flying Tigers," *Liberty,* July 25, 1942-August 8, 1942.

Pentecost, Walter and James J. Sloan. "Advance of the Flying Tigers," *American Aviation Historical Society Journal,* Summer 1970.

Wagner, Ray. "The Chinese Air Force, 1931-1940," *American Aviation Historical Society Journal,* Fall 1974.

Whelpley, Don. "Tigers Over Rangoon," *American Legion,* November 1958.

White, Kenneth. "Battle of the Salween Gorge," *San Antonio Monthly,* November 1983.

Unpublished Diaries and Manuscripts

Adair, Claude Bryant. Personal diary, 1 December 1938-17 June, 1939. Burken Collection.

Bailey, George R. *My Military Diary,* unpublished memoir, August 1936-31 December 1962. George and Mabel Bailey

Burgard, George T. Personal diary, 24 September 1941- 4 July 1942. Lee Burgard.

Chennault, Claire Lee. Personal diary, 1937-1941. Burken Collection.

Chennault, Claire Lee *The Flying Instructor's Manual,* written while instructing at Maxwell Field, Alabama.

Donovan, John T. Personal diary and letters, 17 November 1941-9May 1942. Burken 362

Greene, Paul J. Personal diary, 1 January-23 March 1942. Burken Collection.

Greenlaw, Olga. *First American Volunteer Group Diary, 1941-1942.* Hoover Institution.

Jernstedt, Kenneth. Personal diary, 26 August 1941-9 June 1942. Burken Collection.

Keeton, Robert B. Personal diary, 9 September 1941-23 August 1942. Burken Collection.

Klein, Al V. Personal diary, 25 August 1941-24 August 1943. Burken Collection.

McMillan, George B. Personal diary, 21 July 1941-11 March 1942. Burken Collection.

Mott, Charles D. Personal diary, 9 July-24 December 1941. Burken Collection.

Neale, Robert H. Personal diary, I January-23 July 1942. Burken Collection.

Older, Charles H. Personal diary, 26 August 1941-3 May 1942. Burken Collection.

Olson, Arvid E. Personal diary, 5 August 1941-16 July 1942, Burken Collection.

Olson, Arvid E. *Diary of the Third Pursuit Squadron,* 24 August 1941-9 June 1942. Burken Collection.

Overend, Edmund F. Personal diary, 23 September 1941-26 February 1942. Burken Collection.

Paxton, George L. Personal diary, 24 September 1941-11 August 1942. Author's collection.

Prescott. Robert B. Personal diary, 23 September 1941-25 February 1942. Burken Collection.

Raines, Robert J. Personal diary, 14 April 1941-31 December 1942. Burken Collection.

Reed, William, Personal diary, 22 July, 1941-2 September 1943. Ed Reed.

Riffer, Clarence. Personal diary, 24 May 1941- 12 September1942. Author's Collection.

Rodewald, Donald J. Personal diary, 15 June 1941-5 December 1942. Burken Collection.

Sawyer, Charles W. Personal diary, 5 January 1941-13 December 1942. Burken Collection.

Schaper, William. Personal diary, 10 July 1941-8 August 1942, Burken Collection.

Smith, Curtis E., Duty Officer log, M.S. *Boschfontein,* 24 September 1941-12 November 1941. Burken Collection.

Wright, Peter. Personal diary, 26 June 1941-13 December 1942. Burken Collection.

Personal Interviews and Correspondence

Baisden, Charles. Interview, December 26, 1991.

Bond, Charles. Interview, May 12, 1998.

Burken, Mark. Letters, August 29, 1996, March 13, 14, 15, 1997, April 10, 2002.

Cross, James. Interview, May 10, 1988.

Duke, George. Letter, March 17, 1988.

Fausel, Robert W. Letters, December 4, 1991, December 24, 1991.

Harrington, Jasper J. Interview, March 28, 1981; letter June 28, 1981.

Hill, David L. Interviews, September 27, 1976, February 13, 1992.

Hinoki, Yohei. Continuing correspondence from June 15, 1987 to January 13, 1991.

Ichimura, Hiroshi. Letter, February 13, 2002.

Ieda, Toriyuki. Letter, June 1987.

Kawamoto, Kouki. Letter, July 1988.

Kemph, Merlyn P. Interview, date unknown.

Kochiyama, Shigeru. Several letters from July 22, 1981, to October 13, 1990.

Lansdale, James. Letter, February 2, 1991.

Masahiro, Nakayama. Letter, January 10, 1991

Paxton, George L., Jr. Continuing correspondence from September 10, 1976 to January 12, 1979.

Perry, Paul J. Letter, April 25, 1981.

Rector, Edward F. Letter, unknown date.

Riffer, Clarence. Interview, August 18, 1981.

Rossi, John R. Letters, October 8, 1990, January 30, 1991, August 30, 1999.

Rossi, John R. Interview, May 10, 1988.

Shimizu, Masakazu. Continuing correspondence from February 6, 1987 to June 27, 1988.

Shilling, Erik. Letters, July 28, 1991, November 19, 1993.

Smith, Robert T. Letter, January 8, 1987.

Stolet, Irving. Interview, February 4, 1897.

Towery, William H. Interview, July 7, 1977.

Ueda, Kony. Letters from February 1, 1988, to March 3, 1988.

Watanabe, Fumio. Letter, April 15, 1991.

Wright, Peter. Letter, September 9, 1987.

Oral Histories

Bacon, Noel. U.S.A.A.F. interview, April 22, 1942.

Boyington, Gregory. *Aviation History,* 1988.

Blyer, Robert. Columbia interview, August 1962.

Chennault, Anna. Columbia interview, August 21, 1962.

Chennault, Claire L. U.S.A.F. interview, Washington, D.C., April 16, 1948.

Corcoran, Thomas. Columbia interview, August 21, 1962.

Davis, Doreen. Columbia interview, July 27, 1962.

Frillman, Paul. U.S.A.A.F. interview, July 29, 1942.

Hill, David L. Columbia interview, August 1962.
Jordan, Joe. Columbia interview, August 21, 1962.
Kuykendall, Matt. Columbia interview, August 21, 1962.
MacAlister, Gayle. Columbia interview, August 21, 1962.
McGarry, William. Robert Andrade interview, 1986.
McMillan, George. U.S.A.A.F. interview, November 1942.
Mow, P.T. U.S.A.F. interview, April 21, 1948.
Metasavage, Frank. U.S.A.A.F. interview, August 5, 1942.
Neale, Robert H. Columbia interview, August 21, 1962.
Older, Charles. Columbia interview, August 21, 1962.
Paxton, George. U.S.A.A.F. interview, August 25, 1962.
Pentecost, Walter, U.S.A.A.F. interview, April 28, 1942.
Prescott, Robert W. Columbia interview, August 21, 1942.
Rector, Edward F. U.S.A.A.F. interview, February 2, 1943.
Richardson, Doc. Columbia interview, August 1962.
Rodewald, Donald L. Columbia interview, July 31, 1962.
Schaper, William. Columbia interview, July 31, 1962.
Smith, Robert M. Columbia interview, July 28, 1962.
Trumble, Thomas. Columbia interview, July 28, 1962.
Williams, John M. U.S. Air Force Academy interview.
Wirta, Harvey. Columnia interview, August 1962.

Index

A native Texan, Ward Boyce graduated from Texas A&M with a BS in mechanical engineering and a commission in the U.S. Air Force. He spent 28 years in the Air Force, including 12 years as a "backseater" in the F-101B and the RF-4C. During the Vietnam War he flew 234 combat sorties in the Phantom, including 113 over North Vietnam.

Upon retiring from the Air Force as a colonel in 1988, he was hired by the American Fighter Aces Association as their executive director, a position he held for the next 12 years, and still serves as a member of their board of directors.

Returning home to Austin, Texas, in 1998 with Mary Ray Graves, his wife now for 60 years, he is still active in several veterans' organizations, including the Air Force Association, Daedalians, Quiet Birdmen and the Red River Rats. He and Mary have two daughters, Beverly and Elizabeth, and five grandchildren, all of whom live within an hour's drive.

www.ingramcontent.com/pod-product-compliance
Lightning Source LLC
LaVergne TN
LVRC101024220926
848486LV00012B/389

* 9 7 8 1 9 4 3 4 9 2 8 7 9 *